T0221683

Thermodynamics, Gas Dynamics, and Combustion

Henry Clyde Foust III

Thermodynamics, Gas Dynamics, and Combustion

 Springer

Henry Clyde Foust III
University of Houston – Downtown
Houston, TX, USA

ISBN 978-3-030-87386-8   ISBN 978-3-030-87387-5   (eBook)
https://doi.org/10.1007/978-3-030-87387-5

This Springer imprint is published by the registered company Springer Nature Switzerland AG
The registered company address is: Gewerbestrasse 11, 6330 Cham, Switzerland

To Sheila, may she rest in peace.

Preface

There are several good books on thermodynamics, gas dynamics, and combustion, but there is not one book that covers all three of these areas and shows the integral connections at an advance undergraduate or beginning graduate student level.

This book is written to provide a primer on the subject of thermodynamics (and the allied areas of gas dynamics and combustion) and to be both terse and clear; this is the guiding principle of the book. Most chapters are under 30 pages in length and sufficient examples and problems have been given. Along with the exercises, there are many appendices with either additional information or computer programs (spreadsheets) that further demonstrate some concept.

This book is written in three parts: Part I – Chapters 1, 2, 3, 4, 5, and 6 – is on the fundamentals of thermodynamics; Part II – Chapters 7, 8, and 9 – is on gas dynamics; and Part III – Chapters 10, 11, 12, 13, and 14 – is on combustion.

Chapter 1 begins with ideal gases, then shows when gases deviate from ideal behavior (when pressure significantly increases for a fixed temperature or temperature significantly decreases for a fixed pressure). To address these deviations from ideal behavior, cubic equations of state were developed and mimic the behavior of specific volume versus pressure for a pure substance.

Chapter 2 deals with heat and work. This discussion includes understanding the definitions of specific heat and when to use a certain definition; how these definitions relate to internal energy and enthalpy; various processes described as polytropic processes, the work associated with a particular polytropic process; additionally, states, internal energy, and heat associated with a particular polytropic process are presented for an ideal, perfect gas.

In Chaps. 3 and 4, the first and second laws of thermodynamics are explored.

Chap. 5 presents several heat engines and refrigeration cycles. These cycles include Brayton, Rankine, reverse Brayton, and reverse Rankine cycles.

In Chap. 6, Maxwell relationships, their applications, and ideal gas mixtures are presented. The first and second laws of thermodynamics are re-derived for an ideal mixture of gases.

The next three chapters, Chapters 7, 8, and 9, deal with gas dynamics. Chapter 7 deals with conservation laws applied to a gaseous system, derives the speed of sound for an ideal perfect gas, and goes on to derive the equations for normal shocks. Chapter 8 develops the conservation principles necessary for combustion systems and includes the fact that states change with either location or time; Chapter 7 provides the necessary fluid mechanics for Chaps. 9 and 10 and Chap. 8 provides the necessary fluid mechanics for Chaps. 12 and 13.

Chapter 9 deals with isentropic, choked flow (chemical engineers tend to call this critical flow). A comparison is made between ideal, perfect gases and van der Waal gases. The effects of real gas behavior on isentropic, choked flows are clearly delineated. The chapter goes on to address critical flow and sonic velocity for two phase systems.

Chapter 10 presents a physics-based description of combustions using the Hugoniot/Rayleigh theory that graphically represents the three conservation principles: conservation of mass, conservation of energy, and conservation of moment. Where the Hugoniot Curve and Rayleigh lines intersect represents the end points of the combustion before and after an associated shock.

A combustion system can either be a detonation or a deflagration. A detonation is a wave with an attached reaction zone where the speed of the wave is greater than Mach 1, it's known as a shock wave, and a deflagration is a wave where the speed of the wave is less than Mach 1 and there is an associated reaction zone.

Chapters 11, 12, and 13 deal with a chemically based combustion; a physics-based combustion theory provides information about what occurs after the shock at the von Neumann point and the end of the reaction zone, but not what occurs within the reaction zone. A chemically based combustion theory allows us to investigate state changes within the reaction zone. Chapter 11 deals with the necessary combustion chemistry to include stoichiometry, chemical equilibrium, chemical kinetics, and adiabatic flame temperature.

Chapter 12 discusses deflagration systems, which are combustion systems were the wave travels below Mach 1 ($Ma_1 < .01$) and now transport processes such as momentum diffusion, mass diffusion, and heat diffusion become important. Two forms of deflagrations are discussed: deflagrations where the fuel and oxidizer are pre-mixed (a Bunsen burner is an example) and deflagrations where the fuel and oxidizer are not pre-mixed (a candle flame is an example).

Chapter 13 discussed detonations and presents constant pressure and constant volume combustion, an extended example on a rotational detonation engine, solving for states within the reaction zone, and how the detonation structure is much more complicated and richer than the modeling conducted in this chapter.

In the last chapter, Chapter 14, blast waves are discussed. A blast wave is a wave where the strength of the wave (pressure) decreases with distances from the source. The pivotal works in this area are two papers by G.I. Taylor and blast waves associated with a very intense explosion (atomic bomb). The chapter presents the conservation principles as partial differential equations, reduces these equations to ordinary differential equations through similarity methods, and presents two methods to solve the non-dimensionalized conservation principles. One method is numerical analysis of the ODEs and the other is approximate forms (algebraic equations) to match the numerical results.

Many useful appendices are included. These appendices include worksheets for various heat engines and refrigeration cycles, normal shock waves, various models for critical flow, Rankine-Hugoniot theory, and programs for detonation dynamic systems (ZND models) and deflagration systems (laminar flame theory and diffusion flames). Additionally, there are worksheets associated with the various blast wave systems.

A first course in thermodynamics (14-week semester) could include Sects. 1.1 to 1.4; Chapters 2, 3, 4, and 5.

A second course in thermodynamics (14-week semester) could include Sect. 1.5, Chapters 6, 7, 8, and 9.

An introductory combustion course (14-week semester) could include Chaps. 10, 11, 12, 13, and 14.

Houston, TX, USA   Henry Clyde Foust

Contents

Part I
Fundamentals of Thermodynamics

Chapter 1
Equations of State

1.1 Preview

The purpose of this chapter is to introduce equations that relate temperature (T), pressure (P), and specific volume (ν) for a given set of conditions where T, P, and ν are states of the system.

Specific volume is defined as

$$\nu = \frac{V}{m} = \frac{1}{\rho} \tag{1.1}$$

where V is volume and m is mass, and specific volume is the inverse of density.

Temperature, pressure, and specific volume are fundamental quantities that describe the state of a system and are defined below. Please note we will often assume a system is homogenous in terms of the states, i.e., the temperature is the same throughout the system.

Temperature is a measure of the molecular activity of a fluid at a particular location within the system.

Pressure is an average of the normal stresses that occur on a unit cube of fluid in the three principal directions $\{x, y, z\}$ within the system.

Specific volume is a measure of the mass of a particular system for a unit volume.

We'll start with the simplest representation, which is the ideal gas law. The ideal gas law is predicated on several assumptions that briefly state molecules are perfect, rigid spheres that only transfer momentum during collisions, are infinitely far apart, and occupy no space. When the density increases enough, then some of these assumptions are no longer valid.

Electronic Supplementary Material: The online version of this chapter (https://doi.org/10.1007/978-3-030-87387-5_1) contains supplementary material, which is available to authorized users.

© Springer Nature Switzerland AG 2022
H. C. Foust III, *Thermodynamics, Gas Dynamics, and Combustion*,
https://doi.org/10.1007/978-3-030-87387-5_1

We'll then look at phase diagrams (V vs. T and V vs. P) and develop cubic equations of state to represent the behavior seen on these phase diagrams; the simplest cubic EOS incorporates more accurate models for attractive and repulsive forces. We'll then incorporate the effects of non-spherical molecules and that at higher pressures the volume becomes constant, which become significant as the density gets very large.

A general understanding is gathered by reading Sects. 1.2, 1.3, and 1.4 and more advanced material is given in Sect. 1.5.

1.2   Ideal Gas Law

Through careful experimentation, several researchers of gas behavior that included Boyle, Charles, and Avogadro made observations using several gases near STP[1] [1]. These observations include Boyle's law, Charles' law, and Avogadro's law.

Boyle's Law
Boyle's law states that the volume of a gas (V) is indirectly proportional to its pressure (P) or mathematically

$$V \propto \frac{1}{P} \tag{1.2}$$

Charles Law
Charles' law states that the volume of a gas is directly proportional (V) to its temperature (T) or mathematically

$$V \propto T \tag{1.3}$$

Avogadro's Law
Avogadro's law states that the volume of a gas (V) is directly proportional to the mass given in moles (n) or mathematically

$$V \propto n \tag{1.4}$$

These three laws can be combined mathematically into the following relationship

$$V \propto \frac{nT}{P} \tag{1.5}$$

It can be shown through experimentation (when temperature and pressure do not deviate too far from STP) that

---

[1] Standard temperature and pressure, which is 0 °C and 101.325 kPa.

$$V = R_u \frac{nT}{P} \tag{1.6}$$

Or

$$PV = nR_u T \tag{1.7}$$

where R_u is the universal gas constant and has the value of 8.3145 Pa-m^3 per mole-K. Another form of Eq. 1.7 is

$$PV = mR_g T \tag{1.8}$$

where R_g is now particular to the substance and includes mass in units of kilograms. The value of R_g for a particular substance can be found by the following formula

$$R_g \left[\frac{kJ}{kg - K} \right] = \frac{R_u \left[\dfrac{kJ}{kmole - K} \right]}{MW \left[\dfrac{kg}{kmole} \right]} \tag{1.9}$$

where MW is molecular weight.

We can rearrange Eq. 1.8 to get

$$\rho = \frac{m}{V} = \frac{P}{R_g T} \tag{1.10}$$

The ideal gas law was later explained by the kinetic theory of gases [2]. This theory includes several assumptions

1. The molecules occupy essentially no space and are infinitely far apart
2. Each molecule has the same mass and is perfectly spherical
3. Molecules only exchange momentum during perfectly elastic collisions
4. Each molecule is in constant, rapid, and random motion
5. There are so many molecules that we can discuss the entire set statistically

Some of these assumptions are no longer valid as the density increases. Referring to Eq. 1.10, it is easily seen that the density increases when pressure increases (for fixed T) or when temperature decreases (for fixed P). These observations can be summarized as

$$P \uparrow \text{ for fixed } T \rightarrow \rho \uparrow \rightarrow \text{Non} - \text{ideal behavior} \tag{1.11}$$

$$T \downarrow \text{ for fixed } P \rightarrow \rho \uparrow \rightarrow \text{Non} - \text{ideal behavior} \tag{1.12}$$

1.3   Ideal Gas Law Applications

When using the ideal gas law remember the following.

The first step is to make sure you use the correct value for the gas constant. If the mass is molar, then use R_u. If the mass is in kilograms or pounds, then use R_m. R_m values are given in tables at the back of the book.

A handout will be given for R_u in various units.

The next step is to make sure that T is in absolute scale

$$K = C + 273.15 \ \left(273 \text{ will be utilized in this book}\right) \tag{1.13}$$

and

$$R = 459.67 + F \ \left(460 \text{ will be utilized in this book}\right) \tag{1.14}$$

Further,

$$F = \frac{9}{5}C + 32 \tag{1.15}$$

where K is Kelvin, C is Celsius, R is Rankine, and F is Fahrenheit.

The third step is to make sure P is in absolute scale. Pressure may be given in terms of gauge pressure. The relationship between gauge pressure (P_g) and absolute pressure (P_{abs}) is

$$P_{abs} = P_{atm} + P_g \tag{1.16}$$

where P_{atm} is atmosphere pressure (14.7 psia or 101,300 Pa).

For now, we will consider three possible processes, where a process is a change of states along a particular path. The three paths are -

$$\{P_1, T_1\} \rightarrow \{P_2, T_2\}, \text{ Constant volume process } \left(\text{isometric process}\right)$$

$$\{P_1, v_1\} \rightarrow \{P_2, v_2\}, \text{ Constant temperature process } \left(\text{isothermal process}\right)$$

$$\{v_1, T_1\} \rightarrow \{v_2, T_2\}, \text{ Constant pressure process } \left(\text{isobaric process}\right)$$

To find the equation for a particular process start with ideal gas law (for state 1) and manipulate the equation where the states that are changing are on the left and the states that are constant are on the right. Using the ideal gas law (for state 2) and manipulate the equation where the states that are changing are on the right and the states that are constant are on the left. Equate these two equations to get a relationship between states 1 and 2.

Constant Volume Process

$$PV = nRT \tag{1.17}$$

and

$$\frac{P_1}{T_1} = \frac{nR}{V} = \frac{P_2}{T_1} \tag{1.18}$$

Constant Temperature Process

$$PV = nRT \tag{1.19}$$

and

$$P_1 v_1 = RT = P_2 v_2 \tag{1.20}$$

Constant Pressure Process

$$PV = nRT \tag{1.21}$$

and

$$\frac{v_1}{T_1} = \frac{R}{P} = \frac{v_2}{T_2} \tag{1.22}$$

1.4   Lee/Kessler Charts

Imagine we conduct an experiment with a given mass of water, constant pressure (1 atmosphere), and have a way of measuring both the temperature and volume (V). We heat up our container and plot various points of $\{V, T\}$ for a liquid. Eventually, the positive slope of the line turns to a horizontal line, which is an indication of the mixture range. Again, the line forms a positive slope and the substance is now a gas (see Fig. 1.1).

We can get a similar curve for V versus P at a constant T (see Fig. 1.2).

We observe that there is a point that demarcates the saturated liquid from the saturated vapor curve and that this point is known as the critical point. Substances under the dome are known as mixtures that include both liquid and gas. When a substance is under the mixture dome, not only do we need to know the temperature and pressure but we also need a third quantity, quality (X), to define the substance. Quality is defined as

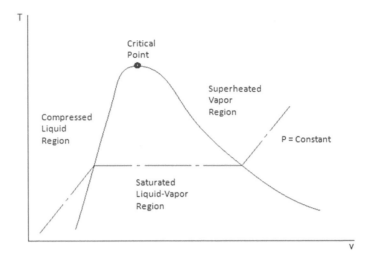

Fig. 1.1 *T* versus *V* for constant *P*

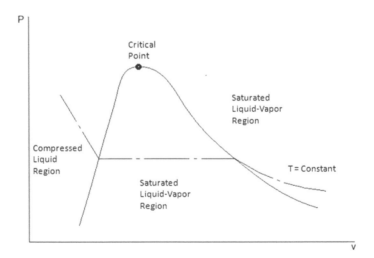

Fig. 1.2 *P* versus *V* for constant *T*

$$X = \frac{m_g}{m_g + m_l} \tag{1.23}$$

where m_g is the mass of gas and m_l is the mass of liquid.

Going back to Figs. 1.1 and 1.2, we can determine for a given pressure and temperature what phase we are in. These rules are provided in Table 1.1.

Table 1.1 Delineating phase from a given T and P

For $T < T_{Sat}$ and $P = P_{Sat} \rightarrow$ Sub-cooled liquid (see V versus T diagram)
For $T > T_{Sat}$ and $P = P_{Sat} \rightarrow$ Super-heated vapour (see V versus T diagram)
For $T = T_{Sat}$ and $P < P_{Sat} \rightarrow$ Super-heated vapour (see V versus P diagram)
For $T = T_{Sat}$ and $P > P_{Sat} \rightarrow$ Sub-cooled liquid (see V versus P diagram)
For $T = T_{Sat}$ and $P = P_{Sat} \rightarrow$ Mixture

We can define a given temperature and pressure relative to the critical pressure (P_c) and critical temperature (T_c).

Reduced pressure is

$$P_R = \frac{P}{P_c} \tag{1.24}$$

and reduced temperature is

$$T_R = \frac{T}{T_C} \tag{1.25}$$

Also define the compressibility factor (Z), which is a measure of the deviation from ideal gas behavior as

$$Z = \frac{PV}{nR_u T} \tag{1.26}$$

We can plot $\{P_R, T_R, Z\}$ for a given pure substance and the result is given as Fig. 1.3. We observe that as T_R decreases (T decreases) for fixed P_R and/or P_R increases (P increases) for fixed T_R that Z decreases, which is consistent with what was said above about when the ideal gas behavior is no longer valid. *We could plot $\{P_R, T_R, Z\}$ for other pure substances and we would find that the points all fall on the same "constant Tr" lines.*

Please note the Lee-Kessler chart is only valid for molecules that are nearly spherical. Below we'll define another parameter acentricity (ω), which is a measure of how much a molecule deviates from being spherical.

Measuring the compressibility factor, Z, from Fig. 1.3 is fraught with error and Appendix 1.C provides a more accurate method. Also note that near the critical point $\{T_r, P_r\} = \{1, 1\}$, the greatest deviation from ideal behavior occurs $\{Z = 0.3\}$ and all models for equation of state notoriously have difficulty predicting the states $\{P, T, \nu\}$ near the critical point.

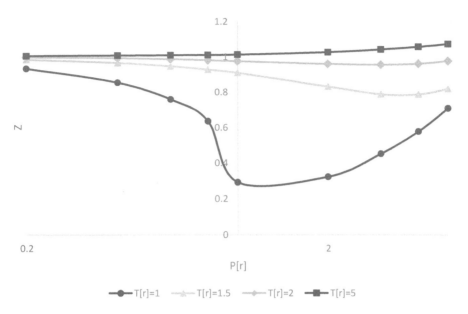

Fig. 1.3 Lee/Kessler compressibility chart [2–4]

1.5   Cubic Equations of State

Three cubic equations of state will be discussed: van der Waals, Redlich-Kwong, and Redlich-Kwong-Soave [3, 5–8]. Each has the following form

$$P = P_{rp} + P_A \tag{1.27}$$

where P_{rp} is the repulsive force and given as

$$P_{rp} = \frac{R_u T}{v - b} \tag{1.28}$$

and P_A is the attractive force and given as

$$P_A = \frac{-a\alpha}{g(v,b)} \tag{1.29}$$

where $a\alpha$ is a measure of the inter-molelcular attractive forces and b is related to the size of the equivalent rigid sphere.

These cubic equations of state are based on the following considerations [2, 3]:

1. Mimic the behavior between v and P given in Fig. 1.2

2. Insure $\dfrac{\partial P}{\partial v} = 0$ and $\dfrac{\partial^2 P}{\partial v^2} = 0$ at the critical point

3. The later cubic models also addressed the fact that all gases take on a constant specific volume at higher pressures, which is stated as $v = 0.26v_c$, where v_c is the specific volume of the gas at the critical point
4. Later models incorporated effects of non-spherical molecules (deviations from the predictions given on the Lee-Kessler chart)

1.5.1 Van der Waal

One of the first attempts to address the deviations from ideal behavior seen as the density increases was the work of van der Waal [4]. The van der Waal equation of state is

$$\left(P + \frac{a}{v^2}\right)(v - b) = R_u T \tag{1.30}$$

or

$$P = P_R + P_A = \frac{R_u T}{v - b} - \frac{a}{v^2} \tag{1.31}$$

These two parameters $\{a, b\}$ can be determined from the following conditions at the critical point

$$\frac{\partial P}{\partial v} = 0 \text{ and } \frac{\partial^2 P}{\partial v^2} = 0 \tag{1.32}$$

The values of $\{a, b\}$ determined from Eqs. 1.30 and 1.32 are

$$a = \frac{27}{64} \frac{R^2 T_c^2}{P_c} \tag{1.33}$$

and

$$b = \frac{RT_c}{8P_c} \tag{1.34}$$

Going back to Fig. 1.2, you'll notice a cubic type of relationship between V and P, which Eq. 1.31 accounts for. Another version of this equation in terms of Z is

$$Z^3 - aZ^2 + bz - c = 0 \tag{1.35}$$

where

$$a = \frac{P_R}{8T_R} + 1, \quad b = \frac{27 P_R}{64 T_R^2}, \quad c = \frac{27 P_R^2}{512 T_R^3} \tag{1.36}$$

It's been found for various reasons that the van der Waal equation of state is inaccurate near and above the critical pressure and other equations of state show higher accuracy in these regions. Van der Waal's equation of state is known as an example of a two-parameter cubic equation of state.

1.5.2   Redlich-Kwong

In 1949, Redlich and Kwong [5, 8] proposed the following model, which provided for a correction to the attractive force term within VW EOS and addresses *Condition 3* given above

$$P = \frac{R_u T}{v - b} - \frac{a\alpha}{v(v + b)} \tag{1.37}$$

Given

$$\alpha = \frac{1}{\sqrt{T}}$$

Using the condition stated as Eq. 1.32, the values for $\{a, b\}$ are

$$a = 0.4748 \frac{R^2 T_c^{2.5}}{P_c}, \quad b = 0.08664 \frac{R T_c}{P_c} \tag{1.38}$$

1.5.3   Acentric Factor

For some substances, P_r and T_r are not enough to specify the compressibility factor (Z). A third factor is needed and is the acentric factor (ω) to account for the molecular structure [2]. In 1955, Pitzer [9] developed a thermodynamic quantity that measures the deviation of a fluid from spherical behavior that would include polar molecules. The acentric factor is defined as

$$\omega = -1.000 - \log_{10}\left(\frac{P_\sigma}{P_C}\right) \tag{1.39}$$

where P_C is the critical pressure for the substance and P_σ is the vapor pressure for T equal to $0.7 T_C$ (the saturated gas curve in Fig. 1.3) and when ω is essentially zero the molecule is considered spherical and has values of $Tr = 0.7$ and $Pr = 0.1$; noble gases have a ω of 0.

The Redlich-Kwong EOS [5, 8] addresses Condition 3 and the Redlich-Kwong-Soave [6] addresses Conditions 3 and 4.

1.5.4   Redlich-Kwong-Soave

Soave in 1972 [6] incorporated into the Redlich-Kwong EOS *Conditions 3 and 4*

$$P = \frac{R_u T}{v - b} - \frac{a\alpha}{v(v + b)} \tag{1.40}$$

where

$$\alpha = \left[1 + S\left(1 - \sqrt{T_R}\right)\right]^2, \quad S = 0.48 + 1.574\omega - 0.177\omega^2 \tag{1.41}$$

And satisfying the condition states as Eq. 1.32, the values for $\{a, b\}$ are

$$a = \frac{0.42748R^2 T_c^2}{P_c}, \quad b = \frac{0.08664RT_c}{P_c} \tag{1.42}$$

In Figs. 1.4, 1.5, and 1.6 the ideal gas law is compared against the van der Waal and Redlich-Kwong Equations of State for methane. For $Tr = 0.79$, we see that as Pr goes up, the density increases, and the deviation from ideal behavior is seen. This trend is countered as the Tr increases and the deviation from ideal behavior becomes less significant (see Figs. 1.5 and 1.6).

For the three cubic equations of state discussed,

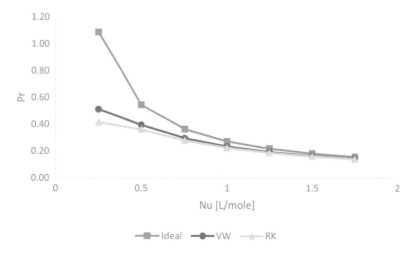

Fig. 1.4 v versus Pr for $Tr = 0.79$

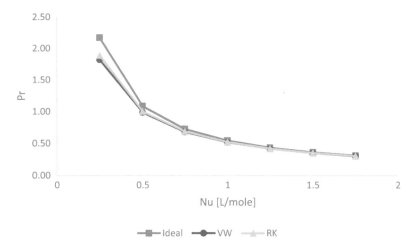

Fig. 1.5 ν versus *Pr* for *Tr* = 1.57

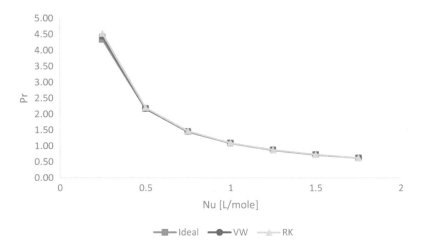

Fig. 1.6 ν versus *Pr* for *Tr* = 3.15

$$Z^3 + \alpha Z^2 + \beta Z + \gamma = 0 \qquad (1.43)$$

where the coefficients are defined in Table 1.2 and $\{a, b\}$ are the coefficients particular to a given equations of state.

Table 1.2 Coefficients for compressibility factor (Z)

	VW	RK	SRK
α	$-1 - B$	-1	-1
β	A	$A - B - B^2$	$A - B - B^2$
γ	$-AB$	$-AB$	$-AB$
A	$\dfrac{aP}{\left(R_uT\right)^{2.5}}$	$\dfrac{aP}{R_u^2T^{2.5}}$	$\dfrac{aP}{\left(R_uT\right)^{2.5}}$
B	$\dfrac{bP}{R_uT}$	$\dfrac{bP}{R_uT}$	$\dfrac{bP}{R_uT}$
Z	$\dfrac{Pv}{R_uT}$	$\dfrac{Pv}{R_uT}$	$\dfrac{Pv}{R_uT}$

1.6   Examples and Problems

1.6.1   Examples

Example 1.1 Ideal Gas Problem, Isothermal Process

An ideal gas goes from 10 bars (absolute) to 20 bars (absolute) for a fixed temperature of 100 F. If the initial specific volume is 20 ft³/lbm, then what's the final specific volume?

Solution

This is a constant temperature (isothermal) process and the appropriate equation is

$$P_1v_1 = P_2v_2$$

Pressure is already in absolute scale and so

$$v_2 = \frac{P_1}{P_2}v_1 = \frac{10}{20}\, 20\,\frac{\text{ft}^3}{\text{lbm}} = 10\,\text{ft}^3/\text{lbm}$$

Example 1.2 Ideal Gas Problem, Isometric Process

An ideal gas goes from {200 °F, 20 Psia} to {400 °F, ??}. What's the final pressure?

Solution

This is an isometric process and the appropriate equation is

$$\frac{P_1}{T_1} = \frac{P_2}{T_2}$$

Pressure is in absolute scale, but temperatures are not and so

$$T_1 = 460 + 200 = 660\,R, T_2 = 460 + 400 = 860\,R$$

Further,

$$P_2 = \frac{T_2}{T_1}P_1 = \frac{860}{660}20\,\text{Psia} = 26.1\,\text{Psia}$$

Example 1.3 Ideal Gas Problem, Isobaric Process
An ideal gas goes from 200 °F to 400 °F for a fixed pressure of 50 Psia. If the initial specific volume is 10 ft³/lbm, then what's the final specific volume?
Solution
This is an isobaric (constant pressure) process and so the appropriate equation is

$$\frac{v_1}{T_1} = \frac{v_2}{T_2}$$

And temperatures are not in absolute so and thus

$$T_1 = 200 + 460 = 660\,R, T_2 = 400 + 460 = 860\,R$$

And

$$v_2 = \frac{T_2}{T_1}v_1 = \frac{860}{660}10\,\frac{\text{ft}^3}{\text{lbm}} = 13\,\text{ft}^3/\text{lbm}$$

Example 1.4 Lee/Kessler Chart Problem
Find the Z factor for CO_2 at 300 K and 5 MPa.
Solution
Find the critical pressure and temperature for CO_2 and then determine the reduced pressure and temperature

$$T_r = \frac{300}{304.1} \approx 1, \ P_r = \frac{5}{7.38} \approx 0.7$$

We then look this point up on the Lee-Kessler chart by first finding Pr equal to 0.7 and then going up the Tr equal to 1 curve and then going to the left to find the Z factor, which is

$$Z = 0.7$$

Example 1.5 Lee/Kessler Chart Problem

Find the Z factor for CO_2 at 300 K and 1 MPa.

Solution

Find the critical pressure and temperature for CO_2 and then determine the reduced pressure and temperature

$$T_r = \frac{300}{304.1} \approx 1, \ P_r = \frac{1}{7.38} \approx 0.1$$

We then look this point up on the Lee-Kessler chart by first finding Pr equal to 0.1 and then going up the Tr equal to 1 curve and then going to the left to find the Z factor, which is $Z \approx 0.98$

Example 1.6 Lee/Kessler Chart Problem

Determine the mass of CO_2 in a hemispherical tank with diameter of 30 inches, pressure of 60 psia, and temperature of $-20\ °C$.

Solution

This problem needs to be done in a series of steps

Step 1 – Convert temperature to Fahrenheit and then put in absolute scale
Step 2 – Get volume of tank in the correct units (ft³)
Step 3 – Determine the Z factor for $\{T_r, P_r\}$
Step 4 – Using ideal gas law, determine the mass [lbm]

Step 1

$$F = \frac{9}{5}\left(-20\,C\right) + 32 = -4\,F, T = 460 + \left(-4\right) = 456\,R$$

Step 2

$$V = \frac{1}{2}\frac{\pi}{6}D^3 = 7069\,\text{in}^3\left[\frac{1\,\text{foot}}{12\,\text{inches}}\right]^3 = 4.09\,\text{ft}^3$$

Step 3

$$T_r = \frac{456}{547.4} \approx 0.8, \ P_r = \frac{60}{1070} \approx 0.06$$

For these conditions, the Z factor is essentially 1 and we treat the substance as an ideal gas

Step 4

$$PV = R_g mT$$

and

$$\frac{PV}{R_gT} = m = \frac{60\,\text{psia}\dfrac{144\,\text{in}^2}{1\,\text{ft}^2}*4.09\text{ft}^3}{35.10\dfrac{\text{ft}-\text{lbf}}{\text{lbm}-R}*456\,R} = 2.21\,\text{lbm}$$

Example 1.7 van der Waal Gas
Derive $\{a, b\}$ for the van der Waal equation of state
Solution
At the critical point,

$$P_c = \frac{R_uT_c}{v_c-b} - \frac{a}{v_c^2} \tag{1.41}$$

and for $\dfrac{\partial P}{\partial v} = 0$

$$\frac{R_uT_c}{\left(v_c-b\right)^2} = \frac{2a}{v_c^3} \tag{1.42}$$

For $\dfrac{\partial^2 P}{\partial v^2} = 0$

$$\frac{R_uT_c}{\left(v_c-b\right)^3} = \frac{3a}{v_c^4} \tag{1.43}$$

If each side of Eq. 1.42 is set equal to X, then Eq. 1.43 in terms of X is

$$\frac{1}{v_c-b}X = \frac{3/2}{v_c}X \tag{1.44}$$

and solving for v_c

$$v_c = 3b \tag{1.45}$$

Substituting Eq. 1.45 into Eq. 1.42 results in

$$\frac{R_uT_c}{4b^2} = \frac{2a}{27b^3} \tag{1.46}$$

or

$$a = \frac{27}{8}bR_uT_c \tag{1.47}$$

Substituting Eq. 1.47 into Eq. 1.41 results in

$$P_c = \frac{R_u T_c}{2b} - \frac{\frac{27}{8} b R_u T_c}{9b^2} = \frac{R_u T_c}{b}\left[\frac{1}{2} - \frac{3}{8}\right] = \frac{R_u T_c}{8b} \tag{1.48}$$

or

$$b = \frac{R_u T_c}{8 P_c} \tag{1.49}$$

and substituting Eq. 1.49 into Eq. 1.47 results in

$$a = \frac{27}{8}\frac{R_u T_c}{8 P_c} R_u T_c = \frac{27}{64}\frac{\left(R_u T_c\right)^2}{P_c} \tag{1.50}$$

Finally, substituting Eq. 1.49 into Eq. 1.45 results in

$$v_c = \frac{3}{8}\frac{R_u T_c}{P_c} \tag{1.51}$$

Example 1.8 Comparing Gas Models

Because of safety concerns, a cylinder with a volume of 20.0 ft³ should not exceed 50 atm. If the tank is filled with 40 lbm of CO_2 at 200 °F, does the pressure exceed the allowable pressure of 50 atm? Check your answer using

(a) Van der Waals EOS
(b) Ideal gas law
(c) RK EOS

Solution
This example was done in Excel and the results are given below.

Van der Waals EOS			Ideal gas law			Redlich-Kwong EOS		
T	[R]	660	T	[R]	660	T	[R]	660
P	[lbf/ft²]	43,779.71	P	[lbf/ft²]	46,332	P	[lbf/ft²]	43,298.82
P	[Psia]	304.03	P	[Psia]	321.75	P	[Psia]	300.69
T[c]	[R]	547.4	nu	[ft³/lbm]	0.5	T[c]	[R]	547.4
P[c]	[lbf/ft²]	154,080	R(g)	[ft-lbf/lbm-R]	35.1	P[c]	[lbf/ft²]	154,080
R(g)	[ft-lbf/lbm-R]	35.1				R(g)	[ft-lbf/lbm-R]	35.1
a		1010.791				a		26,615.87
b	[ft³/lbm]	0.015587				b	[ft³/lbm]	0.010804
nu	[ft³/lbm]	0.5				nu	[ft³/lbm]	0.5

If we use the pressure determine from Redlich-Kwong EOS, which is 301 Psia and convert this number to atmospheres, which is 20.5 atm and less than 50 atm, and so we're good!

1.6.2 Problems

Problem 1.1 Use of Lee-Kessler Chart
Determine the Z factor for air at 10 MPa and 50 F.

Problem 1.2 Use of Lee-Kessler Chart
Determine the Z factor for water at 10 MPa and 50 F.

Problem 1.3 Comparing Z factor for Lee-Kessler Chart and van der Waal Equation
Compare the Z factor for water at the critical conditions using the Lee-Kessler chart and van der Waal equation of state. Do they differ significantly?

Problem 1.4 RK Gas
Derive $\{a, b\}$ for RK EOS

Problem 1.5 SRK Gas
Derive $\{a, b\}$ for SRK EOS

Problem 1.6 van der Waal Gas
Write a program to implement VW EOS

Problem 1.7 RK Gas
Write a program to implement RK EOS

Problem 1.8 SRK Gas
Write a program to implement SRK EOS

Problem 1.9 Comparing Gas Models
Twenty pounds of propane with a volume of 2 ft^3 and a pressure of 300 lbf/in^2. Determine the temperature in F using

(a) Van der Waals EOS
(b) Ideal gas law
(c) Lee-Kessler chart

Problem 1.10 Comparing Gas Models
The pressure within a 20 m^3 tank should not exceed 100 bars. Check the pressure with the tank if filled with 1000 kg of water vapor at a temperature of 350 °C using

(a) Ideal gas law
(b) Lee-Kessler chart
(c) Van der Waals EOS
(d) RK EOS

Problem 1.11 Comparing Gas Models
Ethane gas flows through a pipeline with a volumetric flow rate of 10 ft^3/s at a pressure of 150 atm and a temperature of 50 F. Determine the mass flow rate in lbm/s using

(a) Ideal gas law
(b) Van der Waals EOS
(c) Lee-Kessler chart

Problem 1.12 Comparing Gas Models

A rigid tank contains 1 kg of oxygen (O_2) at a pressure of 50 bars and 300 K. The gas is cooled until the temperature drops to 150 K. Determine the volume of the tank, m^3, and the final pressure (bars) using

(d) Ideal gas law
(e) RK EOS
(f) Lee-Kessler chart

Problem 1.13 Various Equations of State

Calculate the molar volume of propane at 400 K and 300 Bar using the equations of state given below. Compare your values with the experimental value of 0.094 L mol^{-1}.

(a) Ideal gas
(b) VW
(c) RK
(d) SRK

Problem 1.14 RK EOS

Determine the temperature at which the density of methane is 0.183 kg L^{-1} at a pressure of 500.0 bar. Assume that methane obeys the RK equation of state.

The most fundamental equation of state is the virial equation of state, which is expressed as a power series expansion in density or pressure about the ideal gas result. With a sufficient number of terms, the virial equation can give excellent predictions. However, it is usually truncated at the second term due to lack of higher-term coefficient data. In terms of reduced pressure and temperature, the truncated virial equation can be expressed as the following,

$$Z = 1 + B\frac{P_r}{T_r}$$

where the second virial coefficient is determined from the following correlation,

$$B = B_0 + \omega B_1$$

with

$$B_0 = 0.083 - \frac{0.422}{T_r^{1.6}} \qquad B_1 = 0.139 - \frac{0.172}{T_r^{4.2}}$$

Using the truncated virial equation, calculate the temperature of steam at a pressure of 5.0 bar when its specific volume is 0.4744 m^3 kg^{-1}.

Problem 1.15 SRK EOS

The critical temperature and pressure of fluid A are to be determined using a high-pressure laboratory flow device. The molecular weight of A is 28.01 kg kmol^{-1}. The mass flow rate and temperature of fluid A through the device are kept constant at 10.0 g s^{-1} and 300.0 K, respectively. At pressures of 500 bar and 1000 bar the volumetric flow rates of the fluid through the device are measured to be 24.12 cm^3 s^{-1} and 17.11 cm^3 s^{-1}, respectively. Determine the critical temperature and pressure of fluid A, which assumes that A obeys the SRK equation of state.

Appendix 1.1: Table of Ideal Gas Constants

https://en.wikipedia.org/wiki/Gas_constant, accessed on 10/14/2021

Values of R [1]	Units $(V\,P\,T^{-1}\,n^{-1})$
8.3144621(75) [2]	$J\,K^{-1}\,mol^{-1}$
8.31446	$V\,C\,K^{-1}\,mol^{-1}$
5.189×10^{19}	$eV\,K^{-1}\,mol^{-1}$
0.08205746(14)	$L\,atm\,K^{-1}\,mol^{-1}$
1.9872041(18) [3]	$cal\,K^{-1}\,mol^{-1}$
$1.9872041(18) \times 10^{-3}$	$kcal\,K^{-1}\,mol^{-1}$
$8.3144621(75) \times 10^7$	$erg\,K^{-1}\,mol^{-1}$
$8.3144621(75) \times 10^{-3}$	$amu\,(km/s)^2\,K^{-1}$
8.3144621(75)	$L\,kPa\,K^{-1}\,mol^{-1}$
$8.3144621(75) \times 10^3$	$cm^3\,kPa\,K^{-1}\,mol^{-1}$
8.3144621(75)	$m^3\,Pa\,K^{-1}\,mol^{-1}$
8.3144621(75)	$cm^3\,MPa\,K^{-1}\,mol^{-1}$
$8.3144621(75) \times 10^{-2}$	$m^3\,bar\,K^{-1}\,kg\text{-}mol^{-1}$
8.205736×10^{-5}	$m^3\,atm\,K^{-1}\,mol^{-1}$
8.205736×10^{-2}	$L\,atm\,K^{-1}\,kg\text{-}mol^{-1}$
82.05736	$cm^3\,atm\,K^{-1}\,mol^{-1}$
84.78402×10^{-6}	$m^3\,kgf/cm^2\,K^{-1}\,mol^{-1}$
$8.3144621(75) \times 10^{-2}$	$L\,bar\,K^{-1}\,mol^{-1}$
$62.36367(11) \times 10^{-3}$	$m^3\,mmHg\,K^{-1}\,mol^{-1}$
62.36367(11)	$L\,mmHg\,K^{-1}\,mol^{-1}$
62.36367(11)	$L\,Torr\,K^{-1}\,mol^{-1}$
6.132440(10)	$ft\,lbf\,K^{-1}\,g\text{-}mol^{-1}$
1545.34896(3)	$ft\,lbf\,R^{-1}\,lb\text{-}mol^{-1}$
10.73159(2)	$ft^3\,psi\,R^{-1}\,lb\text{-}mol^{-1}$
0.7302413(12)	$ft^3\,atm\,R^{-1}\,lb\text{-}mol^{-1}$
1.31443	$ft^3\,atm\,K^{-1}\,lb\text{-}mol^{-1}$
998.9701(17)	$ft^3\,mmHg\,K^{-1}\,lb\text{-}mol^{-1}$
1.986	$Btu\,lb\text{-}mol^{-1}\,R^{-1}$

Appendix 1.2: Lee/Kessler Chart (website)

The Lee Kessler Chart is provided on the companion website.

Appendix 1.3: Virial Equations of State (website)

An Excel Spreadsheet was developed to illustrate a particular concept and is given on the companion website.

Appendix 1.4: EOS (website)

An Excel Spreadsheet was developed to illustrate a particular concept and is given on the companion website.

References

1. Chan, R. (2010). *Chemistry* (10th ed.). McGraw-Hill.
2. Smith, J. M., Van Ness, H. C., & Abbott, M. M. (2001). *Introduction to chemical engineering thermodynamics* (6th ed.). Mc-Graw Hill.
3. Intermediate thermodynamics, Lecture 5: Equations of state, https://www.scribd.com/doc/171602285/Equation-of-State. Accessed 10/17/2020.
4. Powers, Department of Aerospace and Mechanical Engineering, Notre Dame, Lecture notes, https://www3.nd.edu/~powers/ame.20231/notes.pdf. Accessed 11/15/2018.
5. Redlich, O., & Kwong, J. N. S. (1948). On the thermodynamics of solutions. An equation of state. *114th Meeting of American Chemical Society, Division of Physical and Inorganic Chemistry*, Portland, Oregon, September 13 and 14.
6. Soave, G. (1972). Equilibrium constants from a modified Redlich-Kwong equation of state. *Chemical Engineering Science, 27*, 1197–1203.
7. Peng, D., & Robinson, D. B. (1976). A new two-constant equation of state. *Industrial Engineering Chemistry, Fundamental, 15*(1), 59–64.
8. Wikipedia entry for "Redlich-Kwong Equation of State", https://en.wikipedia.org/wiki/Redlich%E2%80%93Kwong_equation_of_state. Accessed 11/15/2018.
9. Wikipedia entry for "Acentric factor", https://en.wikipedia.org/wiki/Acentric_factor. Accessed 11/15/2018.

Chapter 2
Heat and Work

2.1 Preview

In this chapter, we provide a foundation for what heat and work are and discuss some related topics such as specific heat and polytropic processes. Unlike pressure, temperature, and specific volume (density), both heat and work are not states. What this means essentially is that both heat and work depend on the path taken. We will also see that quantities such as internal energy (e) and enthalpy (h) are independent of path and state functions.

Internal energy (e) is a measure of the thermodynamic energy of a system and is often associated with closed system and enthalpy (h) is a measure of the thermodynamic energy of a system to include changes in "Pv" (where $h = e + Pv$) and is often associated with an open system. Changes in either u or h can be determined by having a specific heat model along either a constant volume or constant pressure path; note though that internal energy and enthalpy are state functions and independent of path.

Work, for an ideal gas, can be determined based on defining the path in terms of a polytropic process and the definition of thermodynamic work.

Quantities such as work and heat for a substance that is either non-ideal or imperfect are generally determined using the thermodynamic tables or a cubic relationship between specific heat and temperature for the working fluid.

Quantities that are state functions and quantities that are not state functions are given in Table 2.1. Note changes in states will be represented as dx and changes in quantities that are not states will be represented as δx.

Electronic Supplementary Material: The online version of this chapter (https://doi.org/10.1007/978-3-030-87387-5_2) contains supplementary material, which is available to authorized users.

H. C. Foust III, *Thermodynamics, Gas Dynamics, and Combustion*,
https://doi.org/10.1007/978-3-030-87387-5_2

Table 2.1 Quantities that are states and not states

Quantity	State	Not States	Comments
Pressure (P)	X		Defined in Chap. 1
Temperature (T)	X		Defined in Chap. 1
Specific volume (ν)	X		Defined in Chap. 1
Internal energy (e)	X		Defined in this chapter
Enthalpy (h)	X		Defined in this chapter
Work (w)		X	Defined in this chapter
Heat (q)		X	Defined in this chapter
Entropy (s)	X		Defined in Chap. 4

2.2   Reversible and Irreversible Processes

A reversible process is defined as "a process that, having taken place, can be reversed, and in so doing, leaves no change in either the system or the surroundings." For a process to be reversible (ideal), the following conditions need to exist

- Quasi-equilibrium
- No friction
- Heat transfer due to an infinitesimal temperature difference only
- Unrestrained expansion does not occur

These conditions are discussed further below in terms of how certain factors can produce an irreversible process.

Four factors that create Irreversible Processes are

- Friction
- Unrestrained expansion
- Heat transfer through a finite temperature difference
- Mixing of two pure substances

Each factor is further discussed below.

Consider a block attached to a pulley where the other end has a weight and this system exists on an inclined ramp. If the weight is great enough, then the block is dragged up the ramp, but due to friction between the block and ramp, pulley and ropes, and internally within the pulley, the work required is greater than it would be without friction. We can then release weights and the block goes down the ramp. Due to the friction between the block and the ramp, the temperature of the ramp is now higher than the temperature of the surrounding and now heat is transferred to the surroundings. The surroundings are no longer back to their original state and this process is irreversible.

Imagine a system where a gas is on one side of a membrane and a vacuum is on the other side. Suddenly, the membrane bursts and the gas freely expands into the vacuum space. In order to restore the system to its original state, work would have to be done to the system and heat transferred out for the gas to go back to its original temperature. Again, the surroundings are no longer back to their original state this process is irreversible.

"Consider as a system a high-temperature body and a low-temperature body, and let heat be transferred from the high-temperature body to the low-temperature body. The only way in which the system can be restored to its initial state is to provide refrigeration, which requires work from the surroundings, and some heat transfer to the surroundings will also be necessary. Because of the heat transfer and the work, the surroundings are not restored to their original state, indicating that the process was irreversible [1]."

Imagine a system where Gas "A" is on one side of a membrane and Gas "B" is on the other side. Suddenly, the membrane bursts. It should be obvious that a certain amount of work and heat will be needed to restore this system to its original state and the surroundings will again not be as they were originally. This is yet another example of an irreversible process.

A reversible process is a process that can maximize the work extracted or minimized the work required and is dependent on the application. An example of a process (device) that extracts work is a turbine and we'd like to maximize the work extracted – we'll call this a "work-out" process; an example of a process (device) that requires work is a pump and we'd like to minimize the work required – we'll call this a "work-in" process.

The following example comes from [2].

For example, suppose we have a thermally insulated cylinder that holds an ideal gas, Fig. 2.1. The gas is contained by a thermally insulated massless piston with a stack of many small weights on top of it. Initially, the system is in mechanical and thermal equilibrium.

This is an example of a "work-out" process.

Our definition of thermodynamic work is

$$\delta w = Pdv \tag{2.1}$$

Fig. 2.1 Frictionless, weightless, and perfectly insulated system [2]

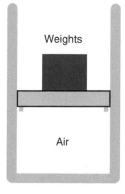

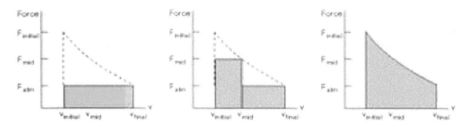

Fig. 2.2 Work related to pathways "*a*", "*b*," and "*c*" [2]

which states the infinitesimal work is equal to the pressure and infinitesimal change in specific volume; because work is dependent on path we use "δ" instead of "*d*."

Consider three different pathways to get from $\{P_1, \nu\}$ to $\{P_2, 4\nu\}$

(a) All of the weights are removed from the piston instantaneously and the gas expands until its volume is increased by a factor of four (a free expansion).
(b) Half of the weight is removed from the piston instantaneously, the system is allowed to double in volume, and then the remaining half of the weight is instantaneously removed from the piston and the gas is allowed to expand until its volume is again doubled.
(c) Each small weight is removed from the piston one at a time, so that the pressure inside the cylinder is always in equilibrium with the weight on top of the piston. When the last weight is removed, the volume has increased by a factor of four.

The resulting graphs in terms of work on a force (P) versus ν graph are given below.

It is obvious from Eq. 2.1 and Fig. 2.2 that the third path maximizes work (maximizes area under the curve), which is the reversible path. This path can equally be termed an isentropic path and we'll discuss this further in Chap. 4.

2.3   Specific Heat, Internal Energy, and Enthalpy

Specific heat, C, of a substance has the following working definition

$$C \approx \frac{1}{m} \frac{\delta Q}{dT} \tag{2.2}$$

It's a measure of the heat released for a given temperature rise and for a unit mass. In a sense, the BTU, British thermal unit, is a measure of specific heat. A 60-degree BTU is defined as the amount of energy released to raise one pound (mass) of water from 59.5 °F to 60.5 °F under atmospheric pressure. The specific heat of water is measured at a particular temperature because the specific heat of water is dependent on temperature.

For liquids and solids, specific heat is independent of path and often considered constant. But for gases, it is dependent on path and so we talk about specific heat (constant pressure) and specific heat (constant volume).

The measure of thermodynamic energy for a closed system is in terms of internal energy (e) where e is related to specific heat along a constant volume path and given as

$$\left(\frac{\partial e}{\partial T}\right)_v = c_v \tag{2.3}$$

And the measure of the thermodynamic energy for an open system is in terms of enthalpy (h) where h is related to specific heat along a constant pressure path and given as

$$\left(\frac{\partial h}{\partial T}\right)_P = c_P \tag{2.4}$$

Gas molecules can store energy in three modes – translational, rotational, and vibrational. For monatomic gas molecules, such as argon, neon, and helium, the only mode of energy is translational (kinetic) energy and in each principal direction

$$E_j = \frac{kNT}{2} \tag{2.5}$$

where k is the Boltzmann constant (specific energy), N is the number of molecules, T is temperature, and j is a principal axis (x, y, z).

And for the three principal directions the total energy is

$$E_{total} = \frac{3kNT}{2} \tag{2.6}$$

And for a diatomic molecule it is

$$E_{total} = \frac{5kNT}{2} \tag{2.7}$$

And more generally for a molecule with n'' atoms per molecule and on a molar basis [4]

$$\frac{E_{total}}{T} = Ak\left(n'' + \frac{1}{2}\right) = R\left(n'' + \frac{1}{2}\right) \tag{2.8}$$

where A is Avogadro's number and represents 1 mole of molecules.

In terms of c_v, the relationship is

$$\frac{c_v}{R} = n'' + \frac{1}{2}$$
(2.9)

In terms of c_p, the relationship is

$$\frac{c_p}{R} = \frac{c_v}{R} + \frac{R}{R} = n'' + \frac{3}{2}$$
(2.10)

where for an ideal gas $c_p - c_v = R$

While for γ, which is defined as $\gamma = \dfrac{c_p}{c_v}$, the relationship is

$$\gamma = \frac{n'' + \dfrac{3}{2}}{n'' + \dfrac{1}{2}}$$
(2.11)

How well does this theory work?

Three classes of molecules to be analyzed are:

Monatomic – neon, argon, helium (noble gases)
Diatomic – hydrogen, oxygen, nitrogen
Triatomic – water, carbon dioxide, nitrogen oxide

Discrepancies will be discussed.
A regressional model c_P for a real gas is

$$c_p = \beta_0 + \beta_1 \theta + \beta_2 \theta^2 + \beta_3 \theta^3$$
(2.12)

where $\theta = \dfrac{T}{1000}$ and $\{\beta_0, \beta_1, \beta_2, \beta_3\}$ are coefficients particular to the substance.

Equation 2.12 will be utilized to make some general observations about specific heat and number of atoms within the molecule.

We see for the monatomic gases (see Fig. 2.3) that the theory and practice are in near perfect agreement and that for noble gases $c_p/R(g)$ versus temperature is constant; this is due to the fact that a monatomic gas only has translational energy, which is the basis to the theory given above. For a diatomic gas (see Fig. 2.4), the theory tends to work at lower temperatures and becomes less and less accurate at higher temperatures. We see a similar trend for the triatomic gases (see Fig. 2.5). The theory tends to work well for molecules with fewer atoms and at lower temperatures.

Powers [5] discusses the modes of energy versus temperature for a diatomic ideal gas and shows

• At very low temperatures – the primary mode of energy is translational

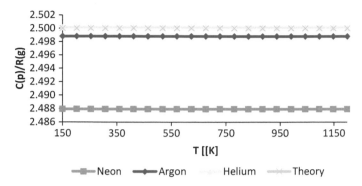

Fig. 2.3 $c_p/R(g)$ versus T for monatomic gases

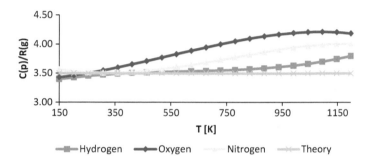

Fig. 2.4 $c_p/R(g)$ versus T for diatomic gases

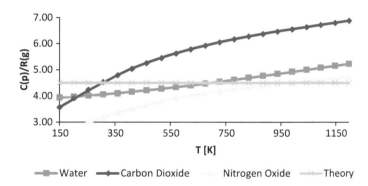

Fig. 2.5 $c_p/R(g)$ versus T for triatomic gases

- At low to normal temperatures (STP) – the primary modes of energy are translational + rotational
- At higher temperatures – all three modes of energy are important

Powers [5] suggests four main methods to calculate enthalpy for an ideal gas

1. Assume a constant c_p at 298 K (often the least accurate method)
2. Assume a constant c_p at some intermediate temperature along the thermodynamic path (often used in engineering analysis)
3. Integrate the differential equation given as Eqs. 2.4 and 2.12 (more accurate)
4. Estimation from thermodynamic tables (most accurate method)

An example is given below to illustrate the four main methods given above, which would equally apply to internal energy.

Example 2.1
Determine the enthalpy change for carbon dioxide that runs from 300 K to 900 K using the four methods provided above.

Given

Substance		CO_2
n''		3
$R(g)$	[kJ/kg-K]	0.1889
$c_p/R(g)$		4.5
c_p	[kJ/kg-K]	0.85005

The betas associated with the regressional model for a real gas are

Beta(0)	0.45
Beta(1)	1.67
Beta(2)	−1.27
Beta(3)	0.39

where c_p has been determined using the regression model at 300 K and 900 K with the average value being 1.03 kJ/kg-K (shown below).

T	c_p
[K]	[kJ/kg-K]
300	0.85
900	1.21
	c_p-bar = 1.03

The integral version of Eq. 2.4 is

$$\int \left[\beta_0 + \beta_1 \frac{T}{1000} + \beta_2 \frac{T^2}{1000^2} + \beta_3 \frac{T^3}{1000^3} \right] dT = \beta_0 T + \frac{\beta_1}{1000} \frac{T^2}{2} + \frac{\beta_2}{1000^2} \frac{T^3}{3} + \frac{\beta_4}{1000^3} \frac{T^4}{4} \quad (2.13)$$

And the solutions at the two endpoints are

	T	$h(T)$
Integral($T1$)	300	199.5
Integral($T2$)	900	836.7

The values of enthalpy at 300 K and 900 K from the thermodynamic tables are

T	h
[K]	[kJ/kg]
300	214.4
900	849.7

The solutions and relative error from the four methods are given below.

Solution methods	Δh [kJ/kg]	RE
1. Using fixed c_p determine dh	510.0	19.7%
2. Using average c_p determine dh	616.8	2.9%
3. Integrating definition of c_p determine dh	637.2	0.3%
4. Using thermodynamic tables determine dh	635.3	

2.4 Polytropic Processes and Work

Definitions:

Control Volume – a boundary around the system that allows us to distinguish the system from the environment and account for forces/energy/mass entering or leaving the system.

Isolated System – mass nor energy passes through the control volume.

Closed System – energy passes, but mass does not pass through the control volume.

Open System – energy and mass pass through the control volume.

Process – a change in states.

Isobaric Process – a process where pressure is constant from state "1" to "2."

Isothermal Process – a process where the temperature is constant from state "1" to "2."

Isentropic Process – a process where there is no heat exchange between the system and environment from state "1" to "2."

Isometric Process – a process where volume is constant from state "1" to "2."

Cycle – a series of processes where the initial and final states coincide.

A polytropic process is defined as

$$PV^n = \text{constant} = K \tag{2.14}$$

where the index indicates the type of process (see Table 2.2).

Table 2.2 Polytropic processes

n	Process	Comments
0	Isobaric	Constant pressure
1	Isothermal	Constant temperature
∞	Isometric	Constant volume
γ	Isentropic	Adiabatic and reversible

Table 2.3 Relationships for different processes [3]

Process	N	Work	State Equation	ΔE	Heat
Constant Pressure (Isobaric)	0	$P(V_2 - V_1)$	$\dfrac{V_1}{T_1} = \dfrac{V_2}{T_2}$	$mc_v(T_2 - T_1)$	$mc_p(T_2 - T_1)$
Constant Volume (Isometric)	∞	0	$\dfrac{P_1}{T_1} = \dfrac{P_2}{T_2}$	$mc_v(T_2 - T_1)$	$mc_v(T_2 - T_1)$
Constant Temperature (Isothermal)	1	$P_1 V_1 \ln\left(\dfrac{V_2}{V_1}\right)$	$P_1 V_1 = P_2 V_2$	0	$P_1 V_1 \ln\left(\dfrac{V_2}{V_1}\right)$
Polytropic	All values	$\dfrac{P_1 V_1 - P_2 V_2}{n-1}$	$\dfrac{T_2}{T_1} = \left(\dfrac{V_2}{V_1}\right)^{n-1}$ $\dfrac{P_2}{P_1} = \left(\dfrac{T_2}{T_1}\right)^{n-\frac{1}{n}}$ $PV^n = K$	$mc_v(T_2 - T_1)$	$Q = W + \Delta U$
Adiabatic	γ	$\dfrac{P_1 V_1 - P_2 V_2}{\gamma - 1}$	$\dfrac{T_2}{T_1} = \left(\dfrac{V_2}{V_1}\right)^{\gamma-1}$ $\dfrac{P_2}{P_1} = \left(\dfrac{T_2}{T_1}\right)^{\gamma-\frac{1}{\gamma}}$ $PV^\gamma = K$	$mc_v(T_2 - T_1)$	0

Using Eqs. 2.1 and 2.14, we can determine the work associated with a given process. There will be two cases we'll explore. The first case is when $n \neq 1$ and when $n = 1$.

Case one ($n \neq 1$)

$$\int_1^2 \delta W = \int_1^2 \frac{K}{V^n} dV = K \left.\frac{V^{1-n}}{1-n}\right|_1^2 = \frac{P_2 V_2^n V_2^{1-n} - P_1 V_1^n V_1^{1-n}}{1-n} \tag{2.15}$$

where

$$K = P_1 V_1 = P_2 V_2 \tag{2.16}$$

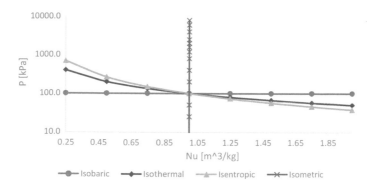

Fig. 2.6 ν versus P for polytrophic processes

and

$$W = \frac{P_2 V_2 - P_1 V_1}{1-n} \tag{2.17}$$

Case two ($n = 1$)

$$W = P_1 V_1 \ln \frac{V_2}{V_1} \tag{2.18}$$

Table 2.3 and Fig. 2.6 below provide a relationship between states, work associated with various processes, internal energy change associated with various processes, and heat associated with various processes. Note – Table 2.3 is only valid for an ideal, perfect gas. This defines a gas that obeys the ideal gas law and has a specific heat that is independent of temperature.

2.5 Examples and Problems

2.5.1 Examples

Determining Δh for Water

Example 2.2 Determine enthalpy change for water at 200 kPa where the temperature runs from 150 °C to 500 °C. Solve by using a constant c_p.

$$\Delta h \approx c_p \Delta T = 1.872 * [500 - 150] = 655.2 \, \text{kJ} / \text{kg}$$

Example 2.3 Determine enthalpy change for water at 200 kPa where the temperature runs from 150 °C to 500 °C. Solve by using an average c_p.

$$\Delta h \approx \frac{c_p^{150} + c_p^{500}}{2} \Delta T = \frac{1.92 + 2.13}{2}[500 - 150] = 709.7\,\text{kJ}\,/\,\text{kg}$$

where each c_p is determined from the cubic relationship between T and c_p.

Example 2.4 Determine enthalpy change for water at 200 kPa where the temperature runs from 150 °C to 500 °C. Solve by using the cubic relationship between T and c_p.

$$\Delta h \approx \int_{150+273}^{500+273} \left\{ \beta_0 + \beta_1\theta + \beta_2\theta^2 + \beta_3\theta^3 \right\} dT = \left[\beta_0 T + \frac{\beta_1}{1000}\frac{T^2}{2} + \frac{\beta_2}{1000^2}\frac{T^3}{3} + \frac{\beta_3}{1000^3}\frac{T^4}{4} \right]_{150+273}^{500+273}$$

$$\Delta h = [h_{500} + h] - [h_{150} + h] = 1488.0 - 779.9 = 708.1\,\text{kJ/kg}$$

Example 2.5 Determine enthalpy change for water at 200 kPa where the temperature runs from 150 °C to 500 °C. Solve by using the thermodynamic tables.

$$\Delta h = h_{500} - h_{150} = 3487.0 - 2676.8 = 718.2\,\text{kJ/kg}$$

The results for Examples 2.2 through 2.5 with relative error are given in the table below.

Determining Δh for carbon dioxide

Method	Δh [kJ/kg]	RE
Constant c_p	655.2	8.78%
Average c_p	709.7	1.19%
Cubic relationship	708.1	1.41%
Table	718.2	

Example 2.6 Determine enthalpy change for carbon dioxide at 400 kPa where the temperature runs from −40 °C to 60 °C. Solve by using a constant c_p.

$$\Delta h \approx c_p \Delta T = .842 * \left[60 - (-40) \right] = 84.2\,\text{kJ/kg}$$

Example 2.7 Determine enthalpy change for carbon dioxide at 400 kPa where the temperature runs from −40 °C to 60 °C. Solve by using an average c_p.

$$\Delta h \approx \frac{c_p^{-40} + c_p^{60}}{2} \Delta T = \frac{0.78 + 0.88}{2} 100 = 82.74\,\text{kJ/kg}$$

where each c_p is determined from the cubic relationship between T and c_p

Example 2.8 Determine enthalpy change for carbon dioxide at 400 kPa where the temperature runs from -40 °C to 60 °C. Solve by using the cubic relationship between T and c_p.

$$\Delta h \approx \int_{40}^{60} \left\{ \beta_0 + \beta_1\theta + \beta_2\theta^2 + \beta_3\theta^3 \right\} dT = \left[\beta_0 T + \frac{\beta_1}{1000}\frac{T^2}{2} + \frac{\beta_2}{1000^2}\frac{T^3}{3} + \frac{\beta_3}{1000^3}\frac{T^4}{4} \right]_{-40+273}^{60+273}$$

$$\Delta h = \left[h_{60} + h \right] - \left[h_{-40} + h \right] = 228.0 - 145.1 = 82.9 \, \text{kJ/kg}$$

Example 2.9 Determine enthalpy change for carbon dioxide at 400 kPa where the temperature runs from -40 °C to 60 °C. Solve by using the thermodynamic tables.

$$\Delta h = h_{60} - h_{-40} = 421.1 - 334.46 = 86.64 \, \text{kJ/kg}$$

The results for Examples 2.6 through 2.9 with relative error are given in the table below.

Method	Δh [kJ/kg]	RE
Constant c_p	84.20	2.82%
Average c_P	82.74	4.50%
Cubic relationship	82.90	4.32%
Table	86.64	

Example 2.10 Determine the work associated with a general polytropic process ($PV^n = K$)

(a) P_1 as 1 atm, V_1 as 10 ft³; P_2 as 10 atm and n equal to 1.5
(b) P_1 as 10 atm, V_1 as 10 ft³; P_2 as 1 atm and n equal to 1.5

For (a), you will find the work is negative and the process is a contraction; for (b), you will find the work is positive and the process is an expansion. The sign associate with work (or heat) determines whether the work (or heat) is directed toward the system or toward the surroundings:

(a) Negative Work → Work on System
(b) Positive Work → Work on Surroundings

The solutions for (a) and (b) are given below.

$P(1)$	[atm]	1	$P(1)$	[atm]	10
$V(1)$	[ft³]	10	$V(1)$	[ft³]	10
$P(2)$	[atm]	10	$P(2)$	[atm]	1
$V(2)$	[ft³]	2.2	$V(2)$	[ft³]	46.4
n		1.5	n		1.5
K		31.6	K		316.2
K		31.6	K		316.2
$W(1,2)$	[Atm-ft³]	-23.1	$W(1,2)$	[Atm-ft³]	107.2

Example 2.11 Determine the work and heat for a series of processes that form a cycle, which establishes first Law for a cycle (*the sum of the work equal the sum of the heat*)

Given

Substance		Air
Mass	[lbms]	1
c_p	[lbf-ft/lbm-R]	186.7608
c_v	[lbf-ft/lbm-R]	133.0671
R	[lbf-ft/lbm-R]	53.34
Gamma		1.4

Complete the following table by using Table 2.1 and the ideal gas law.

State	Process	P [psia]	T [Rankine]	V [ft³]	W [ft-lbf]	Q [ft-lbf]	n
1		100		10			
	Isobaric						0
2		100		1			
	Isometric						Infinity
3		1000		1			
	Isothermal						1
1		100		10			

The solution is given below and we see the sum of the work does equal the sum of the heat (more/less).

State	Process	P [psia]	T [Rankine]	V [ft³]	W [ft-lbf]	Q [ft-lbf]
1		100	2700	10		
	Isobaric				−129,600	−453,772
2		100	270	1		
	Isometric				0	323,313
3		1000	2700	1		
	Isothermal				331,572	331,572
1		100	2700	10		
					Sum of work is	Sum of heat is
					201,972	**201,113**

2.5.2   Problems

Problem 2.1 Consider a process that involves superheated CO_2 that runs from 0 °C to 100 °C at a constant pressure of 400 kPa. Assuming c_p is constant, determine the change in enthalpy.

Problem 2.2 Consider a process that involves superheated CO_2 that runs from 0 °C to 100 °C at a constant pressure of 400 kPa. Assuming a cubic model for c_p, determine the change in enthalpy.

Problem 2.3 Consider a process that involves superheated CO_2 that runs from 0 °C to 100 °C at a constant pressure of 1000 kPa. Using the thermodynamic tables determine the change in enthalpy.

Problem 2.4 Consider a process that involves superheated CO_2 that runs from 0 °C to 100 °C at a constant pressure of 1000 kPa. Assuming c_p is constant, determine the change in enthalpy.

Problem 2.5 Consider a process that involves superheated CO_2 that runs from 0 °C to 100 °C at a constant pressure of 1000 kPa. Assuming a cubic model for c_p, determine the change in enthalpy.

Problem 2.6 Consider a process that involves superheated CO_2 that runs from 0 °C to 100 °C at a constant pressure of 1000 kPa. Using the thermodynamic tables determine the change in enthalpy.

Problem 2.7 Determine the work and heat for a series of processes that form a cycle, which established first Law for a cycle (Air). Assume the mass of the system is 1 kg.

State	Process	P [kPa]	T [K]	V [m³]	W [kJ]	Q [kJ]	n
1		100		5.18			
	Isobaric						0
2		100		1			
	Isometric						Infinity
3		1000		1			
	Isentropic						1.4
1		100		5.18			

Problem 2.8 Determine the work and heat for a series of processes that form a cycle, which established first Law for a cycle. The substance is water and the states are

State 1 – 20 kPa and $X = 0\%$
State 2 – 1 MPa and $v_1 = v_2$
State 3 – 1 MPa and 350 °C
State 4 – 20 kPa and $s_3 = s_4$

Problem 2.9 A piston-cylinder device contains one mole of an ideal gas initially at 30 °C and 1 bar. The gas undergoes the following reversible process: compressed adiabatically to 5 bar, then cooled at a constant pressure of 5 bar to 30 °C, and finally expanded isothermally to its original state. Assuming that for this gas

$$c_P = \frac{7}{2}R$$, calculate Q, W, ΔU, and ΔH for each step of the process and for the entire cycle.

Problem 2.10 One mole of air, initially at 25.0 °C and 100 kPa, undergoes the following mechanically reversible changes: it expands isothermally to a pressure such that when it is cooled at constant volume to 50 °C its final pressure is 3 bar. Assuming that for this gas $c_P = \frac{7}{2}R$, calculate Q, W, ΔU, and ΔH for each step of the process and for the entire cycle.

Problem 2.11 A piston-cylinder device contains 0.150 kg of air at a temperature of 300.0 K. The initial volume of air is 100.0 L. The air is then compressed isothermally that requires 20.0 kJ of work. If air is modeled as a RK gas, compute the final volume and pressure of air in the device.

Problem 2.12 30.0 g s^{-1} of nitrogen gas is flowing through a tube wrapped with a resistance heater. The gas enters the tube at 300.0 K and 100.0 kPa. The resistance heater is turned on and continuously passes a current of 50.0 A from a 240-V source, thereby isobarically heating the gas. Assuming that nitrogen behaves as an ideal gas with a temperature-dependent heat capacity given as a cubic polynomial, determine its exit temperature.

Problem 2.13 1.0 kg of methane initially at 330 K and 50 bar is isothermally compressed to a final pressure that requires 220 kJ of work. If methane behaves as a SRK gas, determine this final pressure.

Appendix 2.1: Cycle Worksheet (Fig. 2.7)

Knowns
Air
$m = 1$ lbm
$R_g = 53.34$ ft-lbf per lbm-R
$c_v = 0.171$ BTU per lbm-R
$c_p = 0.240$ BTU per lbm-R

Fig. 2.7 Cycle on *P* versus *V* graph

State	Process	*P* [psia]	*T* [Rankine]	*V* [ft³]	*W* [ft-lbf]	*Q* [ft-lbf]
1		100		10		
	Isobaric					
2		100		1		
	Isometric					
3		200		1		
	Isothermal					
1						
					Sum =	Sum =

Process

1. Find all states {*P*,*T*,*V*} using relationships for polytropic processes and ideal gas law (make sure to put *T* and *P* in absolute scale)
2. Find *Q* and *W* associated with each process
3. Sum them up and check!

Isobaric Process

Isometric Process

Isothermal

References

1. Borgnakke, & Sonntag. (2009). *Fundamentals of thermodynamics* (7th ed.). Wiley.
2. http://web.mit.edu/16.unified/www/FALL/thermodynamics/notes/node34.html. Accessed 1 Jan 2019.
3. Granet, & Bluestein. (2000). *Thermodynamics and heat power* (6th ed.). Prentice Hall.
4. Mayhew, J. W. (2017). A new perspective for kinetic theory and heat capacity. *Progress in Physics, 13*, 166–173.
5. Powers, J. (2020). *Lecture notes for thermodynamics, AME 20231*. Notre Dame University.

Chapter 3
First Law of Thermodynamics

3.1 Preview

When a system is steady state in terms of mass, which it almost always is, the amount of mass entering is equal to the amount of mass leaving. For a closed system, this accounting is expressed [1, 3] as

$$\delta q - \delta w = \Delta e \qquad (3.1)$$

where e is the internal energy and a measure of the thermodynamic energy of a closed system.

And for an open system where both energy and mass transfer into and out of the control volume [1, 3]

$$\delta q - \delta w = \Delta(\mathrm{KE}) + \Delta(\mathrm{PE}) + \Delta(h) \qquad (3.2)$$

where KE is the kinetic energy, PE is the potential energy, and h is the enthalpy and a measure of the thermodynamic energy of an open system.

The devil is in the details!

3.2 Linear Interpolation

How do we determine the value of a state associated with points between entries on a thermodynamic table? Here's an example.

Given

Electronic Supplementary Material: The online version of this chapter (https://doi.org/10.1007/978-3-030-87387-5_3) contains supplementary material, which is available to authorized users.

© Springer Nature Switzerland AG 2022
H. C. Foust III, *Thermodynamics, Gas Dynamics, and Combustion*,
https://doi.org/10.1007/978-3-030-87387-5_3

For water and at 200 kPa, the following values for u (see Table 3.1). What is the value of u at the same pressure, but temperature equal to 180 °C?

This can be determined by putting the data into Excel, graphing the data, and getting the equation of a line as shown in Fig. 3.1.

Similarly, we can solve for an equation that allows us to solve for y.

$$y_1 = \frac{\Delta y}{\Delta x} x_1 + b \tag{3.3}$$

And

$$\frac{x_2 - x_1}{x_2 - x_1} y_1 - \frac{y_2 - y_1}{x_2 - x_1} x_1 = b = \frac{x_2 y_1 - x_1 y_2}{x_2 - x_1} \tag{3.4}$$

And

$$y = \frac{y_2 - y_1}{x_2 - x_1} x + \frac{x_2 y_1 - x_1 y_2}{x_2 - x_1} = \frac{y_2 - y_1}{x_2 - x_1} (x - x_1) + y_1 \tag{3.5}$$

Incidentally, the answer is $u(180) = 1.5504(180) + 2344.3 = 2623.37$ kJ/kg.

Table 3.1 Temperature versus u

Temperature [°C]	u [KJ/Kg]
150	2576.87
200	2654.39

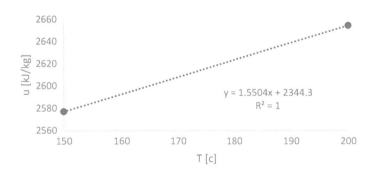

Fig. 3.1 T versus u

3.3   Using Thermodynamic Tables or NIST Chemistry Webbook

Before you can start to use the tables, you need to understand "what is the phase of the substance?", which was previously discussed in Sect. 1.4.

Please note a pure substance is a substance made of only one kind of molecule such as O_2, H_2O, etc. The one exception will be air, which is made of many molecules, but we can ignore this fact for the discussion to follow.

Pure substances (the only kind we deal with in Chaps. 1, 2, 3, 4, and 5) typically have the following ν versus T and ν versus P behavior (see Figs. 3.2 and 3.3). Let's use water and assume we're trying to determine the state of water with the following conditions {100 °C, 200 kPa}. We know water boils at 100 °C and 101.3 kPa. Go to Fig. 3.3.

We see that going toward the left the pressure increases and we're now in the sub-cooled region. This observation is confirmed in two ways. We can go to the "Saturated Water" tables and see that for a pressure of 200 kPa the saturated temperature is 120.23 °C, which is higher than 100 °C. We can also go to the "Superheated Vapor Water" table and see for 200 kPa the lowest temperature is 120.23 °C.

Under the mixture dome, the substance is at 100 °C and 101.3 kPa. We don't know where we are under the mixture dome unless we specify quality (X), which is defined as

$$X = \frac{m_g}{m_g + m_l} \tag{3.6}$$

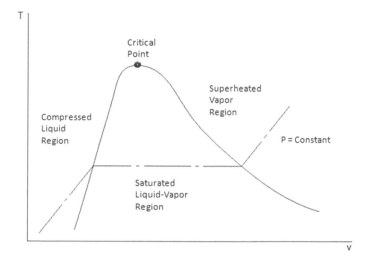

Fig. 3.2 T versus V

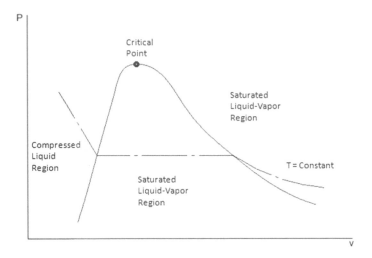

Fig. 3.3 *P* versus *V*

Fig. 3.4 Data type from
NIST Chemistry Webbook

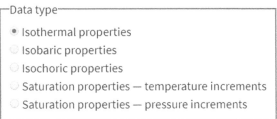

where m_g is the mass of gas and m_l is the mass of liquid. Both gases and liquids are considered fluids.

Also note that the state of a mixture (i.e., specific volume) is a combination of that state as a liquid and a gas and understood through the following equation

$$V_{mix} = v_l + X\left(v_g - v_l\right) \tag{3.7}$$

Always start with the saturated table for the substance this guides you to the correct table. Additionally, water has compressed liquid tables, but most substances do not. We just use the saturated properties of the liquid at the same temperature. States such as specific volume for a liquid are affected by temperature, but not generally by pressure.

Another means to gather thermodynamic data is the National Institute of Standards and Testing (NIST), which runs a website called the NIST Chemistry Webbook [2]. It is required to provide the substance, units and in Fig. 3.4, the type

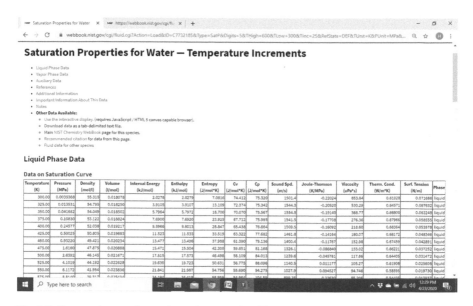

Saturation Properties for Water — Temperature Increments

- Liquid Phase Data
- Vapor Phase Data
- Auxiliary Data
- References
- Additional Information
- Important Information About This Data
- Notes
- **Other Data Available:**
  - Use the interactive display. (requires JavaScript / HTML 5 canvas capable browser).
  - Download data as a tab-delimited text file.
  - Main NIST Chemistry WebBook page for this species.
  - Recommended citation for data from this page.
  - Fluid data for other species

Liquid Phase Data

Data on Saturation Curve

Temperature (K)	Pressure (MPa)	Density (mol/l)	Volume (l/mol)	Internal Energy (kJ/mol)	Enthalpy (kJ/mol)	Entropy (J/mol*K)	Cv (J/mol*K)	Cp (J/mol*K)	Sound Spd. (m/s)	Joule-Thomson (K/MPa)	Viscosity (uPa*s)	Therm. Cond. (W/m*K)	Surf. Tension (N/m)	Phase
300.00	0.0035388	55.319	0.018078	2.0278	2.0279	7.0816	74.412	75.320	1501.4	-0.22024	853.84	0.61028	0.071686	liquid
325.00	0.013531	54.795	0.018250	3.9105	3.9107	13.109	72.374	75.342	1544.3	-0.20520	530.29	0.64571	0.067632	liquid
350.00	0.041682	54.049	0.018502	5.7964	5.7972	18.700	70.070	75.567	1554.8	-0.19140	388.77	0.66800	0.063248	liquid
375.00	0.10830	53.122	0.018824	7.6900	7.6920	23.925	67.712	75.985	1541.5	-0.17708	276.36	0.87966	0.058555	liquid
400.00	0.24577	52.038	0.019217	9.9966	9.6013	28.847	65.438	76.864	1508.5	-0.16092	218.60	0.68364	0.053578	liquid
425.00	0.50025	50.805	0.019683	11.523	11.533	33.519	63.322	77.682	1461.8	-0.14164	180.07	0.68172	0.048346	liquid
450.00	0.95220	49.421	0.020234	13.477	13.496	37.988	61.390	79.136	1400.4	-0.11767	152.98	0.67459	0.042891	liquid
475.00	1.6160	47.875	0.020888	15.471	15.504	42.300	59.651	81.168	1326.1	-0.086840	133.02	0.66221	0.037252	liquid
500.00	2.6392	46.145	0.021671	17.515	17.573	46.498	58.109	84.013	1239.6	-0.045781	117.86	0.64405	0.031472	liquid
525.00	4.1019	44.192	0.022628	19.630	19.723	50.631	56.775	88.096	1140.5	0.011177	105.27	0.61908	0.025608	liquid
550.00	6.1172	41.954	0.023836	21.841	21.987	54.756	55.690	94.275	1027.9	0.094527	94.746	0.58595	0.019730	liquid

Fig. 3.5 Saturated water thermodynamic properties [2]

of data to request must be specified. Once this is done, the results are given either as a graph of one quantity versus another or as a table such as given in Fig. 3.5.

3.4   Conservation of Mass

There are three conservation principles associated with mechanics that we can utilize to solve problems associated with fluid mechanics (open systems). These principles are

- Conservation of mass
- Conservation of momentum
- Conservation of energy (first law of thermodynamics)

Conservation of mass will be discussed in this section and conservation of energy will be discussed in Sect. 3.5.

Essentially conservation of mass states "mass can neither be created, nor destroyed." With this working assumption,

$$\frac{DM_{\text{syst}}}{Dt} = 0 \tag{3.8}$$

and

$$M_{\text{syst}} = \int_{\text{sys}} \rho dV \tag{3.9}$$

Such that

$$\frac{D}{Dt} \int_{\text{sys}} \rho dV = \frac{\partial}{\partial t} \int_{\text{cv}} \rho dV + \int_{\text{cs}} \rho \vec{u} \cdot \hat{n} dA = 0 \tag{3.10}$$

where the first term on the right-hand side accounts for changes in mass of the system (control volume, cv) and the second term accounts for mass entering or leaving the control surface (cs).

When the density of the system is constant, the first term and second term simplify

$$\frac{dm_{\text{sys}}}{dt} + \rho \int_{\text{cs}} \vec{u} \cdot \hat{n} dA = 0 \tag{3.11}$$

and

$$\frac{dV}{dt} + \int_{\text{cs}} \vec{u} \cdot \hat{n} dA = 0 \tag{3.12}$$

Normally, the velocity distribution represented by $\vec{u}$ (see Fig. 1) and $\hat{n}$ which is a normal vector associated with dA (itself a vector) are at an angle of either 0 or 180 degrees and such

$$\vec{u} \cdot \hat{n} = u \quad \text{or} \quad \vec{u} \cdot \hat{n} = -u \tag{3.13}$$

Such that

$$\frac{dm_{\text{sys}}}{dt} + \rho \left\{ \sum \dot{m}_{\text{out}} - \sum \dot{m}_{\text{in}} \right\} = 0 \tag{3.14}$$

where $\dot{m}$ is the mass rate into or out of the control surface (cs).

And when the system is both incompressible (density is the same throughout the control volume) and steady state in mass then

$$\sum Q_{\text{in}} = \sum Q_{\text{out}} \tag{3.15}$$

where Q is volumetric discharge and defined as

$$Q = \frac{dV}{dt} \tag{3.16}$$

When density is change, but the mass of the system is steady state then

$$\sum \dot{m}_{in} = \sum \dot{m}_{out} \qquad (3.17)$$

3.5 First Law

The first law of thermodynamics states that "energy can be neither created, nor destroyed." We naturally exclude special situations such as reactions within the sun or a nuclear reactor where particles are moving appreciable toward the speed of light.

3.5.1 Control Volumes

We use the concept of a control volume to delineate a system from the environment. There are three systems we'll consider: isolated system, closed system, and open system. An isolated system does not allow the transfer of mass or energy to the environment; in a closed system, energy is transferred to the environment, but not energy; in an open system, both energy and mass can be transferred with the environment.

3.5.2 Closed Systems

The first law can be expressed as "the heat added to a system minus the work extracted from a system equal the change in internal energy." This naturally leads to questions such as "what is internal energy"?, "what is work"?, and "what is heat"? We'll define work and heat below; for now, we'll define internal energy as "the thermal energy stored within a system" where there are no changes in the system density (specific volume). This is usually expressed as the container is closed and rigid.

The first law applied to a closed system is given [1, 3] as

$$\delta q - \delta w = de \qquad (3.18)$$

And can be represented graphically in Fig. 3.6.

There is a sign convention on heat (q) and work (w), which is given in Table 3.2.

Fig. 3.6 First law, closed system

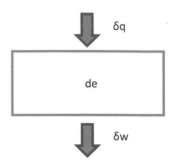

Table 3.2 Heat and work sign convention

When heat is toward system → the sign on δq is (+)
When heat is away from the system → the sign on δq is (−)
When work is toward system → the sign on − δw is (−)
When work is away from the system → the sign on − δw is (+)

3.6   First Law, Open System

The most general form of the equation for first law, open system that will be utilized [1, 3] is

$$\delta q - \delta w = \Delta KE + \Delta PE + \Delta h = E_{in} - E_{out} \tag{3.19}$$

Or

$$\delta \dot{q} - \delta \dot{w} = \dot{m}_{in} \left\{ h_{in} + \frac{v_{in}^2}{2} + z_{in} \right\} - \dot{m}_{out} \left\{ h_{out} + \frac{v_{out}^2}{2} + z_{out} \right\} = \dot{E}_{in} - \dot{E}_{out} \tag{3.20}$$

where ΔKE is the change in kinetic energy, ΔPE is the change in potential energy, Δh is the change in enthalpy and enthalpy is a measure of the total thermodynamic energy of a system and defined as

$$h = u + Pv \tag{3.21}$$

And can be represented graphically in Fig. 3.7.

Fig. 3.7 First law, open system

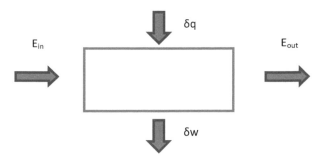

Fig. 3.8 Control volume for a pump

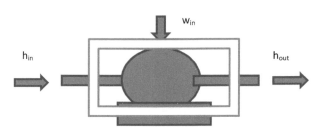

3.7   Engineering Devices

An energy device is some constructed machine that either converts heat to work, work to heat or affects one of the forms of energy shown in Eq. 3.22. Examples of energy devices are pumps, turbines, compressors, and heat exchangers. The governing equation (first law) for a particular device is determined from Eq. 3.22, but usually some simplifying assumptions can be made. An example of these ideas can be applied to a pump.

Example 3.1 First Law, Pump
We can assume the pump is adiabatic ($\delta q = 0$), the inner diameter on intake and discharge are equal ($\Delta KE = 0$), and the pipes are at the same elevation ($\Delta PE = 0$). Therefore, Eq. 3.19 takes the form of Eq. 3.22 for the system of concern.

$$\delta q - \delta w = \Delta KE + \Delta PE + \Delta h = e_{in} - e_{out} \qquad (3.22)$$

Simplifies to

$$h_{in} + w_{in} = h_{out} \qquad (3.23)$$

which is shown graphically below (the sum of what goes into the control volume = the sum of what leaves the control volume) (Fig. 3.8).

Other engineering devices, the respective control volume, and resulting equation are given as Table 3.3.

Table 3.3 First law, engineering devices

Engineering device Assumptions	Control volume	Equation
Turbine $q = 0$ $\Delta KE = 0$ $\Delta PE = 0$	h_i q_o h_e	$w_0 = h_i - h_e$
Pipe flow $\delta w = 0$ $\Delta KE = 0$ $\Delta PE = 0$	q_o h_i h_e	$q_0 = h_i - h_e$
Boilers $\delta w = 0$ $\Delta KE = 0$ $\Delta PE = 0$	h_i h_e q_i	$q_{in} = h_e - h_i$
Condensers $\delta w = 0$ $\Delta KE = 0$ $\Delta PE = 0$	h_i h_e q_o	$q_0 = h_i - h_e$
Nozzles $\delta q = 0$ $W = 0$ $\Delta KE \neq 0$ $\Delta PE = 0$		$h_i + \dfrac{1}{2} v_i^2 = h_e + \dfrac{1}{2} v_e^2$
Diffusers $\delta q = 0$ $\delta w = 0$ $\Delta KE \neq 0$ $\Delta PE = 0$		$h_i + \dfrac{1}{2} v_i^2 = h_e + \dfrac{1}{2} v_e^2$

(continued)

Table 3.3 (continued)

Engineering device Assumptions	Control volume	Equation
Throttling device $\delta q = 0$ $\delta w = 0$ $\Delta KE = 0$ $\Delta PE = 0$		$h_i = h_e$
Pump $\delta q = 0$ $\Delta KE = 0$ $\Delta PE = 0$	h_i $\qquad$ h_e w_i	$-w_i = h_e - h_i$
Compressor $\delta q = 0$ $\Delta KE = 0$ $\Delta PE = 0$	h_i $\qquad$ h_e w_i	$-w_i = h_e - h_i$
Heat exchanger $\delta q = 0$ $\delta w = 0$ $\Delta KE = 0$ $\Delta PE = 0$	$\sum \dot{m}_i h_i$ $\sum \dot{m}_e h_e$	$\sum \dot{m}_i h_i = \sum \dot{m}_e h_e$

3.8 Cycles

We first learn first law for a closed system that helps us to understand how to use the thermodynamics tables and then this leads to first law for an open system and engineering devices. As part of learning first law for closed and open systems, we'll see a method to solve thermodynamic problems, which is given in Sect. 3.10.

Once we've mastered engineering devices, we begin to learn about cycles where

$$\sum \delta q_i = \sum \delta w_i \qquad (3.24)$$

Equation 3.24 is a consequence of a cycle beginning and ending in the same state. Thus the right-hand side of Eq. 3.25 is equal to zero.

$$\delta q - \delta w = \Delta h + \Delta PE + \Delta KE \qquad (3.25)$$

Given in the section on examples and problems are two examples of cycles. The first is an air cycle, which will become important when we discuss Brayton Cycles, which are a form of gas cycle (Sect. 5.3); the second is a water cycle, which will become important when we discuss Rankine Cycles, which is a form of vapor cycle

(Sect. 5.2). An introduction to Rankine Cycles is given below. Cycles will be more fully discussed in Chap. 5.

3.9 Rankine Cycle

When we arrange several engineering devices in a particular order, see Figs. 3.9 and 3.10, we create a series of processes that collectively start and end in the same state, which is the very definition of a cycle. The cycle shown in Figs. 3.9 and 3.10 is a Rankine Cycle.

You'll notice that on the *x*-axis for Fig. 3.9 the quantity is entropy, which we have not formally introduced. A working definition is that entropy change is a measure of the irreversibilities of a system and we would always rather an entropy change of zero, which is a measure of no irreversibilities. But all processes generate some entropy.

In this analysis, we assume the pump and turbine are isentropic, which means there are no irreversibilities. Later, some of the assumptions made will be relaxed.

The Rankine Cycle involves the following four processes and the device that creates a particular process

1 → 2, Pressure Increase, Isentropic Pump
2 → 3, Heat Added, Boiler
3 → 4, Work Extracted, Isentropic Turbine
4 → 1, Heat Dumped, Condenser

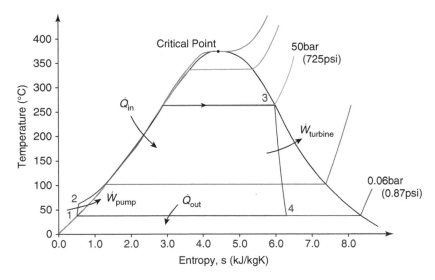

Fig. 3.9 *T* versus *S* for Rankine Cycle

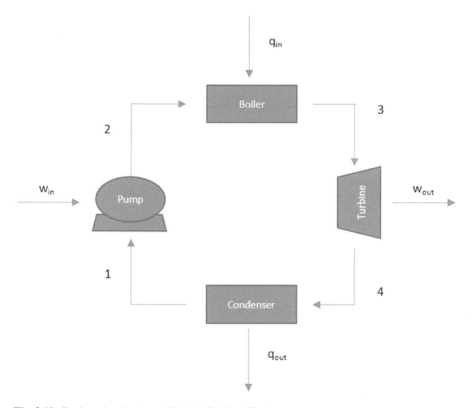

Fig. 3.10 Engineering devices within the Rankine Cycle

In order to determine Eq. 3.2 particular to a Rankine Cycle, an energy balance must be conducted around each engineering device and all states, heat, and work accounted for in a table. What goes into the control volume will be on the left-hand side of the equal sign and what goes out of the control volume will be on the right-hand side of the equal sign.

The following simplifying assumptions and conditions were made

1. Isentropic pump → $s_1 = s_2$
2. Isentropic turbine → $s_3 = s_4$
3. State 1 is on the saturated liquid line @ pressure P_1
4. State 2 is a compressed liquid @ pressure P_2
5. State 3 is a saturated vapor @ pressure P_2
6. State 4 is a mixture @ pressure P_1

Note – the sum of the work equals the sum of the heat for a cycle.
The energy balance around the pump in Fig. 3.11 is

$$h_1 + w_{in} = h_2 \qquad (3.26)$$

Fig. 3.11 Energy balance
for a pump

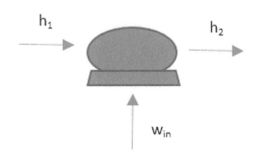

Fig. 3.12 Energy balance
for a boiler

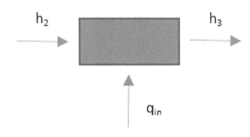

And solving for w_{in}

$$w_{in} = h_2 - h_1 \tag{3.27}$$

The energy balance around the boiler in Fig. 3.12 is

$$h_2 + q_{in} = h_3 \tag{3.28}$$

And solving for q_{in}

$$q_{in} = h_3 - h_2 \tag{3.29}$$

The energy balance around the turbine in Fig. 3.13 is

$$h_3 = w_{out} + h_4 \tag{3.30}$$

The value of h_4 is determined by utilizing the assumption that $s_3 = s_4$ and determining X_4.

And solving for w_o

$$w_{out} = h_4 - h_3 \tag{3.31}$$

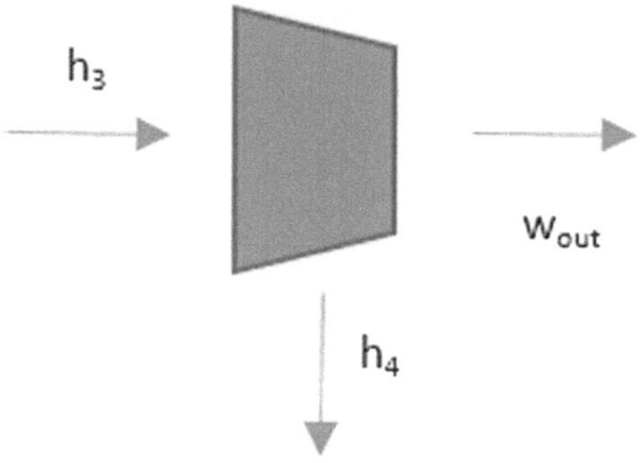

Fig. 3.13 Energy balance
for a turbine

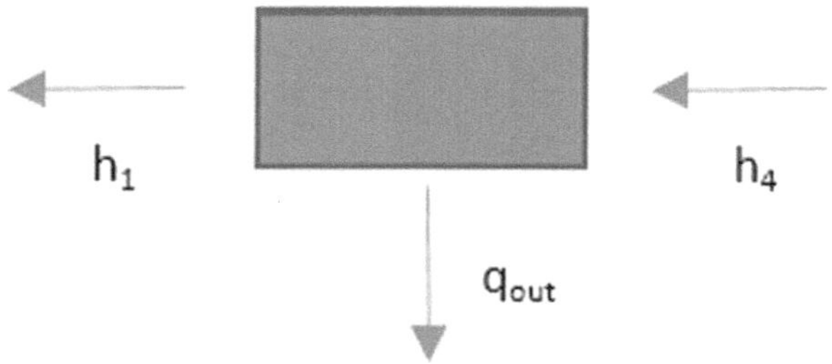

h_3

w_{out}

h_4

Fig. 3.14 Energy balance
for a condenser

h_1 h_4

q_{out}

The energy balance around the condenser in Fig. 3.14 is

$$h_4 = h_1 + q_{out} \tag{3.32}$$

And solving for q_{out}

$$q_{out} = h_4 - h_1 \tag{3.33}$$

Once all enthalpies, works, and heats are determined, then a table (see Table 3.4) is completed for the states given in Fig. 3.9 and the overall thermal efficiency for the cycle is calculated, which is given as

$$\eta_{overall} = \frac{w_o - w_{in}}{q_{in}} = \frac{h_3 - h_4 + h_1 - h_2}{h_3 - h_2} \tag{3.34}$$

The final step is to check the first law of thermodynamics for a cycle

$$\Sigma \delta q = \Sigma \delta w \rightarrow 952.05 \frac{kJ}{kg} = 952.05 \, kJ/kg \tag{3.35}$$

The left-hand side won't exactly match the right-hand side, but if it's off by too much, then check your work. There's probably a mistake!

Table 3.4 Various states, heat, and work for Rankine Cycle

State	Process/Device	T [°C]	P [kPa]	s [KJ/Kg-K]	X	h [kJ/kg]	δw [kJ/kg]	δq [kJ/kg]
1		35.84	6	0.52		149.78		
	Isentropic pump						−6.87	
2		36.33	5000	0.52		156.65		
	Boiler							+2637.68
3		263.99	5000	5.9733		2794.33		
	Isentropic turbine						+958.92	
4		35.84	6	5.9733		1835.41		
	Condenser							−1685.63
1		35.84	6			149.78		

3.10   Problem Solving Procedure for Thermodynamics Problems

Step 1 – Identify the substance
Step 2 – Identify the system

- Isolated system
- Closed system (thermodynamic energy is measured in terms of internal energy)
- Open system (thermodynamic energy is measured in terms of enthalpy)

Step 3 – Can we assume ideal and/or perfect gas behavior?
Step 4 – Fill out as completely as possible the following table, which may involve unit conversions. Please note there may be more than one process involved and it may be useful to plot the states on a pressure versus specific volume or entropy versus temperature graph.

State 1	State 2
Process	
Phase	Phase
Temperature	Temperature
Pressure	Pressure
Specific volume (volume)	Specific volume (volume)
Internal energy (enthalpy)	Internal energy (enthalpy)
Mass	Mass

Step 5 – Identify the appropriate equation/table or simplifying assumption for the following

- Heat
- Work
- Internal energy (enthalpy)
- Kinetic energy, potential energy, and mass rates (if necessary)

Step 6 – Determine the first law for the system
Step 7 – Determine the second law for the system (if necessary)
Step 8 – Solve the problem

Given below in Table 3.5 are the first law and second laws for a particular case where second law will be discussed in Chap. 4.

Table 3.5   First law and second law for various systems

Case	First Law	Second Law	Comments
Closed System Ideal Gas Perfect Gas	$Q - W = \Delta E$	$\Delta s = C_v \ln\left(\dfrac{T_2}{T_1}\right) + R_g \ln\left(\dfrac{v_2}{v_1}\right)$ $\Delta s = C_p \ln\left(\dfrac{T_2}{T_1}\right) - R_g \ln\left(\dfrac{P_2}{P_1}\right)$	Specific heat is constant Work is determined by the path $\Delta h = c_p \Delta T, \Delta e = c_v \Delta T$
Closed System Ideal Gas Non-Perfect	$Q - W = \Delta E$	$\Delta s = s_{T2}^0 - s_{T1}^0 - R_g \ln\left(\dfrac{P_2}{P_1}\right)$ $\Delta s = S_{T2}^0 - S_{T1}^0 + R_g \ln\left(\dfrac{v_2}{v_1}\right)$ + Ideal Gas Thermodynamics Table	Specific heat is NOT constant Work is determined by the path $e = f(T), s = f(T,P), s = f(T,v)$
Closed System Non-Ideal Non-Perfect	$Q - W = \Delta E$	Use Appropriate Thermodynamics Table	Specific heat is NOT constant Work is determined by the path
Open System Ideal Gas Perfect Gas	$Q - W = \Delta H + \Delta KE + \Delta PE$	$\Delta s = C_v \ln\left(\dfrac{T_2}{T_1}\right) + R_g \ln\left(\dfrac{v_2}{v_1}\right)$ $\Delta s = C_p \ln\left(\dfrac{T_2}{T_1}\right) - R_g \ln\left(\dfrac{P_2}{P_1}\right)$	Specific heat is constant Work is determined by the path $\Delta h = c_p \Delta T, \Delta e = c_v \Delta T$
Open System Ideal Gas Non-Perfect	$Q - W = \Delta H + \Delta KE + \Delta PE$	$\Delta s = s_{T2}^0 - s_{T1}^0 - R_g \ln\left(\dfrac{P_2}{P_1}\right)$ $\Delta s = S_{T2}^0 - S_{T1}^0 + R_g \ln\left(\dfrac{v_2}{v_1}\right)$ + Ideal Gas Thermodynamics Table	Specific heat is NOT constant Work is determined by the path $h = f(T), s = f(T,P), s = f(T,v)$
Open System Non-Ideal Non-Perfect	$Q - W = \Delta H + \Delta KE + \Delta PE$	Use Appropriate Thermodynamics Table	Specific heat is NOT constant Work is determined by the path

3.11   Examples and Problems

3.11.1   Examples

Example 3.1 Double Linear Interpolation
Determine the enthalpy for water at 12,000 kPa and 360 °C. This will involve more than one linear interpolation. The raw data from the thermodynamic tables is given in the first table and solving for A, B, and C determines the value to be 2874.41 kJ/kg.

		Pressure [kPa]		
		10,000	12,000	15,000
T [°C]	350	2923.39		2692.41
	360	A	C	B
	400	3096.46		2975.44

$$y = \frac{(y_2 - y_1)}{(x_2 - x_1)}(x - x_1) + y_1$$

$$A = \frac{3096.46 - 2923.39}{400 - 350}(360 - 350) + 2923.39$$

$$B = \frac{2975.44 - 2692.41}{400 - 350}(360 - 350) + 2692.41$$

$$C = \frac{2749.02 - 2957.00}{15,000 - 10,000}(12,000 - 10,000) + 2958.00$$

		Pressure [kPa]		
		10,000	12,000	15,000
T [°C]	350	2923.39		2692.41
	360	**2958**	**2874.41**	**2749.02**
	400	3096.46		2975.44

Example 3.2 Closed System
Saturated vapor R-410a at 0 °C in a rigid tank is cooled to −40 °C. Find the specific heat transfer.

Substance, R-410a
Treat as a non-ideal, non-perfect gas → Use the Thermodynamics Tables
Process, Isometric → No Work

State 1 – Saturated Vapor

T_1 is 0 °C
ν_1 is 0.03267 m³/kg and $\nu_1 = \nu_2$
e_1 is 253.02 kJ/kg

State 2 – Mixture

It's a mixture because
$\nu_l < \nu_1 < \nu_g$

Find X

$$X = \frac{\nu_{mix} - \nu_l}{\nu_g - \nu_l} = \frac{0.03267 - 0.000762}{0.14291 - 0.000762} = 22.45\%$$

Find e_2

$$e_2 = e_l + X\left(e_g - e_l\right) = -0.13 + 22.45\% * \left(237.81 + 0.13\right) = 53.28\,\text{kJ/kg}$$

Heat transfer is

$$\delta q - \delta w = \Delta e$$

And

$$\delta q = -199.72\,\text{kJ/kg}$$

Example 3.3 Closed System

A 100-L rigid tank contains N_2 at 1000 K and 3 MPa. The tank is now cooled to 80 K. What are the work and heat transfer for the process?

Volume is 100 L, Substance is N_2 and the process is Isometric
Use the thermodynamic table for N_2

State 1 – SHV

T_1 is 1000 K
P_1 is 3000 kPa
ν_1 is 0.0996 m³/kg
e_1 is 777.85 kJ/kg

State 2 – Mixture

It's a mixture because
$$v_l < v_1 < v_g$$

Find X

$$X = \frac{V_{mix} - v_l}{v_g - v_l} = \frac{0.03267 - 0.000762}{0.14291 - 0.000762} = 22.45\%$$

Find u_2

$$e_2 = e_l + X\left(e_g - e_l\right) = -0.13 + 22.45\% * \left(237.81 + 0.13\right) = 53.28\,\text{kJ/kg}$$

Heat transfer is

$$\delta q - \delta w = \Delta e$$

And

$$\delta q = -199.72\,\text{kJ/kg}$$

Example 3.4 Reactor Explosion

A water-filled reactor with a volume of 1 m³ is at 10 MPa and 260 °C is placed inside a containment room. The room is well insulated and initially evacuated. Due to a failure, the reactor ruptures and the water fills the containment room. Find the minimum room volume so that the final pressure does not exceed 400 kPa.

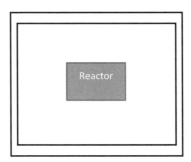

Volume is 1 m³, Substance is Water and the process is that $u_1 = u_2$
The two internal energies are equal because there is neither work nor heat transfer!
Use thermodynamic table for Water

State 1 – SCL

T_1 is 260 °C
P_1 is 10,000 kPa

ν_1 is 0.001265 m³/kg

e_1 is 1121.03 kJ/kg

State 2 – Mixture

It's a mixture because

$\nu_l < \nu_1 < \nu_g$

Find X

$$X = \frac{u_{mix} - u_l}{u_g - u_l} = \frac{1121.03 - 604.29}{2553.55 - 604.29} = 26.5\%$$

Find ν_2

$$\nu_2 = u_l + X\left(\nu_g - \nu_l\right) = 0.001084 + 26.5\% * \left(0.46246 - 0.001084\right) = 0.12 \frac{m^3}{kg}$$

Find mass

$$m = \frac{V_1}{\nu_1} = \frac{1}{0.001265} = 790.5\,kg$$

Find V_2

$$\frac{V_2}{m} = \nu_2 \rightarrow V_2 = 98.86\,m^3$$

Example 3.5 Nozzle

Superheated vapor ammonia enters an insulated nozzle at 30 °C and 1000 kPa with a low velocity and a steady rate of 0.01 kg/s. The ammonia exits at 300 kPa with a velocity of 450 m/s. Determine the temperature (or quality) and the exit area of the nozzle.

$h_1, u_1 \approx 0$ $\qquad\qquad\qquad\qquad$ $h_2, u_2 = 450\ m/s$

Substance is ammonia, Nozzle, Use Thermodynamic Tables

State 1	State 2
Insulated nozzle	
SHV	Phase
20 C	T_2

State 1	State 2
Insulated nozzle	
1000 kPa	300 kPa
$V \approx 0$	450 m/s
h_1	X and h_2
$h_1 + \dfrac{u_1^2}{2} = h_2 + \dfrac{u_2^2}{2}$	

Solve for h_2

$$h_2 = h_1 - \frac{450^2}{2000} = 1377.85 \, \text{kJ/kg}$$

And for 300 kPa in Saturated Ammonia Tables

T	P	$h(\text{l})$	$h(\text{g})$
[°C]	[kPa]	[kJ/kg]	[kJ/kg]
−10	290.9	134.41	1430.8
−5	354.9	157.31	1436.7
−9.29	300	137.67	1431.64

And solving for X

$$X = \frac{u_2 - u_l}{u_g - u_l} h = \frac{1377.85 - 137.67}{1431.64 - 137.67} = 95.84\%$$

Example 3.6 Turbine

A dam along the Green River is 50 m higher than the river that it discharges into. The electric generators driven by the water-powered turbines deliver 250 MW of power. If the discharge water is 20 °C, find the minimum amount of water running through the turbines.

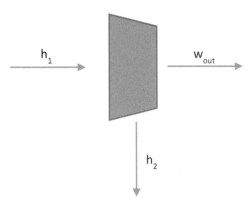

$z_1 = 50$ m, $z_2 = 0$; $\delta \dot{w} = 250\,\text{MW} = 250{,}000\,\text{kW}$; $T = 20\,°\text{C}$; $v = 0.001002\,\text{m}^3/\text{kg}$
Find

$$Q \equiv \frac{dV}{dt}$$

$$\delta \dot{q} - \delta \dot{w} = \dot{e}_2 - \dot{e}_1$$

$$\dot{e}_1 = \dot{m}\left[h_1 + KE_1 + PE_1\right]$$

And

$$\dot{e}_2 = \dot{m}\left[h_2 + KE_2 + PE_2\right]$$

$$h_1 = h_2, \text{Far Upstream}, \text{Far Downstream} \rightarrow KE_1 = KE_2$$

$$PE_1 = \frac{gZ_1}{1000} = \frac{9.81 * 50}{1000}, \ PE_2 = 0$$

$$-250{,}000 = -\dot{m}\frac{9.81 * 50}{1000}$$

$$\dot{m} = 509.84\frac{\text{kg}}{\text{s}}, \ \rho Q = \dot{m} = \frac{Q}{v}, \ Q = 510.7\,\text{m}^3/\text{s}$$

Example 3.7 Boiler and Super-Heater
Saturated liquid nitrogen at 800 kPa enters a boiler at the rate of 0.01 kg/s and exits as saturated vapor. It then flows into a super-heater also at 800 kPa, where it exits at 800 kPa and 280 K. Find the rate of heat transfer in the boiler and the super-heater.

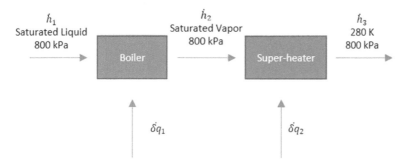

Find $\delta \dot{q}_1, \delta \dot{q}_2$

T	P	h(l)	h(g)
[°C]	[kPa]	[kJ/kg]	[kJ/kg]
100	779.2	−73.2	87.48
105	1084.6	−61.24	87.35
100.34	800	−72.39	87.47

$$h_1 = -72.39 \frac{kJ}{kg}, \ h_2 = 87.47 \frac{kJ}{kg}, \ h_3 = 288.52 \frac{kJ}{kg}$$

And

$$\dot{h}_1 + \delta \dot{q}_1 = \dot{h}_2$$

$$\dot{h}_2 + \delta \dot{q}_2 = \dot{h}_3$$

Further

$$\delta \dot{q}_1 = \dot{m} \left[h_2 - h_1 \right]$$

$$\delta \dot{q}_2 = \dot{m} \left[h_3 - h_2 \right]$$

And

$$\delta \dot{q}_1 = 1.6 \, kW, \ \delta \dot{q}_2 = 2.0 \, kW$$

Example 3.8 Air Cycle

For the cycle shown in the figure below, find the work and heat transfer for 1 lbm of air contained in a cylinder with T_1 at 800 °F assuming the process from 3 to 1 is (a) an isothermal process and (b) an adiabatic process.

We'll assume an ideal, perfect gas and use the work/heat table given at the end of Chap. 2.

$$m = 1 \ \text{lbm}, \quad T_1 = 800°F = 1260 \, R, \quad \gamma = 1.4, \quad R_g = 53.34 \ \frac{ft \ lbf}{lbm \ R},$$

$$c_p^0 = 0.24 \ \frac{BTU}{lbm \ R} = 186.77 \ \frac{ft \ lbf}{lbm \ R}, \quad c_v^0 = 0.171 \ \frac{BTU}{lbm \ R} = 133.1 \ \frac{ft \ lbf}{lbm \ R}$$

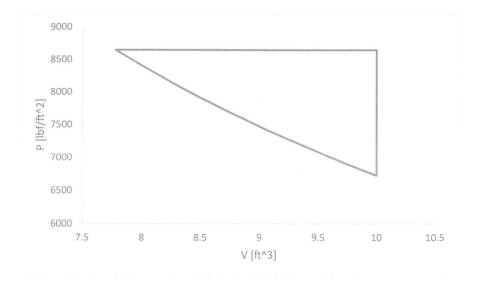

Plan of Attack

(a) Find $\{T, P, V\}$ for all states
(b) Determine work and heat associated with each process
(c) Sum work and sum heat, they should equal

What we are given

State	Process	T [R]	P [lbf/ft^2]	V [ft^3]	Work [ft-lbf]	Heat [ft-lbf]
1		1260	8640	V_1		
	Isobaric				$\delta w_{1 \to 2}$	$\delta q_{1 \to 2}$
2		T_2	8640	10		
	Isometric				0	$\delta q_{2 \to 3}$
3		T_3	P_3	10		
	Isothermal				$\delta w_{3 \to 1}$	$\delta q_{3 \to 1}$
1		1260	8640	V_1		
					X	Y

Find All States
Solve for V_1

$$PV = mR_g T, \quad V_1 = \frac{mR_g T_1}{P_1} = \frac{1 * 53.34 * 1260}{8640} = 7.78\,\text{ft}^3$$

Solve for T_2

$$T_2 = \frac{P_2 V_2}{m R_g} = \frac{8640 * 10}{1 * 53.34} = 1640\,\text{K}$$

Solve for P_3

$$P_3 V_3 = P_1 V_1, \quad P_3 = \frac{P_1 V_1}{V_3} = \frac{8640 * 7.78}{10} = 6721\,\text{lbf} / \text{ft}^3$$

Solve for T_3
$3 \rightarrow 1$ is isothermal and $T_3 = T_4$, $T_3 = 1260$ K

Find Work and Heat for All Processes
$1 \rightarrow 2$, Isobaric

$$\delta w = P(V_2 - V_1) = 60 * 144 * (10 - 7.78) = 19,192\ \text{ft}\quad\text{lbf}$$

$$\delta q = m C_p (T_2 - T_1) = 1 * 187 * (1640 - 1260) = 67,199\ \text{ft}\quad\text{lbf}$$

$2 \rightarrow 3$, Isometric

$$\delta w = 0$$

$$\delta q = m C_v (T_3 - T_2) = 1 * 133 * (1260 - 1640) = -47,889\ \text{ft}\quad\text{lbf}$$

$3 \rightarrow 1$, Isothermal

$$\delta w = P_3 V_3 \ln\left(\frac{V_1}{V_3}\right) = 8640 * 7.78 * \ln\left(\frac{7.78}{10}\right) = -16,882$$

$$\delta q = \delta w$$

The results are

State	Process	T [R]	P [lbf/ft²]	V [ft³]	Work [ft-lbf]	Heat [ft-lbf]
1		1260	8640	7.77875		
	Isobaric				19,191.60	67,199.38
2		1619.798	8640	10		
	Isometric				0.00	−47,889.05
3		1260	6720.84	10		
	Isothermal				−16,882.04	−16,882.04
1		1260	8640	7.77875		
					2309.56	2428.29

Example 3.9 Water Cycle

For the cycle with the following conditions, show that the sum of the work equal the sum of the heat transfer

State 1 – Saturated Liquid
4000 kPa, $h1 = 1087.29$ kJ/kg, $nu1 = 0.001252$ m³/kg
State 2 – Saturated Vapor
4000 kPa, $h2 = 2801.38$ kJ/kg, $nu2 = 0.04978$ m³/kg, $V_2 = 0.5$ m³

$$\frac{0.5\,\text{m}^3}{m} = 0.05\frac{\text{m}^3}{\text{kg}}, m = 10\,\text{kg}$$

State 3 – Mixture
$T = 80\ °C$
$\nu_2 = \nu_3$

$$v_3 = v_l + X\left(v_g - v_l\right) \text{and } X_3 = \frac{0.04978 - 0.001029}{3.40715 - 0.001029} = 1.44\%$$

And

$$h_l = 334.88\frac{\text{kJ}}{\text{kg}} \text{and } h_g = 2643.66\frac{\text{kJ}}{\text{kg}}, h_3 = 334.88 + 1.44\% *\left(2643.66 - 334.88\right)$$
$$= 368.07\,\text{kJ/kg}$$

State 4 – Mixture
$\nu_1 = \nu_4$

$$v_4 = v_l + X\left(v_g - v_l\right) \text{and } X_4 = \frac{0.001252 - 0.001029}{3.40715 - 0.001029} = 0.01\%$$

And

$$h_l = 334.88\frac{\text{kJ}}{\text{kg}} \text{and } h_g = 2643.66\frac{\text{kJ}}{\text{kg}}, h_4 = 334.88 + 0.01\% *\left(2643.66 - 334.88\right)$$
$$= 335.0\,\text{kJ/kg}$$

Determination of Work and Heat

$1 \rightarrow 2$, "Boiler", $\delta q = h_2 - h_1$
$2 \rightarrow 3$, "Turbine", $\delta w = h_2 - h_3$
$3 \rightarrow 4$, "Condenser", $-\delta q = h_3 - h_4$
$4 \rightarrow 1$, "Pump", $-\delta w = h_1 - h_4$

The results are given below.

State	Process	T [°C]	P [kPa]	V [m³]	Comments	h [kJ/kg]	Work [kJ/kg]	Heat [kJ/kg]
1			4000		Saturated liquid	1087.29		
	Isobaric							1714.09
2			4000	0.5	Saturated vapor	2801.38		
	Isometric						2433.31	
3		80		0.5		368.07		
	Isobaric							−33.07
4		80				335		
	Isometric						−752.29	
1						1087.29		
							1681.02	1681.02

3.11.2   Problems

Problem 3.1 Air at 60 °C, 150 kPa flows in a 100 mm by 150 mm rectangular duct in a heating system. The volumetric flow rate is 0.01 m³/s. What is the velocity of the air flowing in the duct and what is the mass flow rate?

Problem 3.2 A boiler receives a constant flow of 6000 kg/h liquid water at 5 MPa and 20 °C, and it heats the flow such that the exit state is 425 °C and 4.5 MPa. Determine the necessary minimum pipe flow area in both the inlet and exit pipe(s) if there should be no velocities larger than 20 m/s. Also, determine the heat rate [BTU/hr.].

Problem 3.3 Nitrogen gas flows into a convergent nozzle at 200 kPa, 400 K, and very low velocity. It flows out of the nozzle at 100.3 kPa, 1000 m/s. If the nozzle is insulated, find the exit temperature.

Problem 3.4 In a jet engine a flow of air at 1000 K, 200 kPa, and 40 m/s enters a nozzle, where the air exits at 500 m/s, 101.3 kPa. Assuming no heat loss, what is the exit temperature?

Problem 3.5 Helium is throttled from 1 MPa, 20 °C to a pressure of 100 kPa. The diameter of the exit pipe is so much larger than that of the inlet pipe that the inlet and exit velocities are equal. Find the exit temperature of the helium and the ratio of the pipe diameters.

Problem 3.6 A compressor in a commercial refrigerator receives R-410a at −20 °C and $X = 100\%$. The exit is at 1200 kPa and 60 °C. Neglect kinetic energies and find the specific work.

Problem 3.7 A rigid 177-ft³ tank contains saturated steam at 73 psia. Initially, 10% of the volume is occupied by liquid and the rest by vapor. Heat transfer takes place until the pressure in the tank reaches 58 psia. Determine the following:

(a) Amount and direction of heat transfer
(b) Initial and final temperatures
(c) Final volumes of liquid and vapor

Problem 3.8 A 4.25-ft³ rigid tank contains saturated liquid R-134a at 116 psia. A valve at the bottom of the tank is now opened, and liquid is withdrawn from the tank. Heat is transferred to the refrigerant such that the pressure inside the tank remains constant. Determine the amount of heat that must be transferred by the time 75% of the total mass has been withdrawn.

Problem 3.9 150 kg of a saturated mixture of steam at 200 kPa is stored in a 15 m³ container. A valve on the container is opened and 15 kg of steam at 500 kPa and 300 °C is gradually added into the container. During this addition process, heat exchange occurs with the surroundings. The final pressure in the container is 300 kPa.

(a) What is the quality of the initial contents of the container?
(b) What is the quality of the final contents of the container?
(c) How much heat was transferred? State whether heat was added or removed.

Appendix 3.1: Rankine Cycle Worksheet

A tableau for the Rankine Cycle is given on the companion website.

Appendix 3.2: Linear Interpolation (website)

An Excel Spreadsheet was developed to illustrate a particular concept and is given on the companion website.

References

1. Borgnakke, C., & Sonntag, R. E. (2009). *Fundamentals of thermodynamics* (7th ed.). Wiley.
2. National Institute of Standards and Testing; https://webbook.nist.gov/chemistry/. Accessed 23 Apr 2020.
3. Potter, M., & Scott, E. P. (2004). *Thermal sciences* (1st ed.). Brooks/Cole – Thomson Learning.

Chapter 4
Entropy and the Second Law of Thermodynamics

4.1 Preview

We'll begin by defining reversible and irreversible processes and give three examples. Example 4.1 defines reversible and irreversible processes; Example 4.2 provides details of a reversible heat engine (Carnot cycle); and Example 4.3 provides details of an irreversible heat engine (Rankine cycle).

We'll move on to defining Carnot cycles, which are reversible heat engines, heat pumps, or refrigeration cycles. A **heat engine** extracts work, w(out), from two reservoirs where one is at a high temperature and the other is at a low temperature; a **heat pump** reverses a heat engine to keep a warm space warm when the surroundings are cold and requires w(in); a **refrigeration cycle** is a reversed heat engine where work is added, w(in), to keep a cold space cold when the surroundings are hot.

We'll then discuss Clausius inequalities, which were illustrated by the three examples discussed above and go on to define entropy and show that it is another state. We then determine entropy changes associated with processes that occur in liquids, solids, and gases.

We will then formally give the second law of thermodynamics for both closed and open systems, and apply the idea of entropy to determine the efficiency of engineering devices.

4.2 Reversible and Irreversible Systems

We'll look at three examples that explore reversible and irreversible systems and tie this to the definition of entropy, which is given as

$$ds = \left(\frac{\delta q}{T} \right)_{rev} \tag{4.1}$$

© Springer Nature Switzerland AG 2022
H. C. Foust III, *Thermodynamics, Gas Dynamics, and Combustion*,
https://doi.org/10.1007/978-3-030-87387-5_4

where ds is the change in entropy (to be defined shortly), T is the temperature of the process, and δq is the heat released assuming a reversible process (to be defined shortly).

Example 4.1 Imagine a system of two ramps with equal angles, a cart, and a connection horizontal section (see Fig. 4.1a), but no friction. No friction due to air, no internal friction within the cart, no friction between the cart and the ramps, and connecting horizontal section. The cart is released at "A", goes through "B", and stops at "C." Because there is no friction, it falls and goes through "B", and stops at "A." Because there is no friction, it does this endlessly.

We know the world doesn't work this way (see Fig. 4.1b). The cart falls away from "A" and likely lands somewhere in the horizontal section dependent on the coefficient of friction, length of each part, and the angle of the ramp. The system in Fig. 4.1b is an example of a reversible system and the system given as Fig. 4.1b is an example of an irreversible system.

Briefly stated, a reversible system is one where there is no additional energy (work) needed to get the system back to the original state, which in this case is "A."

Example 4.2 Let's look at another example.

Imagine a reversible heat engine that contains the four following processes

$$1 \rightarrow 2, Isothermal$$

$$2 \rightarrow 3, Adiabatic$$

$$3 \rightarrow 4, Isothermal$$

$$4 \rightarrow 1, Adiabatic$$

Fig. 4.1 (a) Reversible system. (b) Irreversible system

Fig. 4.1 (continued)

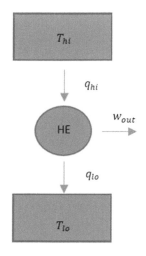

If we assume the substance is an ideal, perfect gas with no phase changes, then the efficiency of this cycle is given as

$$\eta = \frac{w_{net}}{q_{in}} = \frac{q_{in} - q_{out}}{q_{in}} = 1 - \frac{q_{out}}{q_{in}} \tag{4.2}$$

We'll show the efficiency for a reversible heat engine can also be defined as

$$\eta = \frac{w_{net}}{q_{in}} = 1 - \frac{T_{out}}{T_{in}} \tag{4.3}$$

Additionally,

$$\frac{q_{out}}{q_{in}} = \frac{T_{out}}{T_{in}} \tag{4.4}$$

Equally, we can divide $q_{rev}\big/T_{abs}$ which gives us a relative measure of heat transfer and for a reversible heat engine it is

$$\frac{q_{in}}{T_{in}} - \frac{q_{out}}{T_{out}} = 0 \qquad (4.5)$$

Example 4.3 For a real heat engine (Rankine cycle), we have the following conditions [1].

1. Saturated liquid, 0.7 MPa.
2. Saturated vapor, 0.7 MPa.
3. 90% Quality, 15 kPa
4. 10% Quality, 15 kPa.

For these conditions,

$$q_{in} = 2066.3\frac{kJ}{kg}, T_{in} = 164.9^{\circ}C, q_{out} = 1898.4\frac{kJ}{kg}, T_{out} = 53.97^{\circ}C$$

$$\frac{q_{in}}{T_{in}} - \frac{q_{out}}{T_{out}} = \frac{2066.3}{273.15+164.9} - \frac{1898.4}{273.15+53.97} = -1.9\frac{kJ}{kg-K} < 0 \qquad (4.6)$$

Which is consistent for an irreversible heat engine and will be discussed again when we discuss Clausius' inequality.

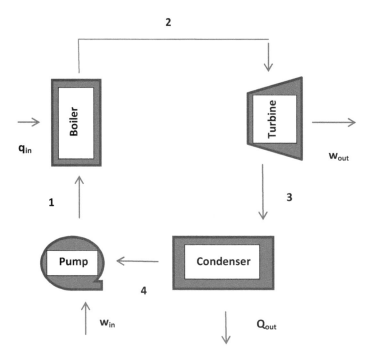

Fig. 4.2 Rankine cycle

Some observations

1. $\dfrac{q_{in}}{T_{in}} - \dfrac{q_{out}}{T_{out}}$ provides a useful measure of whether or not a cycle is reversible or irreversible
2. When the quantity given in (1) is negative, this implies the cycle is irreversible.
3. When the quantity given in (1) is zero, this implies the cycle is reversible.

We'll soon see that the quantity defined as Eq. 4.5 or 4.6 is in fact a state and has several uses to include

1. Provides another state given in the thermodynamic tables.
2. Will provide a measure of whether or not a given process (or cycle) can occur.
3. Provides a means to improve some process (or cycle).
4. Can be utilized to determine the heat transfer for a process (or cycle).

4.3   Carnot Heat Engine and Carnot Heat Pump

The reversible heat engine given above as **Example 2** is actually known as the Carnot heat engine and when the cycle is reversed it either becomes a heat pump or a refrigeration cycle (see Fig. 4.3a, b, and c).

Before we get into the Carnot heat engine there are four concepts that need to be clearly defined – heat engine, heat pump, refrigeration, thermal efficiency, and coefficient of performance (β).

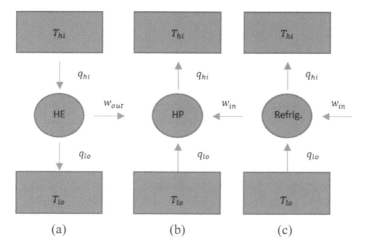

Fig. 4.3 (**a**) Heat engine, (**b**) Heat pump, and (**c**) refrigeration cycle

Heat Engine
An example of a heat engine is given as Fig. 4.3a where heat naturally transfers from a high-temperature reservoir to a low-temperature reservoir and some portion of this heat transfer is extracted as useful work.

Heat Pump
A heat pump (see Fig. 4.3b) would be to add work to the system thereby changes the direction of heat transfer where the T_h (T_{high}) reservoir would be the space you're trying to keep warm. Heat does not naturally want to go from low temperature to high temperature and only does so with the addition of work.

Refrigeration
A refrigeration cycle (see Fig. 4.3c) is another example of a system where work is added in order for the heat transfer to go from a lower temperature to a higher temperature. The only difference between a refrigeration cycle and a heat pump cycle is that now the system is T_c (T_{low}).

Thermal efficiency (η)
Is a measure of how much of the q_{hi} is utilized as useful work and is mathematically defined as

$$\eta_{HE} = \frac{w_o}{q_{hi}} \tag{4.7}$$

Coefficient of Performance (β)
For heat pumps or refrigeration cycles is a measure of efficiency, but is different from thermal efficiency. For a *heat pump*, which is a reverse heat engine, the β is defined as

$$\beta_{HP} = \frac{q_{hi}}{w_o} \tag{4.8}$$

And for *refrigeration*, the β is defined as

$$\beta_{Refrig} = \frac{q_{lo}}{w_o} = \frac{q_{hi} - w_o}{w_o} = \frac{q_{hi}}{w_o} - 1 \tag{4.9}$$

The processes involved in a Carnot heat engine (see Figs. 4.4 and 4.5) are.
A ->B is an isothermal expansion,
B ->C is an adiabatic reversible expansion,
C ->D is an isothermal contraction, and.
D ->A is an adiabatic reversible contraction.
There are a series of postulates invoked when considering a Carnot cycle [1].

Postulate 1 It is impossible to construct an engine that is more efficient than the Carnot heat engine.

Fig. 4.4 Heat engine

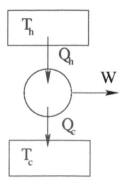

Fig. 4.5 P versus V for a Carnot cycle

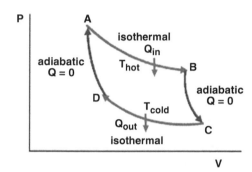

Postulate 2 The efficiency of a Carnot engine is solely a function of the temperature reservoirs.

Postulate 3 All reversible engines, operating between two reservoirs, have the same overall thermal efficiency as a single Carnot engine operating between the same reservoirs.

Prove that for a reversible engine (Carnot heat engine) that the thermal efficiency is defined as

$$\eta = 1 - \frac{q_{lo}}{q_{hi}} = 1 - \frac{T_{lo}}{T_{hi}} \tag{4.10}$$

$$1 \rightarrow 2, Q_{hi} = \int_{V_1}^{V_2} PdV = mRT_{hi} \ln\left(\frac{V_2}{V_1}\right) \tag{4.11}$$

$$2 \rightarrow 3, Q_{2\rightarrow3} = 0 \tag{4.12}$$

$$3 \rightarrow 4,, Q_{lo} = \int_{V_3}^{V_4} PdV = -mRT_{lo} \ln\left(\frac{V_4}{V_3}\right) \tag{4.13}$$

$$4 \rightarrow 1, Q_{3 \rightarrow 4} = 0 \tag{4.14}$$

Substituting Eqs. 4.11 and 4.13 into Eq. 4.10 results in

$$\eta = 1 - \frac{q_{lo}}{q_{hi}} = 1 + \frac{T_{lo} \left(\dfrac{V_4}{V_3} \right)}{T_{hi} \left(\dfrac{V_2}{V_1} \right)} \tag{4.15}$$

Additionally, for the adiabatic process 2->3,

$$\frac{T_3}{T_2} = \left(\frac{V_2}{V_3} \right)^{\gamma - 1} = \frac{T_{lo}}{T_{hi}} \tag{4.16}$$

And for the adiabatic process 4->1,

$$\frac{T_1}{T_4} = \left(\frac{V_4}{V_1} \right)^{\gamma - 1} = \frac{T_{hi}}{T_{lo}} \tag{4.17}$$

Note $T_1 = T_2$ and $T_3 = T_4$.
Further,

$$\frac{V_2}{V_3} = \frac{V_1}{V_4} \text{ and } \frac{V_4}{V_3} = \frac{V_1}{V_2} = X \tag{4.18}$$

Therefore,

$$\eta = 1 - \frac{q_{lo}}{q_{hi}} = 1 + \frac{T_{lo} \left(\dfrac{V_4}{V_3} \right)}{T_{hi} \left(\dfrac{V_2}{V_1} \right)} = 1 + \frac{T_{lo}}{T_{hi}} \frac{\ln(X)}{\ln\left(\frac{1}{X}\right)} = 1 - \frac{T_{lo}}{T_{hi}} \tag{4.19}$$

A Carnot cycle is an idealization that cannot physically be realized, but provides an upper limit on actual performance.

A Summary of Carnot Cycles   Our definition of overall thermal efficiency (η) for a **Carnot heat engine** is

$$\eta_{HE} = \frac{w_o}{q_{hi}} = 1 - \frac{T_{lo}}{T_{hi}} = \frac{T_{hi} - T_{lo}}{T_{hi}} \tag{4.20}$$

Reverse it and you have a **Carnot heat pump**, which has a Beta of

$$\beta_{HP} = \frac{q_{hi}}{w_o} = \frac{1}{1 - \dfrac{T_{lo}}{T_{hi}}} = \frac{T_{hi}}{T_{hi} - T_{lo}} \tag{4.21}$$

And for a **Carnot refrigeration cycle**, the Beta is

$$\beta_{Refrig} = \frac{q_{lo}}{w_o} = \frac{q_{hi} - w_o}{w_o} = \frac{q_{hi}}{w_o} - 1 = \frac{T_{hi} - (T_{hi} - T_{lo})}{T_{hi} - T_{lo}} = \frac{T_{lo}}{T_{hi} - T_{lo}} \tag{4.22}$$

A Summary of Actual Cycles Our definition of overall thermal efficiency (η) for a **Carnot heat engine** is

$$\eta_{HE} = \frac{w_o}{q_{hi}} \tag{4.20}$$

Reverse it and you have a **Carnot heat pump**, which has a Beta of

$$\beta_{HP} = \frac{q_{hi}}{w_o} \tag{4.21}$$

And for a **Carnot refrigeration cycle**, the Beta is

$$\beta_{Refrig} = \frac{q_{lo}}{w_o} = \frac{q_{hi} - w_o}{w_o} = \frac{q_{hi}}{w_o} - 1 \tag{4.22}$$

4.4 Clausius Inequality

We'll show

$$\oint \frac{\delta Q}{T} \leq 0 \tag{4.23}$$

where Eq. 4.23 for a reversible cycle takes the form

$$\oint \frac{\delta Q}{T} = 0 \tag{4.24}$$

And for an irreversible cycle it takes the form

$$\oint \frac{\delta Q}{T} < 0 \qquad\qquad (4.25)$$

4.4.1   Reversible Heat Engines

$$\oint \delta Q = Q_{hi} - Q_{lo} > 0 \qquad\qquad (4.26)$$

And using Eqs. 4.13 and 4.11 for Q_{hi} and Q_{lo} and dividing Q_{hi} by T_{hi} and Q_{lo} by T_{lo} results in

$$\oint \frac{\delta Q}{T} = \frac{Q_{hi}}{T_{hi}} - \frac{Q_{lo}}{T_{lo}} = mRln\left(\frac{V_2}{V_1}\right) + mRln\left(\frac{V_4}{V_3}\right) = 0 \qquad\qquad (4.27)$$

4.4.2   Irreversible Heat Engines

For an irreversible heat engine,

$$W_{irrev} < W_{rev} \qquad\qquad (4.28)$$

And $Q_{hi} - Q_{lo} = W$ therefore,

$$\left(Q_{hi} - Q_{lo}\right)_{irrev} < \left(Q_{hi} - Q_{lo}\right)_{rev} \qquad\qquad (4.29)$$

Since Q_{hi} is the same for both systems,

$$Q_{lo,\,irrev} > Q_{lo,rev} \qquad\qquad (4.30)$$

which results in

$$\oint \frac{\delta Q}{T} = \frac{Q_{hi}}{T_{hi}} - \frac{Q_{lo}}{T_{lo}} < 0 \qquad\qquad (4.31)$$

4.5   Definition of Entropy, Entropy as a State Function, and Area under T Vs. Ds Graphs

4.5.1   Definition of Entropy

Clausius inequality can also be defined as

$$\oint \frac{\delta Q}{T} + \sigma_{cycle} = 0 \tag{4.32}$$

where σ_{cycle} are irreversibilities of the cycle and the first term is a measure of the heat release per temperature and the second term is a measure of the irreversibilities of the cycle.

Another form of Eq. 4.32 is

$$\Delta s = \frac{\delta q}{T} + \sigma_{system} \tag{4.33}$$

which is a measure of the entropy change of a system and when the system is **adiabatic**, the first term on the right-hand side is zero

$$\Delta s = \sigma_{cycle} \tag{4.34}$$

And when the system is **reversible** the second term on the right-hand side is zero.

$$\Delta s = \frac{\delta q}{T} \tag{4.35}$$

A system that is both reversible and adiabatic is **isentropic** and thus

$$\Delta s = 0, s_1 = s_2 \tag{4.36}$$

And for a closed system, the first law is

$$de = \delta q - \delta w \tag{4.37}$$

And work is defined as

$$\delta w = Pdv \tag{4.38}$$

Substituting Eqs. 4.37 and 4.38 into Eq. 4.33 results in

$$ds = \frac{\delta q}{T} + \sigma_{sys} = \frac{du}{T} + \frac{Pdv}{T} + \sigma_{sys} = \frac{C_v(T)}{T} dT + \frac{Pdv}{T} + \sigma_{sys} \tag{4.39}$$

And when the system is reversible

$$ds = \frac{\delta q}{T} = \frac{du}{T} + \frac{Pdv}{T} = \frac{C_v(T)}{T}dT + \frac{Pdv}{T}$$

(4.40)

Additionally, enthalpy is defined as

$$h = e + Pv$$

(4.41)

And the derivative of Eq. 4.41 is

$$dh = de + Pdv + vdP$$

(4.42)

Solving Eq. 4.42 for "du" and substituting into Eq. 4.37 results in

$$dh - Pdv - vdP = \delta q - Pdv$$

(4.43)

Or

$$dh = \delta q + vdP$$

(4.44)

Solving Eq. 4.44 for "δq" and substituting the resultant equation into Eq. 4.40 results in

$$ds = \frac{\delta q}{T} = \frac{dh}{T} - \frac{vdP}{T} = \frac{C_p(T)}{T}dT - \frac{vdP}{T}$$

(4.45)

Note Eqs. 4.35 and 4.45 are known as the Gibbs Equations in honor of Wilfred Gibbs [1, 2].

4.5.2 Entropy as State Function

Given below is a graph of P versus V along several reversible paths
   We've seen for a reversible cycle that

$$\oint \frac{\delta Q}{T} = 0$$

(4.46)

And so

$$\left(\int_1^2 \frac{\delta Q}{T}\right)_A + \left(\int_2^1 \frac{\delta Q}{T}\right)_B = 0$$

(4.47)

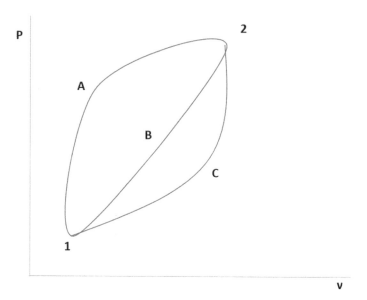

Fig. 4.6 Entropy as a State Function

And

$$\left(\int_1^2 \frac{\delta Q}{T}\right)_A + \left(\int_2^1 \frac{\delta Q}{T}\right)_C = 0 \tag{4.48}$$

Subtracting Eq. 4.48 from 4.47 results in

$$\left(\int_2^1 \frac{\delta Q}{T}\right)_B = \left(\int_2^1 \frac{\delta Q}{T}\right)_C \tag{4.49}$$

Equally, we can say

$$\left(\int_1^2 \frac{\delta Q}{T}\right)_B = \left(\int_1^2 \frac{\delta Q}{T}\right)_C \tag{4.50}$$

This shows that S is independent of path and is a state property of thermodynamics and can be cast as

$$S_2 - S_1 = \int_1^2 \frac{\delta Q}{T} \tag{4.51}$$

Or

$$\delta Q = \int_{1}^{2} T dS \tag{4.52}$$

Which is analogous to

$$\delta W = \int_{1}^{2} P dV \tag{4.53}$$

4.5.3 Graph of T Versus S

Given in Fig. 4.7 is the T versus S graph for a reversible heat engine (Carnot heat engine) where q(in) is the area enclosed by 1->2->S(h)->S(l)->1, q(out) is the area enclosed by S(l)->4->3->S(h)->S(l), and the w(o) is the area enclosed by 1->2->3->4->1.

More generally, the area under a T versus S graph represents δQ.

4.6   Second Law of Thermodynamics

Given

$$\Delta s = \frac{\delta q}{T} + \sigma_{system} \tag{4.54}$$

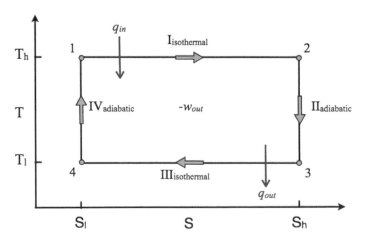

Fig. 4.7 Carnot heat engine

The second law of thermodynamics is as follows
When the $\sigma_{system} > 0 \rightarrow$ *Irreversible System*
When the $\sigma_{system} = 0 \rightarrow$ *Reversible System*
When the $\sigma_{system} < 0 \rightarrow$ *Impossible*
Using the first law of thermodynamics helps us to account for energy sources and sinks and how fast a process or cycle will occur.

Using the second law of thermodynamics helps us to understand "if a process or cycle will ever occur"?

4.7   Entropy of Solids and Liquids

Starting with the entropy balance for a reversible substance

$$ds = \left(\frac{\delta q}{T}\right)_{rev} = \frac{C_v dT}{T} + \frac{P}{T} dv \tag{4.55}$$

And knowing that solids and liquids are treated as incompressible results in two simplifying assumptions

1. dv is zero

$$C_v = C_p = C$$

Results in

$$\Delta s = C ln\left(\frac{T_2}{T_1}\right) \tag{4.56}$$

4.8   Entropy of Gases

For entropy changes of gases, it's not so simple. We'll consider two possible cases

1. Ideal, Perfect Gases.
2. Ideal, Non-Perfect Gases.

4.8.1   Entropy of an Ideal, Perfect Gas

Starting with the entropy balance for a reversible substance

$$Tds = du + Pdv \tag{4.57}$$

and

$$ds = \frac{C_v dT}{T} + \frac{Pdv}{T} \tag{4.58}$$

where $Pv = RT$ such that

$$ds = C_v \frac{dT}{T} + R\frac{dv}{v} \tag{4.59}$$

And integrating both sides of Eq. 4.59 results in

$$\Delta s = C_V \ln\left(\frac{T_2}{T_1}\right) + R ln\left(\frac{v_2}{v_1}\right) \tag{4.60}$$

Or starting from

$$Tds = dh - v dP \tag{4.61}$$

and

$$ds = C_P \frac{dT}{T} - v \frac{dP}{T} \tag{4.62}$$

where $Pv = RT$ such that

$$ds = C_v \frac{dT}{T} - R\frac{dP}{P} \tag{4.63}$$

And integrating both sides of Eq. 4.63 results in

$$\Delta s = C_P \ln\left(\frac{T_2}{T_1}\right) - R ln\left(\frac{P_2}{P_1}\right) \tag{4.64}$$

4.8.2   Entropy Change of an Ideal, Non-Perfect Gas

When the gas is non-perfect and specific heat depends on temperature, then an integration is involved

$$\Delta s = \int_{T_1}^{T_2} \frac{C_v(T)\,dT}{T} + Rln\left(\frac{v_2}{v_1}\right) \qquad (4.65)$$

Or

$$\Delta s = \int_{T_1}^{T_2} \frac{C_P(T)\,dT}{T} - Rln\left(\frac{P_2}{P_1}\right) \qquad (4.66)$$

which is solved by using a combination of equation (second term) and tables (first term) to account for the integration.

See Examples 4.11, 4.12, and 4.13 for further details.

4.9 Engineering Efficiency

In this section, thermal efficiency will be defined in terms of enthalpies associated with a reversible and irreversible engineering device.

4.9.1 Engineering Devices for Work out

For an engineering device where it is work out, the definition of efficiency is

$$\eta = \frac{w_{out}^{Act}}{w_{out}^{iso}} \qquad (4.67)$$

where for an irreversible engineering device $w_{out}^{iso} > w_{out}^{act}$

Reversible Device The energy balance for a reversible engineering device is

$$w_{out}^{iso} = h_1 - h_2^{iso} \qquad (4.68)$$

Irreversible Device The energy balance for an irreversible engineering device is

$$w_{out}^{act} = h_1 - h_2^{act} \qquad (4.69)$$

Using our previous definition of efficiency and Eqs. 4.68 and 4.69, we get

$$\eta_{work\ out} = \frac{h_1 - h_2^{act}}{h_1 - h_2^{iso}} \qquad (4.70)$$

Fig. 4.8 Energy Balance for an Isentropic "Work Out" Device

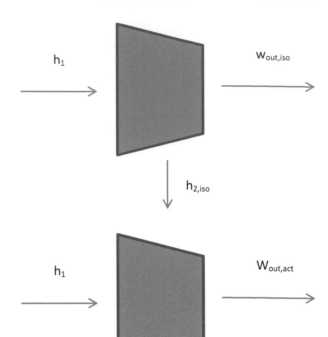

Fig. 4.9 Energy Balance for an Actual "Work Out" Device

4.9.2   Engineering Device for Work in

For an engineering device where it is work out, the definition of efficiency is

$$\eta = \frac{w_{in}^{iso}}{w_{in}^{Act}} \tag{4.71}$$

where for an irreversible engineering device $w_{in}^{act} > w_{in}^{iso}$

Reversible Device   The energy balance for a reversible engineering device is

$$w_{in}^{iso} = h_2^{iso} - h_1 \tag{4.72}$$

Irreversible Device   The energy balance for an irreversible engineering device is

$$w_{in}^{act} = h_2^{act} - h_1 \tag{4.73}$$

Using our previous definition of efficiency and Eqs. 4.72 and 4.73, we get

$$\eta_{work\ in} = \frac{h_2^{iso} - h_1}{h_2^{act} - h_1} \tag{4.74}$$

What are some of the consequences of these definitions in terms of highest and lowest temperatures possible from engineering devices where it's work in or work out?

Example 4.4

Consider a reversible turbine (work out device)

Assume $\eta = 100\%$,

$$S_1 = S_2 \tag{4.75}$$

And

$$\eta = 1 = \frac{h_1 - h_2^{act}}{h_1 - h_2^{iso}} \tag{4.76}$$

Therefore,

$$h_2^{act} = h_2^{iso} \rightarrow T_2 > T_1 \tag{4.77}$$

T_2 can be found through knowing T_1, P_1, which provides s_1 and s_1 is equal to s_2.

Example 4.5

Consider a perfectly irreversible turbine

Assume $\eta = 0$, ideal, and perfect gas

$$S_2 > S_1 \tag{4.78}$$

Therefore,

$$\eta = 0 = \frac{h_1 - h_2^{act}}{h_1 - h_2^{iso}} \tag{4.79}$$

And

$$h_2^{act} = h_1 \rightarrow T_2 = T_1 \tag{4.80}$$

Fig. 4.10 Energy Balance for an Isentropic "Work In" Device

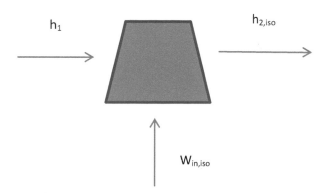

Fig. 4.11 Energy Balance
for an Actual "Work In"
Device

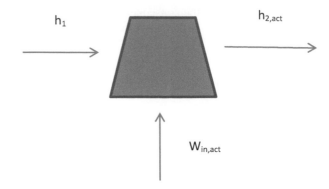

Example 4.6
Consider a reversible pump (work in device)
   Assume $\eta = 100\%$, ideal, and perfect gas

$$S_1 = S_2 \tag{4.81}$$

And

$$\eta = 1 = \frac{h_2^{iso} - h_1}{h_2^{act} - h_1} \tag{4.82}$$

Therefore,

$$h_2^{act} = h_2^{iso} \rightarrow T_2 > T_1 \tag{4.83}$$

T_2 can be found through knowing T_1, P_1, which provides s_1 and s_1 is equal to s_2.

Example 4.7
Consider a perfectly irreversible pump
   Assume $\eta = 0$, ideal, and perfect gas

$$S_2 > S_1 \tag{4.84}$$

And

$$\eta = 0 = \frac{h_2^{iso} - h_1}{h_2^{act} - h_1} \tag{4.85}$$

Therefore,

$$h_2^{act} = h_1 \rightarrow T_2 = T_1 \tag{4.86}$$

4.10   Examples and Problems

4.10.1   Examples

Example 4.8 Heat Engines
A gasoline engine produces 120 HP using 300 kW of heat transfer from burning fuel. What is the thermal efficiency, and how much power is rejected to the environment?

$$\eta = \frac{w_{out}}{q_{hi}} = \frac{89.5\,kW}{300\,kW} = 29.8\%$$

Using first law,

$$\dot{q}_{hi} = \dot{q}_{lo} + \dot{w}_{out}, 300 = \dot{q}_{lo} + 89.5, \dot{q}_{lo} = 210.5\,kW$$

Example 4.9 Carnot Heat Engine Problem
A Carnot heat engine runs between two heat reservoirs where the high-temperature reservoir is at 400 C and the low-temperature reservoir is at 25 C. What's the thermal efficiency of the heat engine? If the heat transfer from the high-temperature reservoir is 5000 BTU/hr., what's the work out [BTU/hr.]? What's the heat rejected [BTU/hr.]?

$$\eta = \frac{\dot{w}_o}{\dot{q}_{hi}} = 1 - \frac{T_{lo}}{T_{hi}} = 1 - \frac{298\,K}{673\,K} = 55.7\%$$

and

$$\dot{w}_o = 2786.0\frac{BTU}{hr}\ and\ \dot{q}_{lo} = 2214.0\frac{BTU}{hr}$$

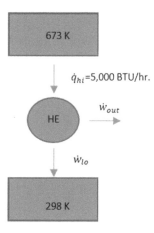

Example 4.10 Carnot Heat Pump and Refrigeration Problem Reverse the Carnot heat engine given in Example 4.2 and determine the beta (coefficient of performance) for a Carnot heat pump and Carnot refrigeration cycle.

$$\beta_{HP} = \frac{T_{hi}}{T_{hi} - T_{lo}} = 1.79, \beta_{Refrig.} = \frac{T_{lo}}{T_{hi} - T_{lo}} = .79$$

Example 4.11 Using Entropy to Determine Heat Transfer for a Given Process
Nitrogen isentropically expands from 300 K to 600 K and the initially specific volume is 0.5 m³/kg, then what's the final specific volume?

$$\Delta s = s_{T_2}^{ref} - s_{T_1}^{ref} + R_g \ln\left(\frac{v_2}{v_1}\right)$$

$$T_1 = 300\,K, T_2 = 600\,K, v_1 = .5\frac{m^3}{kg}, R_g = .2968\frac{kJ}{kg-K}, s_{300\,K}^{ref} = 6.8463\frac{kJ}{kg-K}, s_{600\,K}^{ref}$$

$$= 7.5741\frac{kJ}{kg-K}$$

and

$$0 = 7.7541 - 6.8463 + .2968 * \ln\left(\frac{v_2}{.5}\right), v_2 = .04305\frac{m^3}{kg}$$

Example 4.12 Entropy Change for a Solid
Stainless Steel (304) goes from 300 C to 600 C. What's the change in entropy [kJ/kg-K]?

$$\Delta s = c\ln\left(\frac{T_2}{T_1}\right) = .46\ln\left(\frac{873}{573}\right) = .194\frac{kJ}{kg-K}$$

Example 4.13 Entropy Change for an Ideal, Perfect Gas
Treating air as an ideal perfect gas, determine the entropy change from {25 C, 101.3 kPa} to {200 C, 250 kPa}.

c_p	[kJ/kg-K]	1.004
Rg	[kJ/kg-K]	0.287
T(1)	[C]	25
T(2)	[C]	200
T(1)	[K]	298
T(2)	[K]	571
P(1)	[kPa]	101.3

c_p	[kJ/kg-K]	1.004
P(2)	[kPa]	250
S(0,T1)	[kJ/kg-K]	6.87305
S(0,T2)	[kJ/kg-K]	7.5244
Delta(s)	[kJ/kg-K]	0.393628
Delta(s)	[kJ/kg-K]	0.392082

$$\Delta s = C_p^0 \ln\left(\frac{T_2}{T_1}\right) - R_g \ln\left(\frac{P_2}{P_1}\right) = .3936 \frac{kJ}{kg - K}$$

Example 4.14 Entropy Change for an Ideal, Non-perfect Gas Using Table
Treating air as an ideal non-perfect gas, determine the entropy change from {25 C, 101.3 kPa} to {200 C, 250 kPa}. Use the appropriate equation and thermodynamic tables.

$$\Delta s = s_{T2}^{ref} - s_{T1}^{ref} - R_g \ln\left(\frac{P_2}{P_1}\right) = .3921 \frac{kJ}{kg - K}$$

Example 4.15 Entropy Change for an Ideal, Non-perfect Gas Using Integration
Treating air as an ideal non-perfect gas, determine the entropy change from {25 C, 101.3 kPa} to {200 C, 250 kPa}. Use the appropriate equation and a cubic model for specific heat, integrate the resulting equation.

$$ds = \frac{\delta q}{T}, Tds = C_v dT + Pdv \text{ where } \delta q - \delta w = du$$

and

$$ds = C_v \frac{dT}{T} + R_g \frac{dv}{v}$$

Integrating both sides results in

$$\Delta s = \int_{T_1}^{T_2} c_v(T) \frac{dT}{T} + R_g \ln\left(\frac{v_2}{v_1}\right)$$

where c_p is expressed as

$$c_p = \beta_0 + \beta_1 \theta + \beta_2 \theta^2 + \beta_3 \theta^3 \text{ and } \frac{c_p}{c_v} = \gamma \text{ where } \theta = T[K] \Big/ 1000$$

$$\int_{T_1}^{T_2} \frac{C_p}{T} dT = \beta_0 \ln(T) + \beta_1 \theta + \frac{\beta_2}{2} \theta^2 + \frac{\beta_3}{3} \theta^3$$

and

$$\Delta s = .40 \frac{kJ}{kg-K}$$

Example 4.16 Irreversible Turbine

A steam turbine receives steam at a pressure of 2 MPa and a temperature of 300 C. The steam leaves the turbine at a pressure of 15 kPa. The work output of the turbine is measured and is found to be 600 kJ/kg of steam flowing through the turbine. Determine the efficiency of the turbine.

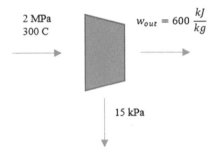

$$h_1 = 3051.15 \frac{kJ}{kg}, s_1 = 7.1228 \frac{kJ}{kg-K}$$

At 15 kPa,

$$h_l = 225.91 \frac{kJ}{kg}, h_g = 2599.06 \frac{kJ}{kg}, s_l = .7548 \frac{kJ}{kg-K}, s_g = 8.0084 \frac{kJ}{kg-K}$$

And

$$s_1 = s_2 \rightarrow X = 82.88\% \text{ and } h_{mix}^{iso} = 2192.68 \frac{kJ}{kg}$$

Therefore,

$$\eta = \frac{600}{3023.50 - 2192.68} = 72\%$$

Example 4.17 Reversible Turbine
What would be the work output of a reversible turbine with an input pressure of
1 MPa and temperature of 300 C and an output pressure of 15 kPa?

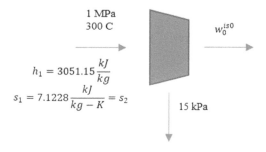

$$X = 87.79\%, h_2^{iso} = 2309.32\frac{kJ}{kg}, and\ w_o^{iso} = 741.83\,kJ\,/\,kg$$

Example 4.18 Adiabatic Turbine
Nitrogen (N_2) enters a turbine where the overall thermal efficiency is 70%. The inlet
temperature and pressure of the fluid are 300 K and 200 kPa and the exit pressure is
100 kPa. Assuming an ideal fluid and assuming the turbine to be adiabatic, determine the exit temperature and work out.

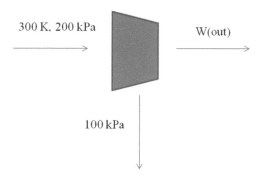

This problem will be solved by two methods

1. Method of engineering efficiency based on enthalpies and steam tables.
2. Analytical solution.

Method 1 Solution
- Assume an ideal gas.
- Adiabatic

$$\eta_{work\ out} = \frac{h_1 - h_2^{Act}}{h_1 - h_2^{iso}} \tag{4.87}$$

and

$$\Delta s = S_{T2}^0 - S_{T1}^0 - R_g \ln\left(\frac{P_2}{P_1}\right) \tag{4.88}$$

and

$$0 = S_{T2}^0 - 6.8463 - .2968 * \ln\left(\frac{1}{2}\right), S_{T2}^0 = 6.641\frac{kJ}{kg-K} \tag{4.89}$$

and using a linear interpolation for S and T

S(1)	6.4250	T(1)	200
S(2)	6.6568	T(2)	250
S(x)	6.6410	T(x)	246.59

T_2(iso) = 247 K and using a linter interpolation for T and h

T(1)	200	h(1)	208
T(2)	250	h(2)	259.7
T(x)	247	h(x)	256.58

h_2(iso) = 256.6 kJ/kg and w(iso) = 55.1 kJ/kg
Additionally,

$$w_{act} = \eta w_{iso} = 70\% * 55.1 = 38.6\frac{kJ}{kg} \tag{4.90}$$

and h_2(act) = 311.67–38.6 = 277.11 kJ/kg and using a linear interpolation for h and T

h(1)	259.70	T(1)	250
h(2)	311.67	T(2)	300
h(x)	273.11	T(x)	262.90

T_2(act) = 263 K

Method 2 Solution
- Assume an ideal, perfect gas.
- Reversible.
- Adiabatic.

First law for the system is

$$w_{out} = h_1 - h_2 = c_p\left(T_1 - T_2\right) \tag{4.91}$$

and second law for the system is

$$\Delta s = \frac{\delta q}{T} + \sigma_{sys} = 0, s_1 = s_2 \tag{4.92}$$

and because the system is isentropic

$$\frac{T_2}{T_1} = \left(\frac{P_2}{P_1}\right)^{\gamma-1/\gamma} \tag{4.93}$$

and substituting Eq. 4.93 into Eq. 4.91 for T_2 results in

$$w_{out}^{iso} = c_p T_1 \left\{ 1 - \left(\frac{P_2}{P_1}\right)^{\gamma-1/\gamma} \right\} = c_p T_1 \left\{ 1 - \left(\frac{P_2}{P_1}\right)^{R_g/c_p} \right\} \tag{4.94}$$

and using the definition of overall thermal efficiency for a work out device

$$\eta = \frac{w_{out}^{act}}{w_{out}^{iso}}, w_{out}^{act} = \eta c_p T_1 \left\{ 1 - \left(\frac{P_2}{P_1}\right)^{R_g/c_p} \right\} \tag{4.95}$$

and

$$\eta c_p T_1 \left\{ 1 - \left(\frac{P_2}{P_1}\right)^{R_g/c_p} \right\} = c_p (T_1 - T_2) \tag{4.96}$$

and solving for T_2 results in

$$T_2 = T_1 - \eta T_1 \left\{ 1 - \left(\frac{P_2}{P_1}\right)^{R_g/c_p} \right\} = T_1 \left[1 - \eta \left\{ 1 - \left(\frac{P_2}{P_1}\right)^{R_g/c_p} \right\} \right] \tag{4.97}$$

Given $c_p = 1.042$ kJ/kg-K and $R_g = 0.2968$ kJ/kg-K
Such that

$$w_{out}^{act} = \eta c_p T_1 \left\{ 1 - \left(\frac{P_2}{P_1}\right)^{R_g/c_p} \right\} = .7 * 1.042 * 300 \left\{ 1 - \left(\frac{1}{2}\right)^{.2968/1.042} \right\} = 39.2 \, kJ/kg \tag{4.98}$$

and

$$T_2 = T_1 \left[1 - \eta \left\{ 1 - \left(\frac{P_2}{P_1} \right)^{R_g / c_p} \right\} \right] = 300 * \left[1 - .7 \left\{ 1 - \left(\frac{1}{2} \right)^{.2968 / 1.042} \right\} \right] = 262 \, K \ (4.99)$$

4.10.2   Problems

Problem 4.1 A room is heated with a 3-kW electric heater. How much power can be saved if a heat pump with a beta of 3 is to be used instead?

Problem 4.2 An air conditioner discards 3 kW to the surroundings with a power input of 1 kW. Find the rate of cooling and the Beta.

Problem 4.3 A Carnot heat engine has a thermal efficiency of 60%. If all else the same but the high temperature is raised 10%, what is the new thermal efficiency?

Problem 4.4 A household freezer operates in a room at 25 C. Heat must be transferred from the cold space at a rate of 4 kW to maintain its temperature at −20 C. What is the theoretically smallest motor required to operate this freezer?

Problem 4.5 A piston/cylinder contains 1 kg of R-134a at 101.3 kPa. It will be compressed in an adiabatic reversible process to 400 kPa and should be at 100 C. What should the initial temperature be?

Problem 4.6 Water at 300 C and 2 MPa is brought to saturated vapor in a rigid container. Find the final temperature and the specific heat transfer.

Problem 4.7 Two 10 kg blocks of copper, one at 300 C and the other at 25 C, come into thermal contact with each other. Find the change in entropy of the system and the final temperature.

Problem 4.8 R-410a at 400 kPa is brought from 20 C to 120 C in a constant pressure process. Evaluate the entropy changes using the constant specific heat approach and the ideal gas tables that evaluate the integral involving specific heat.

Problem 4.9 A rigid tank contains 5 kg of methane at 1000 K and 1 MPa. It is now cooled to 300 K. Find the heat transfer and the entropy change.

Problem 4.10 A steam turbine inlet is at 1 MPa and 300 C. The exit is at 100 kPa. What is the lowest possible exit temperature? Which efficiency does this correspond with?

Problem 4.10 A steam turbine inlet is at 1 MPa and 300 C. The exit is at 100 kPa. What is the highest possible exit temperature? What efficiency does this correspond with?

Problem 4.11 A steam turbine inlet is at 1 MPa and 300 C. The exit conditions are 200 kPa and 200 C. What is the isentropic efficiency of this turbine?

Problem 4.12 Steam enters a turbine at 300 C and 1 MP and exits as saturated vapor at 50 kPa. What is the isentropic efficiency?

Problem 4.13 Two compressors for an agricultural application involving ammonia as a fertilizer are being evaluated in an economic study. The selected compressor will only be utilized for three years, and it will have no market value at the end of the year. The interest rate is 10%, compounded annually. Pertinent data are summarized as follows:

	ABC	XYZ
Investment cost	$2900	$6200
One-year maintenance cost	$170	$510
Efficiency	80%	90%

The compressor will be used to compress 945 kg per hour of ammonia initially at 20 kPa and 300 K to a pressure of 0.1 MPa. If electricity costs 10 cents per kWh and the compressor will be operated 4000 hours per year, which compressor should be chosen? Assume ammonia is an ideal gas with a constant heat capacity.

Problem 4.14 120 kg/min of nitrogen gas is compressed from 1 bar and 25 °C to 10 bar and 25 °C using a two-step process. The first step adiabatically compresses the nitrogen from 1 bar and 25 °C to 10 bar. This adiabatic compression step requires 150 kW of work. The second step involves cooling the nitrogen exiting the compressor to 25 °C in a heat exchanger using water. The water enters the heat exchanger at 25 °C and exits at 40 °C. Determine the temperature of the nitrogen exiting the compressor and the mass flow rate of water needed for the cooling. Use a temperature-dependent heat capacity for nitrogen given as a cubic polynomial.

Problem 4.15 Air enters an adiabatic nozzle steadily at 300 kPa, 200 °C, and 35 m/s. It leaves at 100 kPa. The nozzle inlet has a diameter of 20.0 cm. If the outlet area of the nozzle is 5.60% of its inlet area, determine the exit temperature and velocity of the air. Use a temperature-dependent heat capacity for air given as a cubic polynomial.

Problem 4.16 Ammonia flowing at 1 mol s^{-1} is adiabatically compressed from 0.2 bar and 300 K. The compression requires 6450 J mol^{-1} of work. If the compressor has an efficiency of 75%, determine the outlet pressure and temperature of the ammonia. Assume that ammonia behaves as an ideal gas with a temperature-dependent heat capacity given as a cubic polynomial.

References

1. Borgnakke and Sonntag. (2009). *Fundamentals of thermodynamics* (7th ed.). John Wiley and Sons.
2. Potter and Scott. (2004). *Thermal sciences* (1st ed.). Brooks/Cole – Thomson Learning.

Chapter 5
Various Heat Engines and Refrigeration Cycles

5.1 Preview

Previously, we have seen how four engineering devices can be arranged in a certain order to provide a Rankine cycle, which is given in Sect. 3.8; Rankine cycles form the basis to electrical power generation for much of this country [1]. We have also explored the Carnot heat engine, heat pump, and refrigeration cycles.

In this chapter, we will explore in more depth two classes of heat engines (and refrigeration cycles): the first class is known as a vapor phase cycle and the second class is known as a gas cycle. For vapor phase cycles (Rankine cycle and reverse Rankine cycle), the cycle goes through a series of phases and often involves water. As well we know, when water is involved states change in such a manner that we resort to thermodynamic tables to determine quantities such as heat added or work released. For gas phase cycles, we'll assume the working fluid is air, has constant properties such as c_v, and that the ideal gas law is applicable; in gas phase cycles, there are no changes in phase.

A major aspect of this chapter is the development of an equation for thermal efficiency for heat engines, which is defined as

$$\eta_{overall} = \frac{w_{net}}{q_{in}} = \frac{w_{out} - w_{in}}{q_{in}} \tag{5.1}$$

And for refrigeration cycles, coefficient of performance is determined, and defined as

$$\beta = \frac{q_{lo}}{w_{net}} \tag{5.2}$$

Electronic Supplementary Material: The online version of this chapter (https://doi.org/10.1007/978-3-030-87387-5_5) contains supplementary material, which is available to authorized users.

H. C. Foust III, *Thermodynamics, Gas Dynamics, and Combustion*, https://doi.org/10.1007/978-3-030-87387-5_5

Please note for actual heat engines and refrigeration cycles, finding heat and work are not as easy to determine as for Carnot heat engines and refrigeration cycles and will involve energy and entropy balances around each engineering device. Many of the principles of this chapter will be made clear by examples.

5.2   Vapor Phase Cycles

In this section, we'll review (by examples) the Rankine cycle that is a form of heat engine and then reverse these Rankine cycles to form a refrigeration cycle.

Example 5.1 Rankine Cycle

Determine the overall thermodynamic efficiency of the Rankine cycle given as Table 5.1.

As was shown in Sect. 3.8, the steps in completing the problem are as follows

1. Determine the enthalpy for each state by conducting an energy balance around each engineering device; h_1 is given from the thermodynamic tables; h_2 is determined by knowing the pump work; h_3 is given from the thermodynamic tables; and h_4 is determined by assuming s_3 equals s_4 and finding the enthalpy of the mixture with the same quality as s_4
2. Determine the heat and work associated with each engineering device
3. Check that the sum of the work is equal to the sum of the heat
4. Determine the overall thermal efficiency

Each step is given below
Energy Balance around Pump
Work is considered negative when it's toward the system (Fig. 5.1).
It was determined that $h_1 = 225.91$ *kJ/kg* and $\nu_1 = .00101$ *m³per kg*
The energy balance for the pump is

$$h_1 + \delta w_{in} = h_2 \tag{5.1}$$

Table 5.1 Example 5.1 (Given)

					Given					
State No.	T	P	nu	Quality	Phase	h	s	q	w	Engineering Device
	[C]	[kPa]	[m^3/kg]			[kJ/kg]	[kJ/kg-k]	[kJ/kg]	[kJ/kg]	
1		15		0	Saturated Liquid					
									-w(in)	Isentropic Pump
2		2,000			Compressed Liquid					
								+q(in)		Boiler
3	300	2,000			Superheated Vapor					
									+w(out)	Isentropic Turbine
4		15		X	Mixture					
								-q(out)		Condenser
1		15		0	Saturated Liquid					

Fig. 5.1 Pump

Energy Balance around Pump

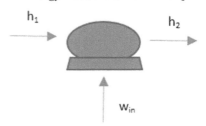

h_1 h_2 w_{in}

Fig. 5.2 Boiler

Energy Balance around Boiler

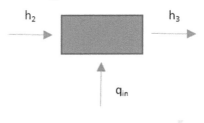

h_2 h_3 q_{in}

Fig. 5.3 Turbine

Energy Balance around Turbine

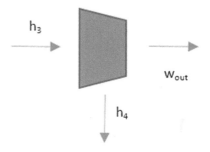

h_3 w_{out} h_4

Fig. 5.4 Condenser

Energy Balance around Condenser

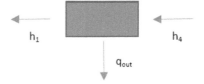

h_1 h_4 q_{out}

Another form of the energy equation for incompressible fluids (assuming $v_1 = v_2$) is

$$PE_1 + (Pv)_1 + KE_1 + frictional\ losses$$
$$+ pump\ work + turbine\ work = PE_2 + (Pv)_2 + KE_2 \tag{5.2}$$

$$gZ_1 + \frac{P_1}{\rho} + \frac{v_1^2}{2} - h_L + h_p - h_T = gZ_2 + \frac{P_2}{\rho} + \frac{v_2^2}{2}$$

where it is assumed $\Delta KE = \Delta PE = h_L = h_T = 0$
Thus,

$$\frac{P_1}{\rho} + h_p = \frac{P_2}{\rho} \tag{5.3}$$

or

$$-\delta w_{in} = v_1 \Delta P = (.00101)(1985) = 2.0\,kJ/kg \tag{5.4}$$

and

$$h_2 = 225.91 + 2 = 227.9\,kJ/kg \tag{5.5}$$

Energy Balance around Boiler
Heat is considered positive when it's toward the system.
It was determined that h_3 is 3023.5 kJ/kg and the energy balance for the boiler is

$$h_2 + \delta q_{in} = h_3, 227.91 + \delta q_{in} = 3023.5, \delta q_{in} = 2795.58\,kJ/kg \tag{5.6}$$

Energy Balance around Turbine
Work is considered positive when it's away from the system.
Because the turbine is isentropic, $s_3 = s_4$ and

$$s_3 = 6.77 \frac{kg}{kg - K} = s_4 = s_l + X(s_g - s_l)$$

Where for 15 kPa

P	s(l)	s(g)	s(mix)	X
[kPa]	[kJ/kg-K]	[kJ/kg-K]	[kJ/kg-K]	
15	0.7584	8.0084	6.77	82.9%
X	h(l)	h(g)	h(mix)	
	[kJ/kg]	[kJ/kg]	[kJ/kg]	
82.9%	225.91	2599.06	2192.69	

and $h_4 = 2192.7\,kJ/kg$ and the energy balance for the turbine is

$$h_3 = \delta w_{out} + h_4, 2023.5 = \delta w_{out} + 2192.7, \delta w_{out} = 830.8\,kJ/kg \tag{5.7}$$

Energy Balance around Condenser
Heat is considered negative when it's away from the system.

The energy balance for the condenser is

$$h_4 = \delta q_{out} + h_1, 2192.7 = \delta q_{out} + 225.9, \delta q_{out} = -1966.7 \, kJ/kg \qquad (5.8)$$

The summary of the solution is given as Table 5.2 where $\sum \delta q_i = \sum \delta w_i$ and the overall thermal efficiency is 29.65%.

5.2.1 Improvements to Rankine Cycle

This section will determine the effects on the overall thermal efficiency of changing the following

1. Increasing the boiler pressure
2. Increasing the boiler temperature
3. Reducing the condenser pressure

These changes are given as Examples 5.2, 5.3, and 5.4.

Example 5.2 Increase Boiler Pressure
The solution is given as Table 5.3.

Example 5.3 Increase Boiler Temperature
The solution is given as Table 5.4.

Example 5.4 Reduce the Condenser Pressure
The solution is given as Table 5.5.

Table 5.2 Example 5.1 (Solution)

					Solution					
State No.	T	P	nu	Quality	Phase	h	s	q	w	Engineering Device
	[C]	[kPa]	[m^3/kg]			[kJ/kg]	[kJ/kg-k]	[kJ/kg]	[kJ/kg]	
1	53.97	15	0.00101	0	Saturated Liquid	225.91	0.75			
									-2.01	Isentropic Pump
2		2,000			Compressed Liquid	227.923	0.75			
								2795.58		Boiler
3	300	2,000	0.12547		Superheated Vapor	3023.5	6.77			
									830.82	Isentropic Turbine
4		15		82.88%	Mixture	2192.68	6.77			
								-1966.77		Condenser
1		15		0	Saturated Liquid	225.91	0.75			
								828.81	828.81	
									Difference =	0
									Eff=	29.65%

				Linear Interpolation			
X(1)	779	Y(1)	87				
X(2)	1,085	Y(2)	87.35	$y = \dfrac{(y_2 - y_1)}{(x_2 - x_1)}(x - x_1) + y_1$	$v_{mix} = v_l + X(v_g - v_l)$		
X	800	Y	87.47				

P	s(l)	s(g)	s(mix)	X		h(l)	h(g)	h(mix)
[kPa]	[kJ/kg-K]	[kJ/kg-K]	[kJ/kg-K]			[kJ/kg]	[kJ/kg]	[kJ/kg]
15	0.7548	8.0084	6.7663	82.88%		225.91	2599.06	2192.68

Table 5.3 Example 5.2 (Solution)

State No.	T	P	nu	Quality	Phase	h	s	q	w	Engineering Device
	[C]	[kPa]	[m^3/kg]			[kJ/kg]	[kJ/kg-k]	[kJ/kg]	[kJ/kg]	
1	53.97	15	0.001014	0	Saturated Liquid	225.91				
									-4.04	Isentropic Pump
2		4,000			Compressed Liquid	229.95				
								2730.73		Boiler
3	300	4,000	0.05884		Superheated Vapor	2960.68	6.36			
									900.47	Isentropic Turbine
4		15		82.88%	Mixture	2060.21	6.36			
								-1834.30		Condenser
1		15		0	Saturated Liquid	225.91				

	896.43	896.43	
		Difference =	0
		Eff=	32.83%

Table 5.4 Example 5.3 (Solution)

State No.	T	P	nu	Quality	Phase	h	s	q	w	Engineering Device
	[C]	[kPa]	[m^3/kg]			[kJ/kg]	[kJ/kg-k]	[kJ/kg]	[kJ/kg]	
1	53.97	15	0.001014	0	Saturated Liquid	225.91	0.75			
									-2.01	Isentropic Pump
2		2,000			Compressed Liquid	227.9228	0.75			
								3129.56		Boiler
3	450	2,000	0.16353		Superheated Vapor	3357.48	7.28			
									995.29	Isentropic Turbine
4		15			Mixture	2362.19	7.28			
								-2136.28		Condenser
1		15		0	Saturated Liquid	225.91	0.75			

	993.28	993.28	
		Difference =	0
		Eff=	31.74%

Table 5.5 Example 5.4 (Solution)

State No.	T	P	nu	Quality	Phase	h	s	q	w	Engineering Device
	[C]	[kPa]	[m^3/kg]			[kJ/kg]	[kJ/kg-k]	[kJ/kg]	[kJ/kg]	
1	32.88	5	0.001005	0	Saturated Liquid	137.79				
									-2.00	Isentropic Pump
2		2,000			Compressed Liquid	139.79				
								2883.71		Boiler
3	300	2,000	0.12547		Superheated Vapor	3023.50	6.77			
									960.54	Isentropic Turbine
4		5		79.43%	Mixture	2062.96	6.77			
								-1925.17		Condenser
1		5		0	Saturated Liquid	137.79	0.00			

	958.54	958.54	
		Difference =	0
		Eff=	33.24%

Table 5.6 Summary

Scenario	Description	q(in)	q(out)	w(in)	w(out)	η	% Change
		[kJ/kg]	[kJ/kg]	[kJ/kg]	[kJ/kg]		
1	Base case	2795.577	−1966.77	−2.013	830.82	29.65%	0.00%
2	Increase T(hi)	3129.56	−2136.28	−2.01	995.29	31.74%	7.05%
3	Decrease P(lo)	2883.705	−1925.17	−2.005	960.54	33.24%	12.12%
4	Increase P(hi)	2730.73	−1834.3	−4.04	900.47	32.83%	10.73%

A summary table is given as Table 5.6. From the table, we see that reducing the pressure in the condenser has the largest positive effect on overall thermal efficiency.

5.2.2 Effects of Engineering Efficiency on Overall Thermal Efficiency

In this section, the effects of engineering efficiency associated with the pump and turbine on overall thermal efficiency are explored. The difference between this section and the previous section is that an additional table is created to determine the enthalpies at states "2" and "4" for actual conditions where the appropriate equations are given below.

$$\eta_{pump} = \frac{h_2^{iso} - h_1}{h_2^{act} - h_1}$$

$$\eta_{turbine} = \frac{h_3 - h_4^{act}}{h_3 - h_4^{iso}}$$

$$h_2^{act} = h_1 + \frac{h_2^{iso} - h_1}{\eta_{pump}}$$

$$h_4^{act} = h_3 - \eta_{turbine}\left(h_3 - h_4^{iso}\right)$$

Example 5.5 90% Efficient Pump and Turbine
The solution is given as Table 5.7.

Example 5.6 80% Efficient Pump and Turbine
The solution is given as Table 5.8.

Example 5.7 70% Efficient Pump and Turbine
The solution is given as Table 5.9.

Table 5.7 Example 5.5 (Solution)

State No.	T [C]	P [kPa]	nu [m^3/kg]	Quality	Phase	h [kJ/kg]	s [kJ/kg-k]	q [kJ/kg]	w [kJ/kg]	Engineering Device
										Solution
1	53.97	15	0.001014	0	Saturated Liquid	225.91				
									-2.24	90% Efficient Pump
2		2,000			Compressed Liquid	228.1464				
								2795.35		Boiler
3	300	2,000	0.12547		Superheated Vapor	3023.5				
									747.74	90% Efficient Turbine
4		15		82.88%	Mixture	2275.762				
								-2049.85		Condenser
1		15		0	Saturated Liquid	225.91				
								745.50	745.50	
								Difference =	0	
								Eff=	26.67%	

Table 5.8 Example 5.6 (Solution)

State No.	T [C]	P [kPa]	nu [m^3/kg]	Quality	Phase	h [kJ/kg]	s [kJ/kg-k]	q [kJ/kg]	w [kJ/kg]	Engineering Device
Solution										
1	53.97	15	0.001014	0	Saturated Liquid	225.91				
									-2.52	80% Efficient Pump
2		2,000			Compressed Liquid	228.426				
								2795.07		Boiler
3	300	2,000	0.12547		Superheated Vapor	3023.5				
									664.66	80% Efficient Turbine
4		15		82.88%	Mixture	2358.844				
								-2132.93		Condenser
1		15		0	Saturated Liquid	225.91				
								662.14	662.14	
								Difference =	0	
								Eff=	23.69%	

Table 5.9 Example 5.7 (Solution)

State No.	T [C]	P [kPa]	nu [m^3/kg]	Quality	Phase	h [kJ/kg]	s [kJ/kg-k]	q [kJ/kg]	w [kJ/kg]	Engineering Device
										Solution
1	53.97	15	0.001014	0	Saturated Liquid	225.91				
									-2.88	70% Efficient Pump
2		2,000			Compressed Liquid	228.7854				
								2794.71		Boiler
3	300	2,000	0.12547		Superheated Vapor	3023.5				
									581.57	70% Efficient Turbine
4		15		82.88%	Mixture	2441.926				
								-2216.02		Condenser
1		15		0	Saturated Liquid	225.91				
								578.70	578.70	
								Difference =	0	
								Eff=	20.71%	

Table 5.10 Summary

Scenario	Description	q(in)	q(out)	w(in)	w(out)	Etta	% Change
		[kJ/kg]	[kJ/kg]	[kJ/kg]	[kJ/kg]		
5	90% efficient pump and turbine	2795.35	−2049.85	−2.24	747.74	26.67%	−10.04%
6	80% efficient pump and turbine	2795.07	−2132.93	−2.52	664.66	23.69%	−20.09%
7	70% efficient pump and turbine	2794.71	−2216.02	−2.88	581.57	20.71%	−30.16%

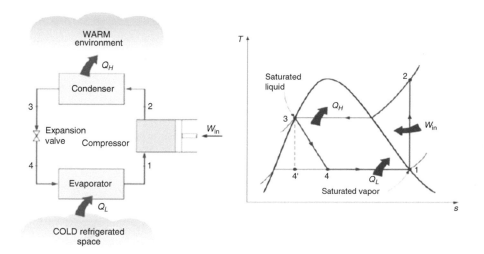

Figs. 5.5 and 5.6 Reverse Rankine cycle [2]

A summary of Examples 5.5, 5.6, and 5.7 is given as Table 5.10. Please note with a 10% decrease in engineering efficiency of the pump and turbine there is an equal decrease in the overall thermal efficiency of 10%; additionally, the increased work(in) due to inefficiencies in the pump are minor compared to the loss of work(out) due to inefficiencies in the turbine and, as such, much of the reduction in overall thermal efficiency is attributable to the turbine inefficiencies.

5.2.3   Reverse Rankine Cycle

The reverse Rankine cycle is shown in Figs. 5.5 and 5.6. The reverse Rankine cycle involves the following four processes and the device that provides undertakes the process

1->2, Pressure Increase (work added), Isentropic Compressor
2->3, Heat Dumped, Condenser
3->4, Pressure Decreased (work extracted), Isenthalpic Valve
4->1, Heat Added, Evaporator

In order to determine Eq. 5.2 particular to a reverse Rankine cycle, an energy balance must be conducted around each engineering device and all states, heat, and work accounted for in a table.

The following simplifying assumptions and conditions were made

$$h_3 = h_4$$

1. Isentropic compressor, $s_1 = s_2$
2. State 1 is a saturated vapor at T_{lo}
3. State 2 is a superheated vapor at T_{hi}
4. State 3 is a saturated liquid at T_{hi}
5. State 4 is a mixture at T_{lo}

Note – the sum of the work equals the sum of the heat for a cycle.

Energy Balance for Compressor The energy balance around the compressor in Fig. 5.7 is

$$h_1 + w_{in} = h_2 \tag{5.9}$$

And solving for w_{in}

$$w_{in} = h_2 - h_1 \tag{5.10}$$

Energy Balance for Condenser The energy balance around the condenser in Fig. 5.8 is

$$h_2 = q_{out} + h_3 \tag{5.11}$$

And solving for q_{out}

$$q_{out} = h_2 - h_3 \tag{5.12}$$

Energy Balance for Expansion Valve The energy balance around the expansion valve in Fig. 5.9 is

$$h_3 = h_4 = h_{sl,Thi} \tag{5.13}$$

Fig. 5.7 Compressor

Energy Balance for Compressor

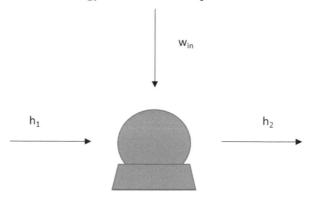

Fig. 5.8 Condenser

Energy Balance for Condenser

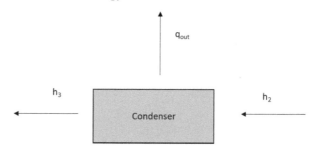

Fig. 5.9 Expansion Valve

Energy Balance for Expansion Valve

Fig. 5.10 Evaporator   **Energy Balance for Evaporator**

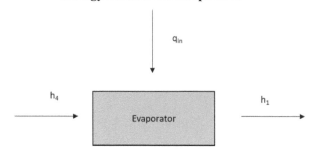

Table 5.11 Various States, heat and work for a reverse Rankine cycle

State	Process/Device	T	P	s	X	h	δw	δq
		[K]	[MPa]	[KJ/kg-K]		[kJ/kg]	[kJ/kg]	[kJ/kg]
1		T_1	P_1			h_1		
	Isentropic compressor						$-w_{in}$	
2		T_2	P_2			h_2		
	Condenser							$+q_{in}$
3		T_3	P_2			h_3		
	Expansion valve						0	
4		T_1	P_1			h_4		
	Evaporator							$-q_{out}$
1		T_1	P_1			h_1		

Energy Balance for Evaporator The energy balance around the evaporator in Fig. 5.10 is

$$h_4 + q_{in} = h_1 \qquad (5.14)$$

And solving for q$_{out}$

$$q_{in} = h_1 - h_4 \qquad (5.15)$$

Once all enthalpies, works, and heats are determined, then a table (see Table 5.11) is completed and the coefficient of performance for the cycle is calculated, which is given as

$$\beta = \frac{q_{lo}}{w_{in}} = \frac{h_1 - h_4}{h_2 - h_1} \qquad (5.16)$$

The final step is to check the first law of thermodynamics for a cycle

$$\sum \delta q = \sum \delta w \qquad (5.17)$$

Example 5.8 Reverse Rankine Cycle
Problem Statement
  A reverse Rankine cycle has a T_{lo} of -10 C, P_{hi} of 1 MPa, and the working fluid is R-134a. Find q_{lo}, q_{hi}, and $Beta_{ref}$.

Strategy
  Determine all enthalpies and using energy balances around each engineering device, determine w_{in}, w_{out}, q_{in}, and q_{out}. Once all this is found, determine beta for the refrigeration cycle.

Solution
State 1 – Saturated Vapor
$h_1 = h$(saturated vapor) $= 392.28$ kJ/kg
$s_1 = 1.7319$ kJ/kg-K $= s_2$

State 2 – Superheated Vapor
Using a linear interpolation, find h_2
1000 kPa

T [C]	s [kJ/kg-K]	h [kJ/kg]
40	1.7148	420.25
50	1.7493	431.24
44.94	1.7319	425.68

  h_2 is 425.68 kJ/kg

State 3 – Saturated Liquid
  At state 3, pressure is 1000 kPa and X = 0%, using saturated R134a tables plus linear interpolations

Pressure [kPa]	T [C]	h [kJ/kg]
877.6	35	249.10
1017	40	256.54
1000	39.34	255.56

  h_3 is 255.56 kJ/kg

State 4 – Mixture
Since $h_3 = h_4$, h_4 is 255.56 kJ/kg
Using the following equations,

$$w_{in} = h_1 - h_2 = 392.29 - 425.68 = -33.40 \, kJ/kg$$

$$q_{hi} = h_3 - h_2 = 255.56 - 425.68 = -170.12 \, kJ/kg$$

$$q_{lo} = h_1 - h_4 = 392.28 - 255.56 = 136.72 \, kJ/kg$$

And beta for the refrigeration cycle is

$$\beta = \frac{136.72}{33.40} = 4.09$$

5.3   Gas Cycles

In this section, the following assumptions will be made (air-standard cycle)

1. Air is the working fluid, acts ideal, and there is no inlet process, nor outlet process
2. Combustion is treated as an external heat added to the system
3. The cycle is completed by heat dumped to the atmosphere
4. All processes are internally reversible
5. Specific heat is assumed constant and takes the value for air at 300 K

5.3.1   Brayton Cycle

The first example of an air cycle we'll look at is the Brayton cycle, which is given in Figs. 5.11 and 5.12.

The following simplifying assumptions and conditions were made

1. Isentropic compressor, $s_1 = s_2$.
2. Isentropic turbine, $s_3 = s_4$
3. The combustion is isobaric, $P_2 = P_3$

Note – the sum of the work always equals the sum of the heat for a cycle.

Energy Balance for Compressor   The energy balance around the compressor in Fig. 5.13 is

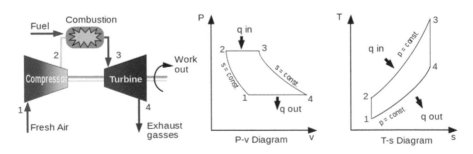

Figs. 5.11 and 5.12   Brayton cycle [3]

Fig. 5.13 Compressor

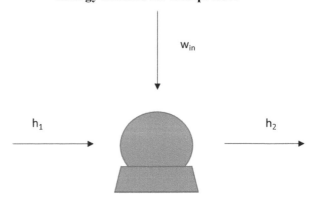

Energy Balance for Compressor

Energy Balance for Combustion Chamber

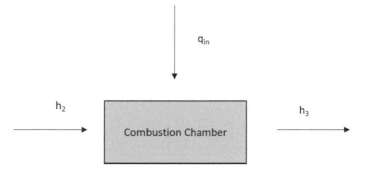

Fig. 5.14 Combustion chamber

$$h_1 + w_{in} = h_2 \qquad (5.18)$$

And solving for w_{in}

$$w_{in} = h_2 - h_1 \qquad (5.19)$$

Energy Balance for Combustion Chamber The energy balance around the combustion chamber in Fig. 5.14 is

$$h_2 + q_{in} = h_3 \qquad (5.20)$$

And solving for q_{in}

$$q_{in} = h_3 - h_2 \qquad (5.21)$$

Fig. 5.15 Turbine

Energy Balance for Turbine

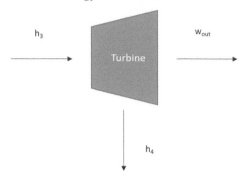

Energy Balance for Turbine The energy balance around the turbine in Fig. 5.15 is

$$h_3 = w_o + h_4 \tag{5.22}$$

And solving for w_o

$$w_o = h_4 - h_3 \tag{5.23}$$

Using Eq. 5.1,

$$\eta_{overall} = 1 - \frac{q_{lo}}{q_{hi}} = 1 - \frac{c_p \left(T_4 - T_1\right)}{c_p \left(T_3 - T_2\right)} = 1 - \frac{T_1}{T_2} \frac{\left(T_4 / T_1 - 1\right)}{\left(T_3 / T_2 - 1\right)} \tag{5.24}$$

Equation 5.24 can be simplified further by putting T_4 / T_1 and T_3 / T_2 in terms of pressures (see Eqs. 5.25, 5.26, and 5.27).
Note

$$\frac{P_2}{P_1} = \frac{P_3}{P_4} \tag{5.25}$$

Using the state equations for the isentropic path from state 1 to 2 results in

$$\frac{P_2}{P_1} = \left(\frac{T_2}{T_1}\right)^{\gamma / \gamma - 1} \tag{5.26}$$

And
Using the state equations for the isentropic path from state 3 to 4 results in

$$\frac{P_3}{P_4} = \left(\frac{T_3}{T_4}\right)^{\gamma/\gamma-1} \tag{5.27}$$

From Eqs. 5.25, 5.26, and 5.27 it is easy to see that

$$\frac{T_3}{T_4} = \frac{T_2}{T_1} \tag{5.28}$$

And

$$\frac{T_3}{T_2} - 1 = \frac{T_4}{T_1} - 1 \tag{5.29}$$

Therefore, Eq. 5.24 simplifies to

$$\eta_{overall} = 1 - \frac{T_1}{T_2}\frac{\left(T_4/T_1 - 1\right)}{\left(T_3/T_2 - 1\right)} = 1 - \frac{T_1}{T_2} \tag{5.30}$$

And from Eq. 5.26

$$\eta_{overall} = 1 - \frac{T_1}{T_2} = 1 - \frac{1}{\left(\frac{P_2}{P_1}\right)^{\gamma-1/\gamma}}$$

And defining

$$r_p = \frac{P_2}{P_1} \tag{5.31}$$

Results in

$$\eta_{overall} = 1 - r_p^{\gamma/\gamma-1} \tag{5.32}$$

Another quantity of interest when dealing with Brayton cycles is to determine how much of the turbine work goes back to the system to drive the compressor, which is defined as back work rate (BWR) and given as

$$BWR = \frac{w_{compressor}}{w_{turbine}} = \frac{h_2 - h_1}{h_3 - h_4} = \frac{T_2 - T_1}{T_3 - T_4} \tag{5.33}$$

Example 5.9 Brayton Cycle
Problem Statement
Given an air standard Brayton cycle where the ambient conditions are 100 kPa and 25 C, the r_p associated with the compressor is 10 and the highest temperature in the system is 1100 C, determine w_{in}, w_{out}, q_{in}, and q_{out}, the overall thermal efficiency, and the back work rate.

Strategy
Determine pressures and temperatures for all states. Once all temperatures are known and assuming an ideal, perfect gas, then the enthalpies are determined. Knowing all enthalpies perform an energy balance around each engineering device to determine work(in), work(out), heat(in), and heat(out). Once all this is determined, find the overall thermal efficiency and back work rate (BWR).

Solution

State 1
This is given as 100 kPa and 25 C (298 K).

State 2
Assuming an isentropic relationship between states 1 and 2, the following equation is appropriate for determining T_2.

$$\frac{T_2}{T_1} = \left(\frac{P_2}{P_1}\right)^{\gamma-1/\gamma} = 10^{\frac{.4}{1.4}} \rightarrow T_2 = 556\,K$$

And P_2 is given as 1000 kPa.

State 3
T_3 is given as 1100 C (1373 K).

State 4

$$\frac{T_4}{T_3} = \left(\frac{P_4}{P_3}\right)^{\gamma-1/\gamma} = \frac{1}{10^{\frac{.4}{1.4}}} \rightarrow T_4 = 711\,K$$

Once all temperatures, compute all enthalpies, $h_i = c_p T_i$, and determine w_{in}, w_{out}, q_{in}, and q_{out} from the following equations:

$$w_{in} = h_1 - h_2 = 289.152\frac{kJ}{kg} - \frac{558.261kJ}{kg} = -269.11kJ/kg$$

$$q_{in} = h_3 - h_2 = 1374.49 - 558.2651 = 820.23\,kJ/kg$$

$$w_{out} = h_3 - h_4 = 1378.49 - 713.99 = 664.51\,kJ/kg$$

$$q_{out} = h_1 - h_4 = 289.15 - 713.99 = -424.83\,kJ/kg$$

Additionally,

$$\eta = \frac{820.23 - 269.11}{820.23} = 48.2\%$$

$$WBR = \frac{269.11}{664.51} = 40.5\%$$

5.3.2 Reverse Brayton Cycle

Using the definition of coefficient of performance for a refrigeration cycle we see

$$\beta = \frac{q_{lo}}{w_{in} - w_{out}} = \frac{h_1 - h_4}{(h_2 - h_1) - (h_3 - h_4)}$$
$$= \frac{c_p(T_1 - T_4)}{c_p(T_2 - T_1) - c_p(T_3 - T_4)} = \frac{T_1 - T_4}{(T_2 - T_1) - (T_3 - T_4)}$$

$$= \frac{T_1 - T_4}{(T_2 - T_4) - (T_1 - T_4)} = \frac{1}{\dfrac{T_2 - T_4}{T_1 - T_4} - 1} \tag{5.34}$$

and

$$\beta = \frac{1}{\dfrac{T_2}{T_1}\left(\dfrac{1 - \dfrac{T_4}{T_2}}{1 - \dfrac{T_4}{T_1}}\right) - 1} \tag{5.35}$$

where

$$\frac{T_3}{T_4} = \frac{T_2}{T_1} \tag{5.36}$$

and

$$1 - \frac{T_3}{T_2} = 1 - \frac{T_4}{T_1} \tag{5.37}$$

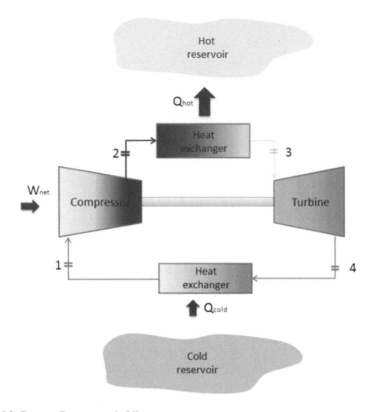

Figs. 5.16 Reverse Brayton cycle [4]

Therefore,

$$\beta = \frac{1}{T_2 / T_1 - 1} = \frac{1}{\left(\dfrac{P_2}{P_1}\right)^{\gamma-1/\gamma} - 1} = \frac{1}{r_p^{\gamma-1/\gamma} - 1} \tag{5.38}$$

5.4   Examples and Problems

5.4.1   Problems

Problem 5.1 Rankine Problem
A Rankine cycle utilizing ammonia as the working fluid runs between two 100 kPa and 1000 kPa with a high temperature of 40 C. Assuming that the pump and turbine are isentropic, what's the overall thermal efficiency?

Problem 5.2 Rankine Problem

Redo Problem 5.1, but with a pump and turbine at 80% isentropic efficiency, what's the overall thermal efficiency?

Problem 5.3 Reverse Rankine Problem

Consider a reverse Rankine cycle utilizing R-134a as the working fluid. The temperature of the refrigerant in the evaporator is −30 C and in the condenser is 40 C. The refrigerant is circulated at the rate of 0.05 kg/s. Determine the beta and the capacity of the plant in rate of refrigeration.

Problem 5.4 Reverse Rankine Problem

Redo Problem 5.3, but with a compressor at 80% isentropic efficiency, what's the new beta?

Problem 5.5 Brayton Problem

A Brayton cycle with r_p of 5 and high temperature of 1200 C has ambient conditions of 0 C and 80 kPa. What's the overall thermal efficiency and BWR?

Problem 5.6 Brayton Problem

Redo Problem 5.5, but with a compressor and turbine at 80% isentropic efficiency, what's the overall thermal efficiency?

Problem 5.7 Reverse Brayton Problem

Consider a reverse Brayton cycle where air enters the compressor at 80 kPa and 0 C and leaves at 400 kPa. Air enters the expander at 20 C. Determine Beta for the cycle.

Problem 5.8 Reverse Brayton Problem

Redo **Problem 5.7**, but with a compressor at 80% isentropic efficiency, what's the overall thermal efficiency?

Problem 5.9 Rankine Cycle

A 210-MW steam power plant operates on a simple ideal Rankine cycle. Steam enters the turbine at 10 MPa and 500 °C and is cooled in the condenser at 10 kPa. If the efficiency of the turbine and pump are 85% and 90%, respectively, determine the following:

(a) The quality of the steam at the turbine exit.
(b) The thermal efficiency of the cycle.
(c) The mass flow rate of the steam.

Problem 5.10 Brayton Cycle

An aircraft engine operates on a simple ideal Brayton cycle with a pressure ratio of 10. Heat is added to the cycle at a rate of 500 kW. Air passes through the engine at a rate of 50 m^3 min^{-1} at STP. Air entering the compressor is at 70 kPa and 0 °C. The efficiency of the compressor and turbine are 80% and 90%, respectively. Assuming that air behaves as an ideal gas with a temperature-dependent heat capacity given as a cubic polynomial, determine the power produced by the engine and its thermal efficiency.

Problem 5.11 Refrigeration Cycle

You are designing a refrigeration cycle that operates on the ideal vapor-compression cycle. The temperature range of the cycle is between 0 °C and 50 °C. The working fluid is R-134a. The cooling load is to be 20 kW. Two compressor options are to be considered: (1) a 70% efficient compressor; or (2) an 80% efficient compressor that is $5000 more expensive. The compressor runs on electricity which costs 10 cents per kWh. Assuming the system runs constantly, how long would the system have to be run in order for the higher-efficiency compressor to be cost-effective?

Appendix 5.1: Rankine Cycle Worksheet (website)

An Excel Spreadsheet was developed to illustrate a particular concept and is given on the companion website.

Appendix 5.2: Reverse Rankine Cycle Worksheet (website)

An Excel Spreadsheet was developed to illustrate a particular concept and is given on the companion website.

Appendix 5.3: Brayton Cycle Worksheet (website)

An Excel Spreadsheet was developed to illustrate a particular concept and is given on the companion website.

References

1. https://energyeducation.ca/encyclopedia/Rankine_cycle
2. https://resources.saylor.org/wwwresources/archived/site/wp-content/uploads/2013/08/BolesLectureNotesThermodynamicsChapter10.pdf
3. https://en.wikipedia.org/wiki/Brayton_cycle
4. https://www.nuclear-power.net/nuclear-engineering/thermodynamics/thermodynamic-cycles/heating-and-air-conditioning/reverse-brayton-cycle-brayton-refrigeration-cycle/

Chapter 6
Thermodynamic Properties and Gas Mixtures

6.1 Preview

The purpose of this chapter is the following. There are certain quantities within the thermodynamic tables that we cannot derive through experimentation but determine through measuring other quantities and what are known as the Maxwell relationships. These relationships relate one set of states to another set of states. An example is

$$\left(\frac{\partial P}{\partial T}\right)_v = \left(\frac{\partial S}{\partial v}\right)_T \tag{6.1}$$

So that if we know the relationship between pressure and temperature at a fixed volume this helps us to understand the relationship between entropy and specific volume at a fixed temperature.

Another facet of the chapter is to utilize the Maxwell relationships to relate states across the mixture dome from saturated liquid to saturated gas; this developed relationship is known as the Clapeyron equation. This process allows us to determine the heat of vaporization solely in terms of $\{T, P, v\}$ through either experimentation or the appropriate equation of state, which was discussed in Chap. 1.

The chapter also provides relationships for the first law and second law solely in terms of temperature and pressure. For these relationships to be complete, we also need to have an appropriate equation of state.

The final part of the chapter provides a method of determining relationships between temperature, pressure, and specific volume for a gas mixture.

Electronic Supplementary Material: The online version of this chapter (https://doi.org/10.1007/978-3-030-87387-5_6) contains supplementary material, which is available to authorized users.

© Springer Nature Switzerland AG 2022
H. C. Foust III, *Thermodynamics, Gas Dynamics, and Combustion*,
https://doi.org/10.1007/978-3-030-87387-5_6

6.2   Maxwell's Equations

For a closed system, the energy of the system is

$$de \equiv \delta q_{rev} - \delta w_{rev} \tag{6.2}$$

And using the definitions of entropy and work

$$de \equiv Tds - Pdv \tag{6.3}$$

For an open system, the energy of the system is

$$dh \equiv de + Pdv + vdP \tag{6.4}$$

Substituting Eq. 6.3 into Eq. 6.4 results in

$$dh \equiv Tds + vdP \tag{6.5}$$

For a closed system, the entropy balance of the system is

$$da \equiv -Pdv - sdT \tag{6.6}$$

where a is a measure of the *Helmholtz* energy.

For an open system, the entropy balance of the system is

$$dg \equiv vdP - sdT \tag{6.7}$$

where g is a measure of the *Gibbs* energy.

All quantities in Eqs. 6.3, 6.5, 6.6, and 6.7 are states and described by exact differentials. The total differential for a quantity F is given as

$$dF = Mdx + Ndy = \left(\frac{\partial F}{\partial x}\right)_y dx + \left(\frac{\partial F}{\partial y}\right)_x dy \tag{6.8}$$

where $M = \left(\dfrac{\partial F}{\partial x}\right)_y$ and $N = \left(\dfrac{\partial F}{\partial y}\right)_x$ and further

$$\left(\frac{\partial M}{\partial y}\right)_x = \frac{\partial^2 F}{\partial y \partial x} \tag{6.9}$$

and

$$\left(\frac{\partial N}{\partial x}\right)_y = \frac{\partial^2 F}{\partial x \partial y} \tag{6.10}$$

Because dF is an exact differential, Eqs. 6.9 and 6.10 are equivalent.

Using the fact that Eqs. 6.3, 6.5, 6.6, and 6.7 are all exact differentials, the following conditions hold for Eq. 6.3

$$\left(\frac{\partial e}{\partial s}\right)_v = T \ and \ \left(\frac{\partial u}{\partial v}\right)_s = -P \tag{6.11}$$

For Eq. 6.5,

$$\left(\frac{\partial h}{\partial s}\right)_P = T \ and \ \left(\frac{\partial h}{\partial P}\right)_s = v \tag{6.12}$$

For Eq. 6.6,

$$\left(\frac{\partial a}{\partial v}\right)_T = -P \ and \ \left(\frac{\partial a}{\partial T}\right)_P = -s \tag{6.13}$$

For Eq. 6.7,

$$\left(\frac{\partial g}{\partial P}\right)_T = v \ and \ \left(\frac{\partial g}{\partial T}\right)_v = -s \tag{6.14}$$

Using Eqs. 6.11 through 6.14, we get a set of equations known as the *Maxwell equations* [1, 2].

$$\left(\frac{\partial e}{\partial s}\right)_v = \left(\frac{\partial h}{\partial s}\right)_P \tag{6.15}$$

$$\left(\frac{\partial h}{\partial P}\right)_s = \left(\frac{\partial a}{\partial v}\right)_T \tag{6.16}$$

$$\left(\frac{\partial e}{\partial v}\right)_s = \left(\frac{\partial g}{\partial P}\right)_T \tag{6.17}$$

$$\left(\frac{\partial a}{\partial T}\right)_P = \left(\frac{\partial g}{\partial T}\right)_v \tag{6.18}$$

Another set of *Maxwell equations* that has proven to be more useful is [1].

$$\left(\frac{\partial T}{\partial v}\right)_s = -\left(\frac{\partial P}{\partial s}\right)_v \tag{6.19}$$

$$\left(\frac{\partial T}{\partial P}\right)_s = \left(\frac{\partial v}{\partial s}\right)_P \tag{6.20}$$

$$\left(\frac{\partial P}{\partial T}\right)_v = \left(\frac{\partial S}{\partial v}\right)_T \tag{6.21}$$

$$\left(\frac{\partial v}{\partial T}\right)_P = -\left(\frac{\partial s}{\partial P}\right)_T \tag{6.22}$$

6.3 Enthalpy and Entropy as Functions of T and P

Given

$$de = \left(\frac{\partial e}{\partial T}\right)_v dT + \left(\frac{\partial e}{\partial P}\right)_T dP \tag{6.23}$$

And

$$ds = \left(\frac{\partial s}{\partial T}\right)_v dT + \left(\frac{\partial s}{\partial v}\right)_T dv \tag{6.24}$$

We want to define Eqs. 6.23 and 6.24 solely in terms of $\{T, P\}$.
By definition,

$$C_V = \left(\frac{\partial e}{\partial T}\right)_v \tag{6.25}$$

And using Eq. 6.21

$$\left(\frac{\partial P}{\partial T}\right)_v = \left(\frac{\partial s}{\partial v}\right)_T \tag{6.26}$$

Substituting Eqs. 6.25 and 6.26 into Eqs. 6.23 and 6.24 results in

$$de = C_V dT + \left(\frac{\partial e}{\partial P}\right)_T dP \tag{6.27}$$

and

$$ds = \left(\frac{\partial s}{\partial T}\right)_v dT + \left(\frac{\partial P}{\partial T}\right)_v dv \qquad (6.28)$$

And using Eqs. 6.3, 6.27, and 6.28 we can get

$$de = C_V dT + \left(\frac{\partial e}{\partial P}\right)_T dP = Tds - Pdv = T\left\{\left(\frac{\partial s}{\partial T}\right)_v dT + \left(\frac{\partial P}{\partial T}\right)_v dv\right\} - Pdv \qquad (6.29)$$

And combining like differentials results in

$$\left\{T\left(\frac{\partial s}{\partial T}\right) - C_{Vv}\right\}dT = \left\{\left(\frac{\partial u}{\partial v}\right)_T + P - T\left(\frac{\partial P}{\partial T}\right)_v\right\}dv \qquad (6.30)$$

where

$$\left(\frac{\partial s}{\partial T}\right)_v = \frac{C_V}{T} \qquad (6.31)$$

and

$$\left(\frac{\partial e}{\partial v}\right)_T + P - T\left(\frac{\partial P}{\partial T}\right) = 0_v \qquad (6.32)$$

Substituting Eq. 6.31 into 6.28 and Eq. 6.32 into Eq. 6.28 results in

$$de = C_V dT + \left[T\left(\frac{\partial P}{\partial T}\right)_v - P\right]dv \qquad (6.33)$$

and

$$ds = \frac{C_V}{T}dT + \left(\frac{\partial P}{\partial T}\right)_v dv \qquad (6.34)$$

Further, du and dh can be related through

$$\Delta h = \Delta e + P_2 v_2 - P_1 v_1 \qquad (6.35)$$

And dh can be shown to be

$$dh = C_p dT + \left[v - T \left(\frac{\partial u}{\partial T} \right)_s \right] dP \tag{6.36}$$

The above discussion was for a single phase. What follows is a discussion that addresses state changes under the mixture dome from the saturated liquid to saturated gas lines.

Starting with Eq. 6.5

$$dh = Tds + vdP \tag{6.37}$$

And for a constant temperature and pressure

$$dh = Tds \tag{6.38}$$

And integrating both sides results in

$$h_g - h_l = T \left(s_g - s_l \right) \tag{6.39}$$

And utilizing the Maxwell relationship (Eq. 6.21)

$$\left(\frac{\partial s}{\partial v} \right)_T = \left(\frac{\partial P}{\partial T} \right)_v \tag{6.40}$$

where the right-hand side is actually independent of specific volume

$$\left(\frac{\partial s}{\partial v} \right)_T = \left(\frac{\partial P}{\partial T} \right)_{Sat} \tag{6.41}$$

And integrating (where $\left(\frac{\partial P}{\partial T} \right)_{Sat}$ is *constant*) results in

$$\frac{s_g - s_l}{v_g - v_l} = \left(\frac{\partial P}{\partial T} \right)_{Sat} \tag{6.42}$$

And substituting Eq. 6.39 results in

$$\frac{1}{T} \frac{h_g - h_l}{v_g - v_l} = \left(\frac{\partial P}{\partial T} \right)_{Sat} \tag{6.43}$$

The beauty of Eq. 6.43 is that $h_g - h_l$ can be determined by knowing the relationship between $\{T, P, v\}$, which can be either determined experimentally or through an appropriate equation of state.

Further, $\nu_g \gg \nu_l$ and assuming $v_g = \dfrac{R_g T}{P}$ results in

$$\frac{1}{T}\frac{h_g - h_l}{\dfrac{R_g T}{P}} = \left(\frac{\partial P}{\partial T}\right)_{Sat} \tag{6.44}$$

Another form of Eq. 6.44 is

$$\frac{dP}{P} = \frac{\Delta h_{Sat}}{R_g}\frac{dT}{T^2} \tag{6.45}$$

And assuming Δh_{sat} is constant results in the Clausius-Clapeyron equation [1].

$$\ln(P) = -\frac{\Delta h_{Sat}}{R_g}\frac{1}{T} + C \tag{6.46}$$

Example 6.1 Application of Clausius-Clapeyron Equation.
We see for water from saturation pressure 5 kPa to 200 kPa has the following relationship between 1/T and P.
   While the Δh_{Sat} does vary as shown in Fig. 6.2, from Fig. 6.1 and using a value of 0.4615 kJ/kg-K for R_g, the calculated value for Δh_{Sat} is 2344 kJ/kg and an acceptable estimation for the given saturated pressure range.

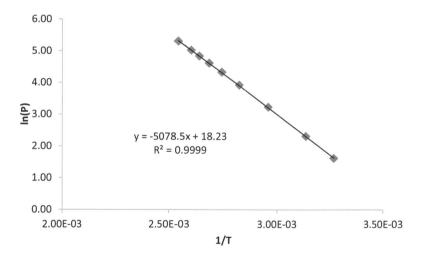

Fig. 6.1 1/T versus ln(P) for Water

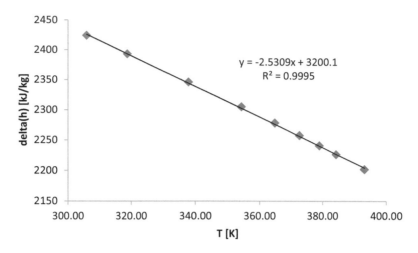

Fig. 6.2 T versus delta(h) for Saturated Water

6.3.1   Enthalpy and Entropy Functions for Ideal, Perfect Gases

For an ideal, perfect gas and utilizing Eqs. 6.34 and 6.36, a relationship will be developed to account for the energy and entropy balance for an ideal gas.

For an ideal gas,

$$\left(\frac{\partial P}{\partial T}\right)_v = \frac{R}{v}, \left(\frac{\partial v}{\partial T}\right)_P = \frac{R}{P}, \text{ and } v = \frac{RT}{P} \tag{6.47}$$

From Eq. 6.36, it is easily shown

$$dh = C_P dT \tag{6.48}$$

Or

$$\Delta h = C_P \Delta T \tag{6.49}$$

And from Eq. 6.34

$$\Delta S = C_V \ln\left(\frac{T_2}{T_1}\right) + R \ln\left(\frac{v_2}{v_1}\right) \tag{6.50}$$

6.3.2   Enthalpy and Entropy Functions for Van der Waal Gas

Using the equation of state for a van der Waal gas and Eqs. 6.34 and 6.36, a relationship will be developed to account for the energy and entropy balance for a non-perfect, van der Waal gas.

$$\left(\frac{\partial P}{\partial T} \right)_v = \frac{R_u}{v - b} \tag{6.51}$$

From Eq. 6.33, it is easily shown

$$de = C_V dT + \frac{a}{v^2} dv \tag{6.52}$$

And from Eq. 6.34

$$ds = C_V \frac{dT}{T} + \frac{R_u}{\upsilon - b} dV \tag{6.53}$$

An enthalpy balance can be determined by relating Δu and Δh through Eq. 6.35.

6.4   Composition of Mixtures

The composition of a mixture can be expressed either on a molar basis or a mass basis; each is discussed below.

6.4.1   Molar Basis

The mole fraction of species "i" is defined as

$$\chi_i = \frac{N_i}{\sum N_i} \tag{6.54}$$

where N_i is the number of moles of species "i".
  By definition

$$\sum \chi_i = 1 \tag{6.55}$$

And

$$\chi_i = Y_i \frac{MW_{mix}}{MW_i} \tag{6.56}$$

6.4.2   Mass Basis

The mass fraction of species "i" is defined as

$$Y_i = \frac{m_i}{\sum m_i} \tag{6.57}$$

where m_i is the mass of species "i".
   By definition

$$\sum Y_i = 1 \tag{6.58}$$

And

$$Y_i = \chi_i \frac{MW_i}{MW_{mix}} \tag{6.59}$$

6.5   Gas Mixtures, Part I

In this section, the question that will be asked is how to determine a specified state $\{T,P,\ v\}$ of a gas system made from two or more substances? Another method applicable to cubic equations of state will be given in Sect. 6.6.
   An example of determining the total mass for a binary mixture is given below.

Example 6.2 Gas Mixture Problem
A 2 kg mixture of 50% argon and 50% nitrogen by mole is in a tank at 2 MPa, 180 K. How large is the volume using (a) Ideal Gas Law and (b) Kay's Rule?
   We'll discuss both methods and then solve this problem. Note there are other methods [1, 2].

6.5.1   Ideal Gas Mixtures

The idea behind an ideal gas mixture is that each component independently contributes to the state $\{n, P, V\}$. This is expressed mathematically for volume as

$$V_{Total} = V_A + V_B + \ldots V_X \tag{6.60}$$

And

$$V_{Total} = \frac{n_A R_u T}{P} + \frac{n_B R_u T}{P} + \ldots + \frac{n_X R_u T}{P} \tag{6.61}$$

Similar expressions can be developed for pressure or molar mass.

6.5.2 Kay's Rule

The idea behind Kay's rule is that the critical pressure and temperature for the mixture is a linear combination of each component and given by

$$T_{critical}^{mixture} = \chi_A T_{critical}^A + \chi_B T_{critical}^B + \ldots + \chi_X T_{critical}^X \tag{6.62}$$

And

$$P_{critical}^{mixture} = \chi_A P_{critical}^A + \chi_B P_{critical}^B + \ldots + \chi_X P_{critical}^X \tag{6.63}$$

where χ_i is defined as

$$\chi_i = \frac{moles\ of\ i}{total\ moles} \tag{6.64}$$

The correction factor to the ideal gas solution is to determine the z factor associated with the Lee Kessler chart using the critical temperature and pressure associated with the mixture.

$$V_X = \frac{n_X R_u T}{P}$$

The solution for Example 6.2 is given below.

Substance(A)		Argon		
Substance(B)		Nitrogen	$mw_A n_A + mw_B n_B = 2\ kg$	
Total mass	[kg]	2.00		
Temperature	[K]	180.00	$39.948 n_A + 28.013 n_B = 2$	
Pressure	[MPa]	2.00		
X(A)		0.50	$39.948\{\chi_A n_{total}\} + 28.013\ \{\chi_B n_{total}\} = 2$	
X(B)		0.50		

Substance(A)		Argon			
Mw(A)	[kg/kmole]	39.95			
Mw(B)	[kg/kmole]	28.01			
R(u)	[kJ/kg-kmole]	8.31			
			$PV = nR_uT$		
Find total volume					
n(total)	[kmoles]	0.0588573			
n(A)	[kmoles]	0.0294286	$PV = nZR_uT$		
n((B)	[kmoles]	0.0294286			
V(A)	[m^3]	0.0220216			
V(B)	[m^3]	0.0220216			
V(Total)	[m^3]	0.0440			

Substance	P[critical]	T[critical]	X(*)	P*[critical]	T*[critical]
	[MPa]	[K]		[MPa]	[K]
Argon	4.87	150.8	50%	2.435	75.4
Nitrogen	3.39	126.2	50%	1.695	63.1
				4.13	138.5

T	P	T[r]	P[r]	Z	V
[K]	[MPa]				[m^3]
180	2	1.30	0.48	0.92	0.0405

Method	V	
	[m^3]	
Ideal gas	0.0440	
Kay's rule	0.0405	

6.6   Gas Mixtures, Part II

We saw in Chap. 1 that when pressure increases for a fixed temperature or temperature decreases for a fixed pressure, then some of the assumptions that ideal gas behavior is based on are no longer valid. To address these concerns, cubic equations of state (EOS) were developed that mimic the behavior of pressure versus ν for real gases; further, these EOSs address other concerns.

The cubic EOS can be extended to gas mixtures through what are called mixing rules; the mixing rule utilized in this section is known as van der Waals one-fluid mixing rules [2]. Here's how it works.

The mixture coefficients $\{a_{mix}, b_{mix}\}$ are defined [2] as

$$a_{mix} = \begin{bmatrix} \chi_1 & \cdots & \chi_n \end{bmatrix} \begin{bmatrix} a_{11} & \cdots & a_{1n} \\ \vdots & \ddots & \vdots \\ a_{n1} & \cdots & a_{nn} \end{bmatrix} \begin{bmatrix} \chi_1 \\ \cdots \\ \chi_n \end{bmatrix} \qquad (6.65)$$

and

$$b_{mix} = \begin{bmatrix} \chi_1 & \cdots & \chi_n \end{bmatrix} \begin{bmatrix} a_1 \\ \cdots \\ a_n \end{bmatrix} \qquad (6.66)$$

where

$$a_{ij} = a_{ji} = \sqrt{a_i a_j} \qquad (6.67)$$

And x_i is the mole fraction for pure component i, $\{a_i, b_i\}$ are the coefficients for pure component i, and n is the number of components.

An example of this method for a binary mixture is given below.

Example 6.3 Gas Mixture for Cubic Equations of State

Determine how the pressure changes for a mixture of methane and air for temperatures of 300 K, 600 K, and 1200 K and $v_{mix} = .1\dfrac{L}{mole}$. Do this by varying the percentage of methane from 0 to 100% and for both the van der Waal and Redlich Kwong EOS.

The results of this analysis are given below.

6.7   Examples and Problems

6.7.1   Examples

Example 6.4 For van der Waal's equation of state, determine

(a) dP,
(b) Show mixed second derivatives are equal

$$\left(\frac{\partial v}{\partial T} \right)_P$$

Solution for (a)

$$dP = \left(\frac{\partial P}{\partial T} \right)_v dT + \left(\frac{\partial P}{\partial v} \right)_T dv = MdT + Ndv \qquad (6.68)$$

where

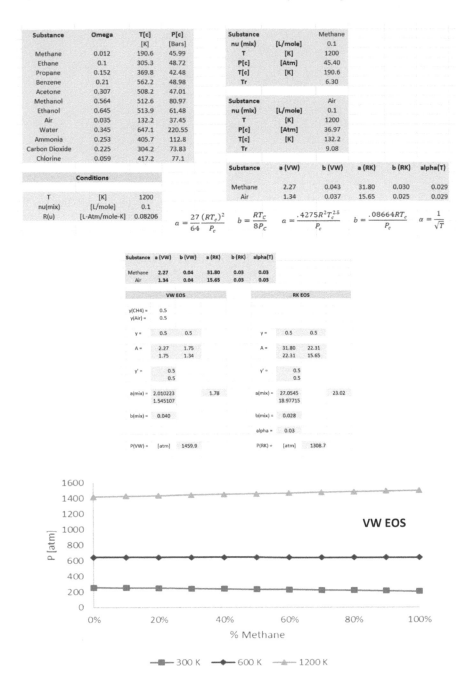

Substance	Omega	T[c] [K]	P[c] [Bars]
Methane	0.012	190.6	45.99
Ethane	0.1	305.3	48.72
Propane	0.152	369.8	42.48
Benzene	0.21	562.2	48.98
Acetone	0.307	508.2	47.01
Methanol	0.564	512.6	80.97
Ethanol	0.645	513.9	61.48
Air	0.035	132.2	37.45
Water	0.345	647.1	220.55
Ammonia	0.253	405.7	112.8
Carbon Dioxide	0.225	304.2	73.83
Chlorine	0.059	417.2	77.1

Substance nu (mix)		Methane 0.1
T	[K]	1200
P[c]	[Atm]	45.40
T[c]	[K]	190.6
Tr		6.30

Substance nu (mix)		Air 0.1
T	[K]	1200
P[c]	[Atm]	36.97
T[c]	[K]	132.2
Tr		9.08

Conditions		
T	[K]	1200
nu(mix)	[L/mole]	0.1
R(u)	[L-Atm/mole-K]	0.08206

Substance	a (VW)	b (VW)	a (RK)	b (RK)	alpha(T)
Methane	2.27	0.043	31.80	0.030	0.029
Air	1.34	0.037	15.65	0.025	0.029

$$a = \frac{27}{64}\frac{(RT_c)^2}{P_c} \qquad b = \frac{RT_c}{8P_c} \qquad a = \frac{.4275R^2T_c^{2.5}}{P_c} \qquad b = \frac{.08664RT_c}{P_c} \qquad \alpha = \frac{1}{\sqrt{T}}$$

Substance	a (VW)	b (VW)	a (RK)	b (RK)	alpha(T)
Methane	2.27	0.04	31.80	0.03	0.03
Air	1.34	0.04	15.65	0.03	0.03

VW EOS

y(CH4) =	0.5
y(Air) =	0.5

y =	0.5	0.5

A =	2.27	1.75
	1.75	1.34

y' =	0.5
	0.5

a(mix) =	2.010223	1.78
	1.545107	

b(mix) =	0.040

P(VW) =	[atm]	1459.9

RK EOS

y =	0.5	0.5

A =	31.80	22.31
	22.31	15.65

y' =	0.5
	0.5

a(mix) =	27.0545	23.02
	18.97715	

b(mix) =	0.028

alpha =	0.03

P(RK) =	[atm]	1308.7

VW EOS

P [atm] versus % Methane

300 K   600 K   1200 K

Fig. 6.3 Pressure versus % Methane (VW EOS)

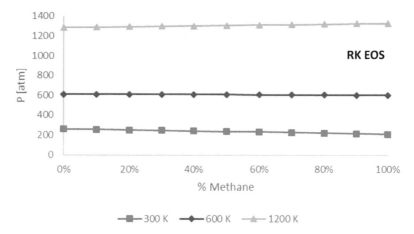

Fig. 6.4 Pressure versus % Methane (RK EOS)

$$M = \frac{R}{v - b} \tag{6.69}$$

and

$$N = \frac{-RT}{(v - b)^2} + 2\frac{a}{v^3} \tag{6.70}$$

such that

$$dP = \left[\frac{R}{v - b}\right]dT + \left[\frac{-RT}{(v - b)^2} + 2\frac{a}{v^3}\right]dv \tag{6.71}$$

Solution for (b)

$$\left(\frac{\partial M}{\partial v}\right)_T = \frac{-R}{(v - b)^2} = \left(\frac{\partial N}{\partial T}\right)_v = \frac{-R}{(v - b)^2} \tag{6.72}$$

Solution for (c).
The triple product states

$$\left(\frac{\partial v}{\partial T}\right)_P \left(\frac{\partial P}{\partial v}\right)_T \left(\frac{\partial T}{\partial P}\right)_v = -1 \tag{6.73}$$

and

$$\left(\frac{\partial v}{\partial T}\right)_P = \frac{-1}{\left(\dfrac{\partial P}{\partial v}\right)_T \left(\dfrac{\partial T}{\partial P}\right)_v} \tag{6.74}$$

Additionally, a double product shows

$$\left(\frac{\partial T}{\partial P}\right)_v = \frac{1}{\left(\dfrac{\partial P}{\partial T}\right)_v} \tag{6.75}$$

Substituting Eq. 6.75 into Eq. 6.74 results in

$$\left(\frac{\partial v}{\partial T}\right)_P = \frac{-\left(\dfrac{\partial P}{\partial T}\right)_v}{\left(\dfrac{\partial P}{\partial v}\right)_T} \tag{6.76}$$

For VW EOS,

$$\left(\frac{\partial P}{\partial T}\right)_v = \frac{R}{v-b}, \left(\frac{\partial P}{\partial v}\right)_T = \frac{-RT}{\left(v-b\right)^2} + 2\frac{a}{v^3} \tag{6.77}$$

Substituting Eq. 6.77 into 6.76 results in

$$\left(\frac{\partial v}{\partial T}\right)_P = \frac{R\left(v-b\right)}{RT - 2\dfrac{a}{v^3}\left(v-b\right)^2} \tag{6.78}$$

Example 6.5 Determine $\left(\dfrac{\partial s}{\partial v}\right)_T$ for water at 240 C and 0.4646 m³/kg using Maxwell relationships and RK EOS.

From Maxwell relationships,

$$\left(\frac{\partial s}{\partial v}\right)_T = \left(\frac{\partial P}{\partial T}\right)_v \tag{6.79}$$

RK EOS is

$$P = \frac{RT}{\bar{v} - b} - \frac{a}{\bar{v}\left(\bar{v}+b\right)} \frac{1}{\sqrt{T}} \tag{6.80}$$

and

$$\left(\frac{\partial P}{\partial T}\right)_v = \frac{R}{\bar{v}-b} + \frac{a}{\bar{v}(\bar{v}+b)} \frac{1}{2T^{1.5}} \tag{6.81}$$

Given

$$\bar{v} = .4646 \frac{m^3}{kg} = 8.372 \frac{m^3}{kmole}, a = 142.59 \, Bar \left(\frac{m^3}{kmole}\right)^2 K^{.5}, b = .0211 \frac{m^3}{kmole} \tag{6.82}$$

and thus

$$\left(\frac{\partial P}{\partial T}\right)_v = 996 \frac{J}{m^3 - K} = \left(\frac{\partial s}{\partial v}\right)_T$$

Example 6.6 Using RK EOS and for an isothermal process determine equations for Δs, Δu, and Δh.

Defining ds as the following

$$ds = \frac{C_P}{T} dT - \left(\frac{\partial v}{\partial T}\right)_P dP = \frac{C_V}{T} dT + \left(\frac{\partial P}{\partial T}\right)_v dv \tag{6.83}$$

and du as

$$dh = C_P dT + \left[v - T\left(\frac{\partial v}{\partial T}\right)_P\right] dP \tag{6.84}$$

And $T_1 = T_2$ implies

$$ds = \left(\frac{\partial P}{\partial T}\right)_v dv \tag{6.85}$$

$$du = \left[T\left(\frac{\partial P}{\partial T}\right)_v - P\right] dv \tag{6.86}$$

and

$$\Delta h = \Delta u + P_2 v_2 - P_1 v_1 \tag{6.87}$$

Further,

$$\left(\frac{\partial P}{\partial T}\right)_v = \frac{R}{\overline{v}-b} + \frac{a}{\overline{v}(\overline{v}+b)} \frac{1}{2T^{1.5}} \tag{6.88}$$

And so

$$ds = \left[\frac{R}{\overline{v}-b} + \frac{a}{\overline{v}(\overline{v}+b)} \frac{1}{2T^{1.5}}\right] dv \tag{6.89}$$

$$du = \left[\left(\frac{RT}{\overline{v}-b} + \frac{a}{\overline{v}(\overline{v}+b)} \frac{1}{\sqrt{T}}\right) - \frac{RT}{\overline{v}-b} + \frac{a}{\overline{v}(\overline{v}+b)} \frac{1}{\sqrt{T}}\right] dv \tag{6.90}$$

$$= \frac{3}{2} \frac{a}{\overline{v}(\overline{v}+b)} \frac{1}{\sqrt{T}} dv$$

Example 6.7 Show that for an ideal gas,

$$C_p - C_v = R_u \tag{6.91}$$

Given

$$ds = \frac{C_v}{T} dT + \left(\frac{\partial P}{\partial T}\right)_v dv = \frac{C_p}{T} dT - \left(\frac{\partial v}{\partial T}\right)_P dP \tag{6.92}$$

and

$$\frac{C_p - C_v}{T} dT = \left(\frac{\partial P}{\partial T}\right)_v dv + \left(\frac{\partial v}{\partial T}\right)_P dP \tag{6.93}$$

And dP can be defined as

$$dP = \left(\frac{\partial P}{\partial T}\right)_v dT + \left(\frac{\partial P}{\partial v}\right)_T dv \tag{6.94}$$

Substituting Eq. 6.94 into 6.93 results in

$$\frac{C_p - C_v}{T} dT = \left(\frac{\partial P}{\partial T}\right)_v dv + \left(\frac{\partial v}{\partial T}\right)_P \left\{\left(\frac{\partial P}{\partial T}\right)_v dT + \left(\frac{\partial P}{\partial v}\right)_T dv\right\} \tag{6.95}$$

Also,

$$\left(C_p - C_v\right) dT = T\left(\frac{\partial P}{\partial T}\right)_v dv + T\left(\frac{\partial v}{\partial T}\right)_P \left\{\left(\frac{\partial P}{\partial T}\right)_v dT + \left(\frac{\partial P}{\partial v}\right)_T dv\right\} \tag{6.96}$$

and

$$\left(C_p - C_v\right)dT - T\left(\frac{\partial v}{\partial T}\right)_P\left(\frac{\partial P}{\partial T}\right)_v dT = \left[T\left(\frac{\partial P}{\partial T}\right)_v + T\left(\frac{\partial v}{\partial T}\right)_P\left(\frac{\partial P}{\partial v}\right)_T\right]dv \quad (6.97)$$

Since T and v are independent variables, there derivates can be set to zero and thus

$$C_P - C_V = T\left(\frac{\partial v}{\partial T}\right)_P\left(\frac{\partial P}{\partial T}\right)_v \quad (6.98)$$

and

$$\left(\frac{\partial P}{\partial T}\right)_v = -\left(\frac{\partial v}{\partial T}\right)_P\left(\frac{\partial P}{\partial v}\right)_T \quad (6.99)$$

And substituting Eq. 6.99 into 6.98 results in

$$C_P - C_V = -T\left(\frac{\partial v}{\partial T}\right)_P^2\left(\frac{\partial P}{\partial v}\right)_T \quad (6.100)$$

Another form of Eq. 6.10 in terms of material science considerations is

$$C_P - C_V = vT\frac{\beta^2}{\kappa} \quad (6.101)$$

where β is volume expansivity and κ is isothermal compressibility where β is defined as

$$\beta = -\frac{1}{v}\left(\frac{\partial v}{\partial T}\right)_P \quad (6.102)$$

And κ is defined as

$$\kappa = -\frac{1}{v}\left(\frac{\partial v}{\partial P}\right)_T \quad (6.103)$$

For an ideal gas,

$$\left(\frac{\partial v}{\partial T}\right)_P = \frac{R_u}{P} = \frac{R_u}{R_uT}v = \frac{v}{T} \quad (6.104)$$

and

$$\left(\frac{\partial P}{\partial v}\right)_T = -\frac{R_u T}{v^2} \tag{6.105}$$

And using the definition for $c_P - c_V$, which is Eq. 6.100

$$C_P - C_V = -T\left(\frac{v}{T}\right)^2 \left(\frac{-R_u T}{v^2}\right) = R_u \tag{6.106}$$

Example 6.8 Determine the error in assuming $c_P = c_V$ for water at 1 atmosphere and 20 C. In Chap. 2, we assumed for liquids and solids that the specific heat of the material is independent of path, temperature, and pressure. How accurate is this assumption?

Starting from Eq. 6.11

$$C_P - C_V = vT\frac{\beta^2}{\kappa} \tag{6.107}$$

And for water at 20 C and 1 atmosphere values are available for κ and β, it is determined that.

T [C]	P [kg/m^3]	β*1e6 [1/K]	K*1e6 [1/Bar]
20	998.21	206.6	45.90

Therefore,

$$C_P - C_V = .027\frac{kJ}{kg - K} \tag{6.108}$$

and $C_P = 4.188 \; {}^{kJ}\!/\!_{kg} - K$ and $C_V = 4.161 \; {}^{kJ}\!/\!_{kg} - K$

The relative error in assuming $c_P = c_V$ is

$$RE = \frac{C_P - C_V}{C_V} = .66\% \tag{6.109}$$

which is acceptable!

Example 6.9 A mixture is 0.18 kmoles as CH_4 and 0.274 kmoles as C_4H_{10}, total volume is 0.241 m^3 at 238 C, and measured pressure is 68.9 Bars. Using a) ideal gas law and b) Kay's rule determine the pressure and compare against measured pressure.

Ideal Gas Solution
Total Mole = 0.454 kmoles, $y_1 = 0.396$, $y_2 = 0.604$

$$\bar{v} = \frac{V_{total}}{n_{total}} = .531 \frac{m^3}{kmole} \tag{6.110}$$

And

$$P = \frac{R_u T}{\bar{v}} = \frac{8314 \frac{N-m}{kmole-K}(238+273)K}{.531 \frac{m^3}{kmoe}} = 80e6\ Pa = 80\ Bars \tag{6.111}$$

Kay's Rule

$$T_c = \sum y_i T_{c,i}, P_c = \sum y_i P_{c,i} \tag{6.112}$$

Substance	y(i)	T[c] [K]	P[c] [Bars]	y(i)T[c,i]	y(i)P[c,i]
Methane	0.40	191.00	46.40	75.64	18.37
Butane	0.60	425.00	38.00	256.70	22.95
	1			T[c] = 332.3	P[c] = 41.3

And

Using the Lee Kessler chart to determine Z where Z is dependent on a changing pressure, the following iterative solution is found.

T	P	T[r]	P[r]	Z	P
[K]	[bars]				[bars]
511	80	1.5	1.9	0.84	67.2
511	67.2	1.5	1.6	0.88	70.4
511	70.4	1.5	1.7	0.88	70.4

6.7.2 Problems

Problem 6.1 Using the saturated nitrogen tables plot five points for $1/T$ versus $\ln(P)$ between 65 K and 100 K and show the slope of this line is appropriately $-\dfrac{\Delta h_{Sat}}{R_g}$.

Problem 6.2 Using the saturated solid-liquid water tables plot five points for $1/T$ versus $\ln(P)$ between 0 and -20 C and show the slope of this line is appropriately $-\dfrac{\Delta h_{Sat}}{R_g}$.

Problem 6.3 Using the following relationship $du = Tds - Pdv$ and the Maxwell relationships to find a relationship for $\left(\dfrac{\partial u}{\partial P}\right)_T$ that is only a function of $\{nu, T, P\}$.

Problem 6.4 A mixture of CO_2 (30% by moles) and nitrogen (70% by moles) with mass 2 kg is within a 1 m^3 container at 400 C. Using both a) ideal gas law and b) Kay's rule determine the pressure in the tank.

Note your solutions lies between the solutions for CO_2 (100% by moles) and nitrogen (100% by moles). Show this!

Problem 6.5 A mixture of CO_2 (70% by moles) and nitrogen (30% by moles) with mass 2 kg is at 2 MPa. Using both a) ideal gas law and b) Kay's rule determine the volume in the tank.

Problem 6.6 Derive the expression $dh = C_p dT + \left[v - T\left(\dfrac{\partial v}{\partial T}\right)_P\right]dv$.

Problem 6.8 Using the equations given in Problem 6.5 find Δh for van der Waal's EOS.

Problem 6.9 Using thermodynamic data for water estimate the freezing temperature of liquid water at a pressure of 4000 lbf/in^2.

Problem 6.10 Using Eq. 6.43 for ds determine Δs for water from 100 C and 50 kPa to 100 C and 200 kPa and compare against the thermodynamic tables.

Appendix 6.1: Thermodynamic Relationships

Maxwell Relationships

$$\left(\frac{\partial u}{\partial s}\right)_v = T \ and \ \left(\frac{\partial u}{\partial v}\right)_s = -P$$

$$\left(\frac{\partial h}{\partial s}\right)_P = T \ and \ \left(\frac{\partial h}{\partial P}\right)_S = v$$

$$\left(\frac{\partial A}{\partial v}\right)_T = -P \ and \ \left(\frac{\partial A}{\partial T}\right)_P = -s$$

$$\left(\frac{\partial g}{\partial P}\right)_T = v \ and \ \left(\frac{\partial g}{\partial T}\right)_V = -s$$

$$\left(\frac{\partial T}{\partial v}\right)_S = -\left(\frac{\partial P}{\partial s}\right)_V$$

$$\left(\frac{\partial T}{\partial P}\right)_s = \left(\frac{\partial v}{\partial s}\right)_P$$

$$\left(\frac{\partial P}{\partial T}\right)_v = \left(\frac{\partial s}{\partial v}\right)_T$$

$$\left(\frac{\partial v}{\partial T}\right)_P = -\left(\frac{\partial s}{\partial P}\right)_T$$

Triple Product Rules

$$\left(\frac{\partial P}{\partial T}\right)_v \left(\frac{\partial v}{\partial P}\right)_T \left(\frac{\partial T}{\partial v}\right)_P = -1$$

$$\left(\frac{\partial T}{\partial P}\right)_v \left(\frac{\partial P}{\partial v}\right)_T \left(\frac{\partial v}{\partial T}\right)_P = -1$$

Double Product Rules

$$\left(\frac{\partial P}{\partial T}\right)_v \left(\frac{\partial T}{\partial P}\right)_v = 1$$

$$\left(\frac{\partial P}{\partial v}\right)_T \left(\frac{\partial v}{\partial P}\right)_T = 1$$

$$\left(\frac{\partial v}{\partial T}\right)_P \left(\frac{\partial T}{\partial v}\right)_P = 1$$

Other Relationships

$$\left(\frac{dP}{dT}\right)_{Sat} = \frac{h_g - h_l}{T\left(v_g - v_l\right)}$$

$$C_P - C_V = -T\left(\frac{\partial v}{\partial T}\right)_P^2 \left(\frac{\partial P}{\partial v}\right)_T$$

$$C_P - C_V = vT\frac{\beta^2}{\kappa}$$

$$\beta = -\frac{1}{v}\left(\frac{\partial v}{\partial T}\right)_P$$

$$\kappa = -\frac{1}{v}\left(\frac{\partial v}{\partial P}\right)_T$$

Equations of State
Ideal Gas Law

$$Pv = RT$$

Van der Waals EOS

$$P = \frac{RT}{v-b} - \frac{a}{v^2}$$

Redlich-Kwong EOS

$$P = \frac{RT}{\bar{v}-b} - \frac{a}{\bar{v}\left(\bar{v}+b\right)}\frac{1}{\sqrt{T}}$$

First and Second Law
First Law

$$du = C_V dT + \left[T\left(\frac{\partial P}{\partial T}\right)_v - P\right]dv$$

$$dh = C_p dT + \left[v - T\left(\frac{\partial u}{\partial T}\right)_s\right]dP$$

Second Law

$$\Delta h = \Delta u + P_2 v_2 - P_1 v_1$$

$$ds = C_V \frac{dT}{T} + \left(\frac{\partial P}{\partial T}\right)_v dv + \sigma_{irr}$$

References

1. Moran and Shapiro. (2008). *Fundamentals of engineering thermodynamics* (6th ed.). John Wiley and Sons.
2. Smith, N., & Abbott. (2001). *Chemical engineering thermodynamics*. McGraw-Hill.

Part II
Fundamentals of Gas Dynamics

Chapter 7
Conservation Principles for a Gaseous System, Part I

7.1 Preview

Before we discuss the theories of combustion, we need to go over some basics. These basic considerations include conservation principles for a gaseous system, the definition of speed of sound, applying this definition to an ideal, perfect gas, and normal shocks. A shock forms when a gas travels faster than the local speed of sound.

Mention needs to be made of what is an ideal, perfect gas and consider the assumption of assuming a substance is ideal and perfect. An ideal gas is simply a gas that obeys ideal gas behavior, which for air near 300 K is acceptable to pressures of 10 to 12 MPa [1]. A thermodynamically perfect gas is one where the specific heat is independent of both pressure and temperature. For air, this is generally true.

7.2 Conservation Principles

Conservation principles for a gaseous system given as Fig. 7.1 are derived below.

The three equations to be derived below for conservation of energy, mass, and momentum will be utilized throughout this book. In this chapter, these principles will form the basis to our definition of speed of sound and equations to relate states on both sides of a normal shock.

Electronic Supplementary Material: The online version of this chapter (https://doi. org/10.1007/978-3-030-87387-5_7) contains supplementary material, which is available to authorized users.

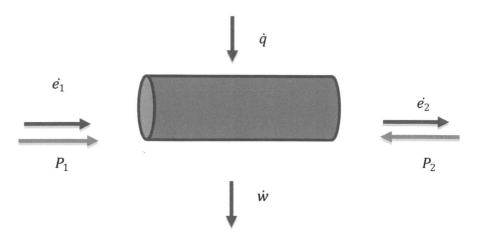

Fig. 7.1 Gaseous fluid system

7.2.1   Conservation of Energy

Accounting for the energy of the system (time invariant) provides the following equation

$$\dot{q} + e_1 = \dot{w} + e_2 \tag{7.1}$$

where $\dot{q}$ is the heat transfer associated with the control volume, $\dot{w}$ is the work associated with the control volume, $\dot{e}_1$ is the energy associated with the fluid entering the control volume, and e_2 is the energy associated with the fluid leaving the control volume.

Another form of Eq. 7.1 is

$$\dot{m}_1\left[h_1 + z_1 + \frac{u_1^2}{2}\right] + \dot{q} = \dot{m}_2\left[h_2 + z_2 + \frac{u_2^2}{2}\right] + \dot{w} \tag{7.2}$$

where $\dot{m}_i$ represents the mass rate (1 for "in" and 2 for "out" of the control volume), h_i is the enthalpy, Z_i is the potential energy above some datum, and $\frac{u_i^2}{2}$ is the kinetic energy.

In this chapter, it will be assumed that there is no heat transfer, work, no change in elevation and the mass rates in and out of the control volume are equal. With these assumptions, Eq. 7.2 becomes

$$h_1 + \frac{u_1^2}{2} = h_2 + \frac{u_2^2}{2} \tag{7.3}$$

7.2.2 Conservation of Mass

Assuming the rate of mass in is equal to the rate of mass out

$$\dot{m}_1 = \rho_1 A_1 u_1 = \dot{m}_2 = \rho_2 A_2 u_2 \tag{7.4}$$

where it is assumed that the areas are equivalent, results in

$$\rho_1 u_1 = \rho_2 u_2 \tag{7.5}$$

7.2.3 Conservation of Momentum

From Newton's second law,

$$+ \rightarrow \Sigma F_x = P_1 A - P_2 A = \frac{d(mV)}{dt} = \dot{m}(u_2 - u_1) = \rho u A(u_2 - u_1) \tag{7.6}$$

And dividing by A

$$P_1 - P_2 = \rho u(u_2 - u_1) \tag{7.7}$$

Or

$$P_1 + \rho_1 u_1^2 = P_2 + \rho_2 u_2^2 \tag{7.8}$$

7.3 Speed of Sound

Figure 7.2 provides the conditions for a substance moving at the speed of sound (see Appendix 7.1 for more details).

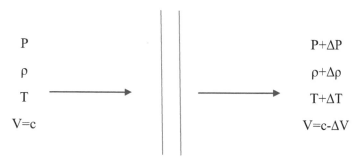

Fig. 7.2 Control Volume for Gas Moving at Speed of Sound [2]

From conservation of mass,

$$\rho A c = (\rho + \Delta\rho) A (c - \Delta V) \tag{7.9}$$

Expanding becomes

$$\rho A c = \rho A c + \Delta\rho A c - \rho A \Delta V - \Delta\rho A \Delta V \tag{7.10}$$

which simplifies to

$$\rho \Delta V + \Delta\rho \Delta V = \Delta\rho c \tag{7.11}$$

And solving for ΔV

$$\Delta V = \frac{c \Delta\rho}{\rho + \Delta\rho} \tag{7.12}$$

From conservation of momentum,

$$PA - (P + \Delta P) A = \dot{m}(V_{out} - V_{in}) = \rho A c (c - \Delta V - c) \tag{7.13}$$

Divide by A and simplify

$$\Delta P = \rho c \Delta V \tag{7.14}$$

Substituting Eq. 7.12 for ΔV results in

$$\Delta P = \rho \frac{c^2 \Delta\rho}{\rho + \Delta\rho} \tag{7.15}$$

And solving for C^2

$$c^2 = \frac{\Delta P}{\Delta\rho}\left(\frac{\rho + \Delta\rho}{\rho}\right) = \frac{\Delta P}{\Delta\rho}\left(1 + \frac{\Delta\rho}{\rho}\right) \tag{7.16}$$

Taking the limit as $\Delta\rho$ goes to zero results in

$$c^2 = \lim_{\Delta\rho \to 0} \frac{\Delta P}{\Delta\rho}\left(1 + \frac{\Delta\rho}{\rho}\right) = \frac{\partial P}{\partial \rho} \tag{7.17}$$

Along an isentropic path.
Equation 7.17 along with the condition of isentropic flow is the definition of speed of sound.

7.3.1 Speed of Sound in an Ideal Gas

Entropy is a state function defined for a reversible process as

$$ds = \frac{\delta q}{T} \tag{7.18}$$

and

$$de = \delta q - \delta w \tag{7.19}$$

Substitution of Eq. 7.19 into 7.18 and applying the definition of δw results in

$$Tds = de + Pd\upsilon \tag{7.20}$$

If $v = \frac{1}{\rho}$, then $dv = -\frac{1}{\rho^2}d\rho$, thus

$$Tds = de - \frac{P}{\rho^2}d\rho \tag{7.21}$$

e is a function of both P and ρ such that

$$de = \frac{\partial e}{\partial P}\bigg|_\rho dP + \frac{\partial e}{\partial \rho}\bigg|_P d\rho \tag{7.22}$$

Substitution of Eq. 7.22 into 7.21 results in

$$Tds = \frac{\partial e}{\partial P}\bigg|_\rho dP + \frac{\partial e}{\partial \rho}\bigg|_P d\rho - \frac{P}{\rho^2}d\rho \tag{7.23}$$

When ds equals zero then

$$\frac{\partial P}{\partial \rho}\bigg|_{ds=0} = \frac{-\dfrac{\partial e}{\partial \rho}\bigg|_P + \dfrac{P}{\rho^2}}{\dfrac{\partial e}{\partial P}\bigg|_\rho} \tag{7.24}$$

For an ideal, perfect gas

$$e = c_v T + \text{constant} \tag{7.25}$$

and

$$e = \frac{1}{\gamma - 1}\frac{P}{\rho} + \text{constant} \tag{7.26}$$

where the relationship $c_p - c_v = R$ has been utilized.

Derivatives for $\dfrac{\partial e}{\partial P}\Big|_\rho, \dfrac{\partial e}{\partial \rho}\Big|_P$, and $\dfrac{\partial P}{\partial \rho}\Big|_s$ are given below

$$\frac{\partial e}{\partial P}\bigg|_\rho = \frac{1}{\gamma - 1}\frac{1}{\rho} \tag{7.27}$$

$$\frac{\partial e}{\partial \rho}\bigg|_P = \frac{-1}{\gamma - 1}\frac{P}{\rho^2} \tag{7.28}$$

and

$$\frac{\partial P}{\partial \rho}\bigg|_s = \frac{\dfrac{1}{\gamma - 1}\dfrac{P}{\rho^2} + \dfrac{P}{\rho^2}}{\dfrac{1}{\gamma - 1}\dfrac{1}{\rho}} = \frac{P}{\rho} + (\gamma - 1)\frac{P}{\rho} = \gamma\frac{P}{\rho} = \gamma RT \tag{7.29}$$

which is the definition for the speed of sound for an ideal, perfect gas.

Naturally, if the conditions dictate another equation of state, then the equation for speed of sound will differ from Equation 7.29. For a van der Waal equation of state, the speed of sound is given in [3].

7.3.2 Speed of Sound in Liquids and Solids

By definition speed of sound [2] is

$$c^2 = \frac{\partial P}{\partial \rho}\bigg|_s \tag{7.30}$$

And for liquids and solids the bulk modulus (K) is defined as

$$K = \rho\frac{\partial P}{\partial \rho}\bigg|_s \tag{7.31}$$

where for liquids, the speed of sound is usually determined from Eq. 7.31.

The bulk modulus (K) is related to Young's modulus (E) through

$$\frac{E}{K} = 3(1 - 2\sigma) \tag{7.32}$$

where σ is Poisson's ratio and often takes on the value $\sigma = \dfrac{1}{3}$, thus

$$E \approx K \tag{7.33}$$

And for many solids

$$c^2 \approx \frac{E}{\rho} \tag{7.34}$$

7.4   Normal Shocks

Just as in incompressible flow Reynold's number, which is a non-dimensional number, is important, in compressible flow, Mach's number is important and defined as

$$Ma = \frac{u}{c} \tag{7.35}$$

where u is the local gas speed and c is the local speed of sound. It is generally acceptable that for a Mach number above 0.3, the fluid is considered compressible and the density on the fluid field varies with location.

Given in Fig. 7.4 are the upstream and downstream conditions for a normal shock where the upstream Mach number is always

$$Ma_1 > 1 \tag{7.36}$$

And the downstream Mach number is always

$$Ma_2 < 1 \tag{7.37}$$

For a given set up upstream conditions to include Mach number, we want to be able to calculate the downstream conditions. This is done by algebraic manipulation of Eqs. 7.38, 7.39, and 7.40.

Fig. 7.3 Shock Wave [2]

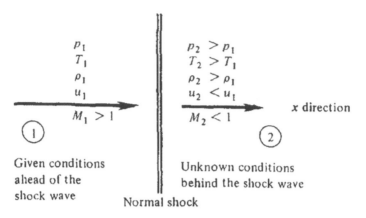

Fig. 7.4 Normal shock conditions upstream (1) and downstream (2)

$$\rho_1 u_1 A_1 = \rho_2 u_2 A_2 \tag{7.38}$$

$$P_1 + \rho_1 u_1^2 = P_2 + \rho_2 u_2^2 \tag{7.39}$$

$$h_1 + \frac{u_1^2}{2} = h_2 + \frac{u_2^2}{2} \tag{7.40}$$

Typically, $A_1 = A_2$ and Eq. 7.39 takes on the simpler form

$$\rho_1 u_1 = \rho_2 u_2 \tag{7.41}$$

Another form of enthalpy can be derived from Eqs. 7.42 and 7.43 when assuming an ideal, perfect gas

$$c_P - c_V = R \tag{7.42}$$

and

$$\gamma = \frac{c_P}{c_V} \tag{7.43}$$

Dividing Eq. 7.42 by c_p results in

$$\frac{c_P - c_V}{c_P} = \frac{R}{c_P} = 1 - \frac{1}{\gamma} \tag{7.44}$$

and

$$\frac{R}{1 - \frac{1}{\gamma}} = c_P = \frac{R\gamma}{\gamma - 1} \tag{7.45}$$

and

$$\Delta h = \frac{R\gamma}{\gamma - 1} \Delta T \qquad (7.46)$$

The strategy is to non-dimensionalize Eqs. 7.39, 7.40, and 7.41 and get the following relations (which will be derived shortly) [4, 5].

$$\frac{Ma_2}{Ma_1} = \frac{\sqrt{1 + \frac{\gamma - 1}{2} Ma_1^2}}{\sqrt{1 + \frac{\gamma - 1}{2} Ma_2^2}} \frac{1 + \gamma Ma_2^2}{1 + \gamma Ma_1^2} \qquad (7.47)$$

$$\frac{P_2}{P_1} = \frac{1 + \gamma Ma_1^2}{1 + \gamma Ma_2^2} \qquad (7.48)$$

and

$$\frac{T_2}{T_1} = \left(\frac{P_2 Ma_2}{P_1 Ma_1} \right) \qquad (7.49)$$

or

$$\frac{T_2}{T_1} = \frac{1 + \frac{\gamma - 1}{2} Ma_1^2}{1 + \frac{\gamma - 1}{2} Ma_2^2} \qquad (7.50)$$

The first equation to be derived will be Eq. 7.50 from Eq. 7.40 and Eq. 7.46.

$$\frac{\gamma - 1}{R\gamma} \left\{ \frac{R\gamma}{\gamma - 1} T_1 + \frac{u_1^2}{2} = \frac{R\gamma}{\gamma - 1} T_2 + \frac{u_2^2}{2} \right\} \qquad (7.51)$$

or

$$T_1 + \frac{(\gamma - 1)}{2R\gamma} u_1^2 = T_2 + \frac{(\gamma - 1)}{2R\gamma} u_2^2 \qquad (7.52)$$

and

$$T_1 \left[1 + \frac{\gamma - 1}{2} Ma_1^2 \right] = T_2 \left[1 + \frac{\gamma - 1}{2} Ma_2^2 \right] \qquad (7.53)$$

Further, Eq. 7.39 can be rewritten as

$$P_1\left[1+\frac{\rho_1}{P_1}u_1^2\right]=P_2\left[1+\frac{\rho_2}{P_2}u_2^2\right] \tag{7.54}$$

and

$$\frac{u_i^2}{P_iv_i}=\frac{u_i^2}{RT_i}=\frac{\gamma u_i^2}{\gamma RT_i}=\gamma Ma_i^2 \tag{7.55}$$

Therefore,

$$P_1\left[1+\gamma Ma_1^2\right]=P_2\left[1+\gamma Ma_2^2\right] \tag{7.56}$$

Lastly, substituting $\rho_i=\dfrac{P_i}{RT_i}$ into Eq. 7.41 results in

$$\frac{P_1}{RT_1}u_1=\frac{P_2}{RT_2}u_2 \tag{7.57}$$

And dividing both sides by γ results in

$$\frac{P_1Ma_1}{\sqrt{T_1}}=\frac{P_2Ma_2}{\sqrt{T_2}} \tag{7.58}$$

Thus from Eq. 7.52 comes

$$\frac{T_2}{T_1}=\frac{1+\dfrac{\gamma-1}{2}Ma_1^2}{1+\dfrac{\gamma-1}{2}Ma_2^2} \tag{7.59}$$

From Eq. 7.56 comes

$$\frac{P_2}{P_1}=\frac{1+\gamma Ma_1^2}{1+\gamma Ma_2^2} \tag{7.60}$$

And Eq. 7.58 comes

$$\frac{T_2}{T_1}=\frac{\left(P_2Ma_2\right)^2}{\left(P_1Ma_1\right)^2} \tag{7.61}$$

Equating Eqs. 7.59 and 7.61 results in

$$\frac{\left(P_2 Ma_2\right)^2}{\left(P_1 Ma_1\right)^2} = \frac{1 + \frac{\gamma - 1}{2} Ma_1^2}{1 + \frac{\gamma - 1}{2} Ma_2^2} \tag{7.62}$$

And substituting for the pressures Eq. 7.60 gives us

$$\frac{Ma_2^2}{Ma_1^2}\left\{\frac{1 + \gamma Ma_1^2}{1 + \gamma Ma_2^2}\right\}^2 = \frac{1 + \frac{\gamma - 1}{2} Ma_1^2}{1 + \frac{\gamma - 1}{2} Ma_2^2} \tag{7.63}$$

or

$$\frac{Ma_2}{Ma_1} = \frac{\sqrt{1 + \frac{\gamma - 1}{2} Ma_1^2}}{\sqrt{1 + \frac{\gamma - 1}{2} Ma_2^2}} \frac{1 + \gamma Ma_2^2}{1 + \gamma Ma_1^2} \tag{7.64}$$

An example of these ideas is given below.

Example 7.1 Normal Shock

A normal shock occurs in air where T_1 is 300 K, P_1 is 100 kPa, and Ma_1 is 3. What are the states on the other side of the shock?

Substance		Air
Rg	[J/kg-K]	287
Gamma		1.4

			Blue = Input
			Yellow = Goal Seek
			Tan = Output

T(1)	[K]	300
P(1)	[kPa]	100
Ma(1)		3
U(1)	[m/s]	1041.57

T(2)	[K]	803.72
P(2)	[kPa]	1033.48
Ma(2)		0.48
U(2)	[m/s]	269.96

$$\frac{Ma_2}{Ma_1} = \frac{\sqrt{1 + \frac{\gamma - 1}{2} Ma_1^2}}{\sqrt{1 + \frac{\gamma - 1}{2} Ma_2^2}} \frac{1 + \gamma Ma_2^2}{1 + \gamma Ma_1^2}$$

$$\frac{P_2}{P_1} = \frac{1 + \gamma Ma_1^2}{1 + \gamma Ma_2^2}$$

$$\frac{T_2}{T_1} = \left(\frac{P_2 Ma_2}{P_1 Ma_1}\right)^2$$

Ma(1)	Ma(2)	Term 1	Term 2	Term 3	Term 4	LHS	RHS	Diff
3	0.48	2.80	1.05	1.32	13.60	0.16	0.16	-2.68E-05

Ma(2)/Ma(1)	0.16
P(2)/P(1)	10.33

Given in Fig. 7.5 are the ratios of pressure, temperature, and Mach number for Ma_1.

We'll now discuss a topic related to standing normal waves, which is moving shock waves that exhibit a different frame of reference (see Fig. 7.6).

where C is equal to u_1 from the normal shock theory and $C-\Delta V$ is equal to u_1-u_2 from the normal shock theory (see Appendix 7.2).

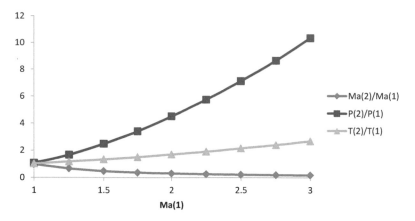

Fig. 7.5 Moving shock waves

Fig. 7.6 Moving normal
shock

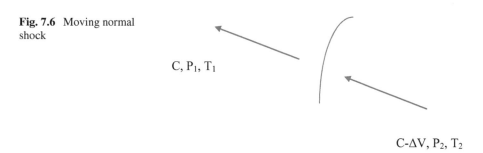

C, P_1, T_1

$C-\Delta V, P_2, T_2$

Example 7.2 Explosions
An atomic explosion generates a pressure of 4500 psia inside the shock. If the ambient conditions are 15 psia and 300 K. What are the states within the shock and both the shock speed (C) and trailing wave ($C-\Delta V$)?

We know the pressure ratio is equal to $P_2 \Big/ P_2 = \dfrac{4500}{15} = 300$.

This problem is solved by using our spreadsheet labeled "normal shock" and going to English unit worksheet. The problem is then solved by using goal seek to solve two problems

1. Force "Diff" to zero by changing Ma$_2$.
2. Force Ma(2)/Ma(1) to 300 by changing Ma$_1$.

This iterative process was done several times and the final results are shown below where T_2 is 26,458, shock speed is 7324 fps, and trailing wave is 6087 fps.

T(1)	[R]	520
P(1)	[psia]	15
Ma(1)		16.02
U(1)	[fps]	7,324

T(2)	[R]	26,458
P(2)	[psia]	4,500
Ma(2)		0.38
U(2)	[fps]	1,237

Ma(1)	Ma(2)	Term 1	Term 2	Term 3	Term 4	LHS	RHS	Diff
16.02243341	0.38	52.34	1.03	1.20	360.41	0.02	0.02	-1.07E-04

Ma(2)/Ma(1)		0.02
P(2)/P(1)		300.00
T(2)/T(1)		50.88

6,087

7.5 Examples & Problems

7.5.1 Examples

Example 7.3 Speed of Sound for an Ideal and Van der Waal Gas
Determine the speed of sound of Argon at 400 K and 1 atm assuming an ideal, perfect gas and then a van der Waal gas. The speed of sound for a van der Waal gas [3] is given as

$$c^2 = \left(1 + \frac{R}{c_V}\right) \frac{v^2 RT}{(v-b)^2} - 2\frac{a}{v}$$

where {a, b} are coefficients associated with van der Waal's EOS and R is the gas constant.
Given

$$c_v = .31\frac{kJ}{kg - K}, R_g = .2081\frac{kJ}{kg - K}, \gamma = 1.667$$

The solution is given below where it is naturally seen for a noble gas that the ideal gas and van der Waal gas speeds of sound are the same.

Cv	[kJ/kg-K]	0.31
R(g)	[kJ/kg-K]	0.2081
R(g)	[J/kg-K]	208.1
Gamma		1.667
T	[K]	400

| c (ideal) | [m/s] | 373 |

$$c^2 = \gamma R_g T$$

| T[c] | [K] | 150.8 |
| P[c] | [Mpa] | 4.87 |

| a | [m^3/kg] | 85.3 |
| b | [m^3/kg] | 8.05E-04 |

$$a = \frac{27}{64}\frac{R^2 T_c^2}{P_c} \qquad b = \frac{RT_c}{8P_c}$$

| nu(ideal) | [m^3/kg] | 0.82 |

$$v = \frac{V}{m} = \frac{R_g T}{P}$$

| c^2 | | 1.39E+05 |
| c (VW) | [m/s] | 373 |

$$c^2 = \left(1 + \frac{R}{C_V}\right)\frac{v^2 RT}{(v-b)^2} - 2\frac{a}{v}$$

Example 7.4 Speed of Sound in a Solid

Determine the speed of sound for aluminum at STP.

The density and Young's modulus can vary dependent on the type of aluminum, but the values used are

$$\rho_{Al} = 2700 \frac{kg}{m^3} \text{ and } E_{Al} = 69\,GPa$$

Thus

$$c^2 \approx \frac{E}{\rho} \text{ and } c \approx 5055\,m/s$$

7.5.2 Problems

Problem 7.1 Speed of Sound for a Van der Waal Gas

The speed of sound for an ideal gas and van der Waal gas should diverge as the compressibility factor (Z) deviates from 1.0. Show this for CO_2 by varying P and T and referring to the work of Lee and Kessler.

Problem 7.2 Normal Shock
Determine the states within a normal shock when $P_2\big/P_1$ is 10 and T_1 is 300 K.

Problem 7.3 Frank White's Atomic Bomb Explosion Problem
An atomic explosion propagates into still air at 15 lbf/in² and 600 R. the pressure just inside the shock is 6000 lbf/in². Assuming $\gamma = 1.4$, what are the speed of the shock and the velocity inside the shock?

Problem 7.4 Speed of Sound for a Gas Mixture (Van der Waal Gas)
Using the speed of sound equation for a van der Waal gas as given above and the data given below for specific heat constant volume determine the speed of sound for a mixture of argon and methane {50% argon by molar mass} at 1 atm and 800 K (Hint: use the method provided in Sect. 6.6 to find $\{a_{mix}, b_{mix}\}$ and Table 1.2 to determine Z).

This table provides critical constants for each gas and the specific heat constant pressure models utilized to create the figure given below.

	T[c]	P[c]	γ	β_0	β_1	β_2	β_3
	[K]	[MPa]					
Argon	150.8	4.87	1.667	0.52			
Methane	190.4	4.6	1.2999	1.2	3.25	0.75	−0.71

Problem 7.5 Speed of Sound for a Gas Mixture (Ideal Gas)
Redo Problem 7.4 but now assume ideal gas behavior. How do the answers differ between Problems 7.4 and 7.5?

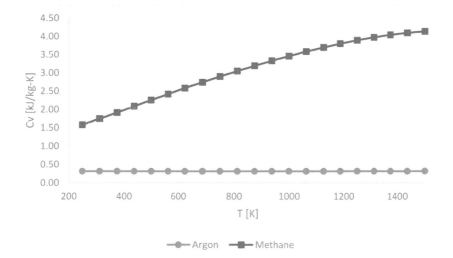

Appendix 7.1: Moving Shock Wave Frame of Reference

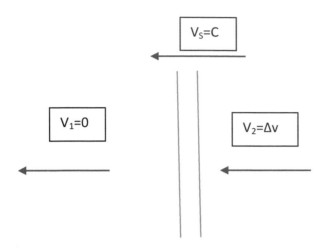

Fixed Wave Frame of Reference If we ride the shock, then we subtract C from each velocity and get

Or

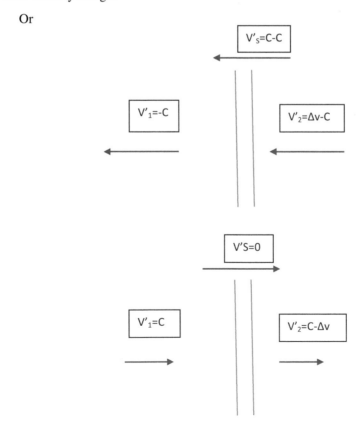

Appendix 7.2: Moving Shock Wave Frame of Reference

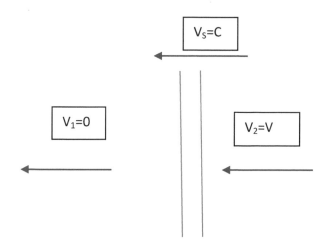

Fixed Wave Frame of Reference If we ride the shock, then we subtract C from each velocity and get.

Or

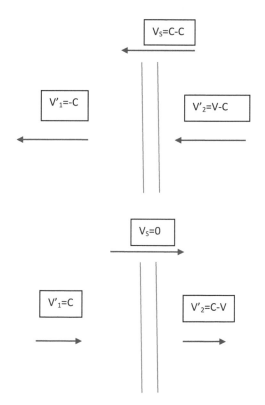

References

1. Borgnakke and Sonntag. (2008). *Fundamentals of thermodynamics* (7th ed.). John Wiley and Sons.
2. White, F. M. (1999). *Fluid mechanics* (4th ed.). Mc-Graw Hill.
3. Donaldson, Coleman duP. (1948). *The importance of imperfect-gas effects and variation of heath capacities on the isentropic flow of gases.* NACA RM No. L8J14.
4. Townend, L. H. (1970). An analysis of oblique and normal detonation waves. *Ministry of Technology, Aeronautical Research Council Reports and Memorandums, R&M No. 3638.*
5. Zucker and Biblarz. (2002). *Fundamentals of gas dynamics* (2nd ed.). John Wiley and Sons.

Chapter 8
Conservation Principles for a Gaseous System, Part II

8.1 Preview

In the previous chapter, conservation principles were developed for a gaseous system and are appropriate for several types of phenomena to include:

- Normal Shocks (Chap. 7).
- Critical Flow (Chap. 9).
- Rankine-Hugoniot Systems (Chap. 10).

But there are times when more general conservation principles are required and involve changes in states with either time or space, which leads to differential equations in a particular coordinate system. The main reason for the more general description when it's a detonation is to incorporate heat released based on some type of Arrhenius reaction rate equation that changes with temperature and results with changes in states with either time or space; for deflagrations non-premixed, the heat release is considered instantaneous, but other transport processes dictate the changes in state with either time or space; and for deflagration premixed, both heat release and transport processes are important.

These transport processes, which are discussed below, include

- Mass transfer.
- Momentum transfer.
- Heat transfer.

Two classes of conservation principles are explored:

- Conservation principles appropriate to detonation systems.
- Conservation principles appropriate to deflagration systems.

Electronic Supplementary Material: The online version of this chapter (https://doi.org/10.1007/978-3-030-87387-5_8) contains supplementary material, which is available to authorized users.

H. C. Foust III, *Thermodynamics, Gas Dynamics, and Combustion*,
https://doi.org/10.1007/978-3-030-87387-5_8

In Sect. 8.2, conservation principles for detonation systems are developed that allows for the following:

- Area changes.
- Shear stress at the wall.
- Heat release (Arrhenius reaction rate).

These equations do not address:

- Mass transfer to include diffusion.
- Momentum transfer.

In Sect. 8.3, relationships will be developed between temperature, pressure, and density for an isentropic, compressible flow. These developed relationships will be utilized to relate static and stagnation conditions for a given Mach number. The developed relationships will be seen again in Chap. 9.

Mass transfer is discussed in Sect. 8.4. In Sect. 8.5, conservation principles appropriate for a deflagration system that involves premixed laminar flames is developed and is in a planar coordinate system. In Sect. 8.6, conservation principles appropriate for a deflagration system that involves non-premixed laminar flames is developed and is in an axi-symmetric coordinate system.

Various sets of conservation principles and the related fluid phenomena are summarized in Table 8.1.

8.2   More General Conservation Principles for Detonations

In Chap. 7, conservation principles were developed for a gaseous system to include heat release. These equations were.

Conservation of Mass

$$\rho_1 u_1 = \rho_2 u_2 \tag{8.1}$$

Conservation of Momentum

$$P_1 + \rho_1 u_1^2 = P_2 + \rho_2 u_2^2 \tag{8.2}$$

Table 8.1 Conservation principles appropriate for a particular fluid phenomenon

Chapter/ Section	Relevant Phenomena	Relevant Chapter/ Section
Chapter 7	Normal shocks, critical flow, Rankine-Hugoniot systems	Chapters 9 and 10
Section 8.2	Detonations	Section 13.5
Section 8.3	Isentropic, compressible flow	Chapter 9
Section 8.5	Premixed laminar flames	Section 12.3
Section 8.6	Non-premixed laminar flames	Section 12.4

Conservation of Energy

$$h_1 + q + \frac{u_1^2}{2} = h_2 + \frac{u_2^2}{2} \tag{8.3}$$

More general equations are now derived that include heat release, area change, and shear stress at the wall (see Fig. 8.1). This derivation is taken from [1].

Where the following assumptions/conditions are made [1]

1. Surfaces 1 and 2 are open and allow fluxes of mass, momentum, and energy.
2. Surface w is closed – no mass flux allowed through the surface.
3. An external heat is applied to through surface w.
4. No longitudinal heat transfer and thermal conductivity is zero.
5. No diffusive viscous stresses are allowed and so viscosity is zero.
6. The cross-section is solely a function of x and so A(x).
7. L_p is the circumferential perimeter length.

8.2.1   Conservation of Mass

$$\left. \bar{\rho}\bar{A}\Delta x \right|_{t+\Delta t} - \left. \bar{\rho}\bar{A}\Delta x \right|_t = \rho_1 A_1 \left(u_1 \Delta t \right) - \rho_2 A_2 \left(u_2 \Delta t \right) \tag{8.4}$$

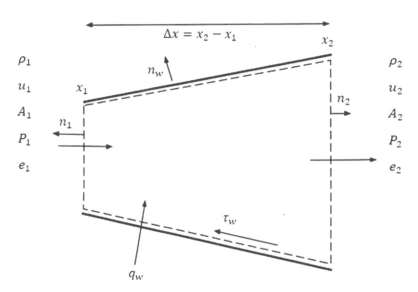

Fig. 8.1 Control volume for more general conservation principles [1]

or

$$\frac{\overline{\rho A}\big|_{t+\Delta t} - \overline{\rho A}\big|_{t}}{\Delta t} - \frac{\rho_1 A_1 (u_1) - \rho_2 A_2 (u_2)}{\Delta x} = 0 \tag{8.5}$$

The limit as $\Delta t \to 0$ implies $\Delta x \to 0$, therefore

$$\frac{\partial}{\partial t}(\rho A) + \frac{\partial}{\partial x}(\rho A u) = 0 \tag{8.6}$$

At steady state,

$$\frac{\partial}{\partial x}(\rho A u) = 0 \tag{8.7}$$

and

$$\rho_1 A_1 u_1 = \rho_2 A_2 u_2 \tag{8.8}$$

8.2.2   Conservation of Momentum

$$\frac{d}{dt}(mu) = u\dot{m} + m\dot{u} = \sum F_x = F_1 + F_2 + F_3 \tag{8.9}$$

where
F_1 = Pressure differential.
F_2 = Reaction at the wall due to pressure differential.
F_3 = Shear stresses at the wall.

or

$$\frac{mu\big|_{t+\Delta t} - mu\big|_{t}}{\Delta t} + \frac{um\big|_{t+\Delta t} - um\big|_{t}}{\Delta t} = \sum F_x \tag{8.10}$$

and

$$mu\big|_{t+\Delta t} = mu\big|_{t} + \left\{ um\big|_{t+\Delta t} - um\big|_{t} \right\} \Delta t + \sum F_x \Delta t \tag{8.11}$$

Further,

$$\left(\bar{\rho}\bar{A}\Delta x\right)\bar{u}\Big|_{t+\Delta t}=\left(\bar{\rho}\bar{A}\Delta x\right)\bar{u}\Big|_{t}+\left[\rho_{1}A_{1}\left(u_{1}\Delta t\right)\right]u_{1}$$
$$-\left[\rho_{2}A_{2}\left(u_{2}\Delta t\right)\right]u_{2}+\left[\rho_{1}A_{1}\right]\Delta t-\left[\rho_{2}A_{2}\right]\Delta t$$
$$+\left[\bar{P}\left(A_{2}-A_{1}\right)\right]\Delta t-\left(\tau_{w}\bar{L}_{p}\Delta x\right)\Delta t \tag{8.12}$$

or

$$\frac{\left(\bar{\rho}\bar{A}\right)\bar{u}\Big|_{t+\Delta t}-\left(\bar{\rho}\bar{A}\right)\bar{u}\Big|_{t}}{\Delta t}+\frac{\left[\rho_{2}A_{2}\left(u_{2}\right)\right]u_{2}-\left[\rho_{1}A_{1}\left(u_{1}\right)\right]u_{1}}{\Delta x}=-\left(\frac{P_{2}A_{2}-P_{1}A_{1}}{\Delta x}\right)$$
$$+\bar{P}\frac{A_{2}-A_{1}}{\Delta x}-\tau_{w}\bar{L}_{p} \tag{8.13}$$

The limit as $\Delta x \to 0$ implies $\Delta t \to 0$

$$\frac{\partial}{\partial t}\left(\rho A u\right)+\frac{\partial}{\partial x}\left(\rho A u^{2}\right)=-\frac{\partial}{\partial x}\left(PA\right)+P\frac{\partial A}{\partial x}-\tau_{w}L_{p} \tag{8.14}$$

And at steady state

$$\frac{d}{dx}\left(\rho A u^{2}\right)=-\frac{d}{dx}\left(PA\right)+P\frac{dA}{dx}-\tau_{w}L_{p} \tag{8.15}$$

Further

$$\rho A u\frac{du}{dx}+u\frac{d}{dx}\left(\rho A u\right)=-P\frac{dA}{dx}-A\frac{dP}{dx}+P\frac{dA}{dx}-\tau_{w}L_{p} \tag{8.16}$$

Where the second term of the left-hand side is zero and the sum of the first and third term on the right-hand side is zero.

$$\rho A u\frac{du}{dx}=-A\frac{dP}{dx}+-\tau_{w}L_{p} \tag{8.17}$$

And when shear stress is zero

$$\rho A u\frac{du}{dx}+A\frac{dP}{dx}=0 \tag{8.18}$$

From Eq. 8.17, when $\frac{dA}{dx}=\tau_{w}=0$, then

$$\frac{\partial}{\partial x}\left(\rho A u^{2}\right)=-\frac{\partial}{\partial x}\left(PA\right) \tag{8.19}$$

or

$$\frac{\partial}{\partial x}\left(\rho A u^2 + PA\right) = 0, \frac{\partial}{\partial x}\left(P + \rho u^2\right) = 0 \tag{8.20}$$

$$P_1 + \rho_1 u_1^2 = P_2 + \rho_2 u_2^2 \tag{8.21}$$

8.2.3   Conservation of Energy

$$\left[\bar{\rho}\bar{A}\Delta x\right]\left[\bar{e} + \frac{\bar{u}^2}{2}\right]\Big|_{t+\Delta t} = \left[\bar{\rho}\bar{A}\Delta x\right]\left[\bar{e} + \frac{\bar{u}^2}{2}\right]\Big|_{t}$$
$$+ \rho_1 A_1 \left(u_1 \Delta t\right)\left(e_1 + \frac{u_1^2}{2}\right) - \rho_2 A_2 \left(u_2 \Delta t\right)\left(e_2 + \frac{u_2^2}{2}\right)$$
$$+ q_w \left(\bar{L}_p \Delta x\right)\Delta t + \left(P_1 A_1\right)\left(u_1 \Delta t\right) - \left(P_2 A_2\right)\left(u_2 \Delta t\right) \tag{8.22}$$

or

$$\frac{\left[\bar{\rho}\bar{A}\right]\left[\bar{e} + \frac{\bar{u}^2}{2}\right]\Big|_{t+\Delta t} - \left[\bar{\rho}\bar{A}\right]\left[\bar{e} + \frac{\bar{u}^2}{2}\right]\Big|_{t}}{\Delta t}$$
$$+ \frac{\rho_2 A_2 \left(u_2\right)\left(e_2 + \frac{u_2^2}{2} + \frac{P_2}{\rho_1}\right) - \rho_1 A_1 \left(u_1\right)\left(e_1 + \frac{u_1^2}{2} + \frac{P_1}{\rho_1}\right)}{\Delta x} = q_w \bar{L}_p \tag{8.23}$$

The limit as $\Delta x \to 0$ implies $\Delta t \to 0$

$$\frac{\partial}{\partial t}\left[\rho A\left(e + \frac{u^2}{2}\right)\right] + \frac{\partial}{\partial x}\left[\rho A u\left(e + \frac{u^2}{2} + \frac{P}{\rho}\right)\right] = q_w L_p \tag{8.24}$$

At steady state

$$\frac{\partial}{\partial x}\left[\rho A u\left(e + \frac{u^2}{2} + \frac{P}{\rho}\right)\right] = q_w L_p \tag{8.25}$$

or

$$\rho u \frac{d}{dx}\left[e + \frac{u^2}{2} + \frac{P}{\rho}\right] = \frac{q_w L_p}{A} \tag{8.26}$$

Further

$$\rho u \left[\frac{de}{dx} + u\frac{du}{dx} + \frac{1}{\rho}\frac{dP}{dx} - \frac{P}{\rho^2}\frac{dP}{dx} \right] = \frac{q_w L_p}{A} \tag{8.27}$$

From the conservation of momentum

$$\rho u^2 \frac{du}{dx} + u\frac{dP}{dx} = -\frac{\tau_w L_p}{A} \tag{8.28}$$

Subtracting Eq. 8.28 from Eq. 8.27 results in

$$\rho u \frac{de}{dx} - u\frac{P}{\rho}\frac{d\rho}{dx} = \frac{q_w L_p}{A} + \frac{\tau_w L_p}{A} \tag{8.29}$$

or

$$\frac{de}{dx} - \frac{P}{\rho^2}\frac{d\rho}{dx} = \frac{(q_w + \tau_w)L_p}{\dot{m}} \tag{8.30}$$

When there is no heat transfer, nor shear stress Eq. 8.30 takes on the simpler form

$$\dot{e} + P\dot{v} = 0 \tag{8.31}$$

And from Eq. 8.30 de can be defined as

$$de = \left.\frac{\partial e}{\partial \rho}\right|_P d\rho + \left.\frac{\partial e}{\partial P}\right|_\rho dP \tag{8.32}$$

And $\dfrac{de}{dx}$ can be defined as

$$\frac{de}{dx} = \left.\frac{\partial e}{\partial \rho}\right|_P \frac{d\rho}{dx} + \left.\frac{\partial e}{\partial P}\right|_\rho \frac{dP}{dx} \tag{8.33}$$

Substitution of Eq. 8.33 into Eq. 8.30 results in

$$\left.\frac{\partial e}{\partial \rho}\right|_P \frac{d\rho}{dx} + \left.\frac{\partial e}{\partial P}\right|_\rho \frac{dP}{dx} - \frac{P}{\rho^2}\frac{d\rho}{dx} = \frac{(q_w + \tau_w u)L_p}{\dot{m}} \tag{8.34}$$

or

$$\frac{dP}{dx} - \left[\frac{\dfrac{P}{\rho_2} - \left.\dfrac{\partial e}{\partial \rho}\right|_P}{\left.\dfrac{\partial e}{\partial P}\right|_\rho}\right]\frac{d\rho}{dx} = \frac{\left(q_w + \tau_w u\right)L_p}{\dot{m}\left.\dfrac{\partial e}{\partial P}\right|_\rho} \tag{8.35}$$

Where the quantity in brackets was shown in Chap. 7 to be equal to c^2

$$\frac{dP}{dx} - c^2 \frac{d\rho}{dx} = \frac{\left(q_w + \tau_w u\right)L_p}{\dot{m}\left.\dfrac{\partial e}{\partial P}\right|_\rho} \tag{8.36}$$

8.2.4　Conservation of Energy for Detonation System

The total derivative for internal energy can be defined as

$$de = \left.\frac{\partial e}{\partial P}\right|_{v,\lambda} dP + \left.\frac{\partial e}{\partial v}\right|_{P,\lambda} dv + \left.\frac{\partial e}{\partial \lambda}\right|_{P,v} d\lambda = 0 \tag{8.37}$$

And taking the time derivative

$$\dot{e} = \left.\frac{\partial e}{\partial P}\right|_{v,\lambda} \dot{P} + \left.\frac{\partial e}{\partial v}\right|_{P,\lambda} \dot{v} + \left.\frac{\partial e}{\partial \lambda}\right|_{P,v} \dot{\lambda} \tag{8.38}$$

And substituting Eq. 8.38 into Eq. 8.31 results in

$$\left.\frac{\partial e}{\partial P}\right|_{v,\lambda} \dot{P} + \left.\frac{\partial e}{\partial v}\right|_{P,\lambda} \dot{v} + \left.\frac{\partial e}{\partial \lambda}\right|_{P,v} \dot{\lambda} + P\dot{v} = 0 \tag{8.39}$$

or

$$\dot{P} + \left[\frac{P + \left.\dfrac{\partial e}{\partial v}\right|_{P,\lambda}}{\left.\dfrac{\partial e}{\partial P}\right|_{v,\lambda}}\right]\dot{v} + \frac{\left.\dfrac{\partial e}{\partial v}\right|_{P,\lambda}}{\left.\dfrac{\partial e}{\partial P}\right|_{v,\lambda}}\dot{\lambda} = 0 \tag{8.40}$$

Where it can be shown that

$$\rho^2 c^2 = \frac{P + \overline{\frac{\partial u}{\partial v}}_{P,v}}{\overline{\frac{\partial u}{\partial P}}_{v,\lambda}} \tag{8.41}$$

and

$$\rho c^2 \sigma = -\frac{\overline{\frac{\partial e}{\partial \lambda}}_{P,\lambda}}{\overline{\frac{\partial e}{\partial P}}_{v,\lambda}} \tag{8.42}$$

In Eq. 8.42 σ is thermicity and is defined as

$$\sigma = \frac{1}{\rho c^2} \frac{\partial P}{\partial \lambda}_{u,v} \tag{8.43}$$

Thermicity is a measure of the pressure rise for a given change in λ [2]. And for an ideal gas

$$\sigma = \frac{\gamma - 1}{\gamma} \frac{\rho}{P} q_w \tag{8.44}$$

Substitution of Eqs. 8.41 and 8.42 into Eq. 8.40 results in

$$\dot{P} + \rho^2 c^2 \dot{v} - \rho c^2 \sigma = 0 \tag{8.45}$$

and

$$\dot{P} = c^2 \dot{\rho} + \rho c^2 \sigma r \tag{8.46}$$

Which essentially shows that changes in pressure with time are a function of density change with time and heat release through a chemical reaction.

Another form of Eq. 8.46 is

$$u \frac{dP}{dx} - uc^2 \frac{d\rho}{dx} = \rho c^2 \sigma r \tag{8.47}$$

Where $u = \dfrac{dx}{dt}$

A summary table is given below (Table 8.2).

Table 8.2 Conservation Principles for a Gaseous System, Part II

	Conservation of Mass	Conservation of Momentum	Conservation of Energy
Unsteady, heat release, shear stress at the wall, and area change	$\frac{\partial}{\partial t}(\rho A) + \frac{\partial}{\partial x}(\rho A u) = 0$	$\frac{\partial}{\partial t}(\rho A u) + \frac{\partial}{\partial x}(\rho A u^2) = -\frac{\partial}{\partial x}(PA) + P\frac{\partial A}{\partial x} - \tau_w L_p$	$\frac{\partial}{\partial t}\left[\rho A\left(e+\frac{u^2}{2}\right)\right] + \frac{\partial}{\partial x}\left[\rho A u\left(e+\frac{u^2}{2}+\frac{P}{\rho}\right)\right] = q_w L_p$
Steady, heat release, shear stress at the wall, and area change	$\frac{\partial}{\partial x}(\rho A u) = 0$	$\frac{d}{dx}(\rho A u^2) = -\frac{d}{dx}(PA) + P\frac{dA}{dx} - \tau_w L_p$	$\frac{dP}{dx} - c^2 \frac{d\rho}{dx} = \frac{(q_w + \tau_w u)L_p}{m\left.\frac{\partial e}{\partial P}\right|_\rho}$
Steady, no area change, no shear stress, and heat release from detonation	$\rho\frac{du}{dx} + u\frac{d\rho}{dx} = 0$	$u\frac{du}{dx} + \frac{1}{\rho}\frac{dP}{dx} = 0$	$u\frac{dP}{dx} - uc^2\frac{d\rho}{dx} = \rho c^2 \sigma r$
Steady, no area change, no shear stress, and no state changes within the control volume	$\rho_1 u_1 = \rho_2 u_2$	$P_1 + \rho_1 u_1^2 = P_2 + \rho_2 u_2^2$	$h_1 + q + \frac{u_1^2}{2} = h_2 + \frac{u_2^2}{2}$

8.3 Reversible, Adiabatic (Isentropic) Compressible Flow

In this section, equations are developed to relate states during isentropic, compressible flow. These relationships will allow us to see how Mach number affects temperature, pressure, and density when comparing static and stagnation conditions [3, 4].

From fluid mechanics [4], static and stagnation pressure are defined in the following manner. "The pressure at a point in a fluid is called the 'static pressure'. The 'stagnation pressure' is the pressure that the fluid would obtain if brought to rest without loss of mechanical energy. The difference between the two is the 'dynamic pressure'. The 'total pressure' is the sum of the static pressure, the dynamic pressure, and the gravitational potential energy per unit volume. It is therefore the sum of the mechanical energy per unit volume in a fluid."

The relationship between static and stagnation pressure (P_0) with no change in potential energy is given as

$$P_0 = \frac{1}{2}\rho u^2 + P_{static} \tag{8.48}$$

or

$$\frac{P_0}{P_{static}} = 1 + \frac{1}{2}\frac{u^2}{P_{static}/\rho} = 1 + \frac{1}{2}\gamma Ma^2 \tag{8.49}$$

Please note the relationship given as Eq. 8.49, which is derived from Bernoulli's law, is no longer valid for a compressible flow.

The first law for an isentropic, compressible flow is

$$h_o + \frac{u_0^2}{2} = h + \frac{u^2}{2} \tag{8.50}$$

where "0" denotes the stagnation conditions and assuming $u_0 \approx 0$ for an ideal gas such that

$$c_p T_0 = c_p T + \frac{u^2}{2} \tag{8.51}$$

or

$$T_0 = T + \frac{u^2}{2c_p} \tag{8.52}$$

and

$$\frac{T_0}{T} = 1 + \frac{u^2}{2c_p T} \tag{8.53}$$

where R_g is equal to

$$R_g = c_p - c_v \tag{8.54}$$

Dividing both sides of Eq. 8.54 by c_p results in

$$\frac{R_g}{c_p} = \frac{c_p - c_v}{c_p} = 1 - \frac{1}{\gamma} = \frac{\gamma - 1}{\gamma} \tag{8.55}$$

or

$$\frac{1}{c_p} = \frac{\gamma - 1}{\gamma} \frac{1}{R_g} \tag{8.56}$$

Substitution of Eq. 8.56 into Eq. 8.53 results in

$$\frac{T_0}{T} = 1 + \frac{u^2}{\gamma R_g T} \frac{\gamma - 1}{2} \tag{8.57}$$

and

$$\frac{T_0}{T} = 1 + \frac{\gamma - 1}{2} Ma^2 \tag{8.58}$$

Additionally, for an isentropic, compressible flow the second law is

$$ds = \frac{c_v dT}{T} + \frac{Pdv}{T} \tag{8.59}$$

And for an ideal gas

$$ds = \frac{c_v dT}{T} + R_g \frac{dv}{v} \tag{8.60}$$

or

$$ds = \frac{c_v dT}{T} + \left(c_p - c_v\right)\frac{dv}{v} \tag{8.61}$$

And for $\Delta s = 0$, the integration is

$$c_v \ln\left(\frac{T}{T_0}\right) + \left(c_p - c_v\right)\ln\left(\frac{v}{v_0}\right) = s_2 - s_1 = 0 \tag{8.62}$$

or

$$\frac{T}{T_0} = \left(\frac{v_0}{v}\right)^{\gamma-1} = \left(\frac{\rho}{\rho_0}\right)^{\gamma-1} = \left(\frac{P}{P_0}\right)^{\frac{\gamma-1}{\gamma}} \qquad (8.63)$$

Where it can be shown that

$$\frac{\rho}{\rho_0} = \left(1 + \frac{\gamma-1}{2} Ma^2\right)^{-\frac{1}{\gamma-1}} \qquad (8.64)$$

and

$$\frac{P}{P_0} = \left(1 + \frac{\gamma-1}{2} Ma^2\right)^{-\frac{\gamma}{\gamma-1}} \qquad (8.65)$$

For air, $\gamma = 1.4$, Eqs. 8.58, 8.64, and 8.65 become

$$\frac{T}{T_0} = \left(1 + \frac{1}{5} Ma^2\right)^{-1} \qquad (8.66)$$

$$\frac{\rho}{\rho_0} = \left(1 + \frac{1}{5} Ma^2\right)^{-2.5} \qquad (8.67)$$

$$\frac{P}{P_0} = \left(1 + \frac{1}{5} Ma^2\right)^{-3.5} \qquad (8.68)$$

Graphically Eqs. 8.66, 8.67, and 8.68 relate static and stagnation states with Mach number, which is given in Fig. 8.2.

The last thing to do in this section is to show the relationship between stagnation properties (0) and the location where Ma = 1 (*), which we'll see in Chap. 9 is the choke point.

$$\frac{T*}{T_0} = \left(1 + \frac{\gamma-1}{2} 1^2\right)^{-1} = \frac{2}{\gamma+1} \qquad (8.69)$$

$$\frac{\rho*}{\rho_0} = \left(1 + \frac{\gamma-1}{2} 1^2\right)^{-\frac{1}{\gamma-1}} = \left(\frac{2}{\gamma+1}\right)^{\frac{1}{\gamma-1}} \qquad (8.70)$$

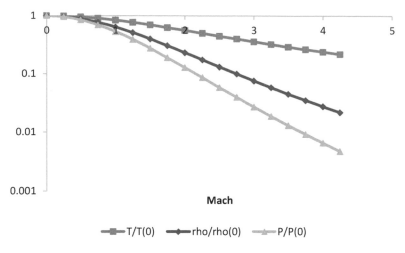

Fig. 8.2 Static/Stagnation States versus Mach Number

and

$$\frac{P*}{P_0} = \left(1 + \frac{\gamma-1}{2}1^2\right)^{-\frac{\gamma}{\gamma-1}} = \left(\frac{2}{\gamma+1}\right)^{\frac{\gamma}{\gamma-1}} \tag{8.71}$$

And for air, $\gamma = 1.4$, Eqs. 8.69, 8.70, and 8.71 become

$$\frac{T*}{T_0} = .8333 \tag{8.72}$$

$$\frac{\rho*}{\rho_0} = .6339 \tag{8.73}$$

And

$$\frac{P*}{P_0} = .5283 \tag{8.74}$$

We'll see in Chap. 9 that when the pressure at the stagnation condition is twice the pressure at the choke point, then the mass rate will be independent of the pressure difference between the two states and this determined mass rate will be the maximum mass rate allowed (critical flow).

8.4   Mass Transfer

Mass diffusion, moment diffusion, and energy diffusion are understood through the following equations.

Mass Diffusion

$$J_a = -D_{ab} \frac{\partial}{\partial y}\left(\rho_a\right) \tag{8.75}$$

Momentum Diffusion

$$\tau_{xy} = -\nu \frac{\partial}{\partial y}\left(\rho v_x\right) \tag{8.76}$$

Energy Diffusion

$$q_y = -\alpha \frac{\partial}{\partial y}\left(\rho c_p T\right) \tag{8.77}$$

where $D_{ab}\left[\dfrac{m^2}{s}\right]$ is molecular diffusion for a binary gas, $\nu\left[\dfrac{m^2}{s}\right]$ is kinematic viscosity, $\alpha\left[\dfrac{m^2}{s}\right]$ is heat diffusion, $J_a\left[\dfrac{N}{m^2}\right]$ is mass diffusion, $\tau_{xy}\left[\dfrac{N}{m^2}\right]$ is shear stress, and $q_y\left[\dfrac{N}{m^2}\right]$ is heat transfer and ρ is constant in Eqs. 8.75 and 8.76 and ρc_p is constant in Eq. 8.77. Please note that D_{ab}, υ, and α all have the same dimensions.

In the above equations, the quantities $\{D_{AB}, \alpha, \upsilon\}$ are related through three non-dimensional numbers given below.

Schmidt Number

$$Sc = \frac{\nu}{D} = \frac{rate\ of\ momentum\ transport}{rate\ of\ mass\ transport} \tag{8.78}$$

Prandtl Number

$$\text{Pr} = \frac{\nu}{\alpha} = \frac{rate\ of\ momentum\ transport}{rate\ of\ energy\ transport} \tag{8.79}$$

Lewis Number

$$Le = \frac{\alpha}{D} = \frac{rate\ of\ energy\ transport}{rate\ of\ mass\ transport} \tag{8.80}$$

And these three non-dimensional numbers are related in the following way

$$Le = \frac{Sc}{Pr} \qquad (8.81)$$

Typically for deflagrations [5],

$$Le \approx 1 \qquad (8.82)$$

And

$$Sc \approx Pr \rightarrow \alpha \approx D \qquad (8.83)$$

Equation 8.83 states that the mass transfer (D) is approximately equal to the heat transfer (α) for deflagration systems; Glassman and Yetter [6] caution assuming $Le \approx 1$ and this assumption must always be established for a particular system of interest.

8.4.1 Fick's Law and Species Conservation Principles

Fick's law of diffusion for a binary gas, which is the type of mixture considered in this book, is defined as the rate at which two gas species diffusion through each other. For one dimensional, binary diffusion (the type considered in the chapter on deflagration) it is

mass flow of A per unit area
= mass flow of A associated with bulk flow per unit area
− mass flow of A associated with molectular diffusion   (8.84)

or

$$\dot{m}_A^{''} = \frac{\dot{m}}{A} = Y_A \left(\dot{m}_A^{''} + \dot{m}_B^{''} \right) - \rho D_{AB} \frac{dY_A}{dx} \qquad (8.85)$$

where gas "A" is transported by two means:

- Bulk motion of the fluid.
- Molecular diffusion.

So that for a mixture of gases "A" and "B" the mixture mass flux is

$$\dot{m}^{''} = \dot{m}_A^{''} + \dot{m}_B^{''} = Y_A \dot{m}^{''} - \rho D_{AB} \frac{dY_A}{dx} + Y_B \dot{m}^{''} - \rho D_{AB} \frac{dY_B}{dx} \qquad (8.86)$$

And for a binary gas mixture, $Y_A + Y_B = 1$, such that Eq. 8.86 is

$$\dot{m}'' = \dot{m}''_A + \dot{m}''_B = \dot{m}'' - \rho D_{AB} \frac{dY_A}{dx} + -\rho D_{AB} \frac{dY_B}{dx} \tag{8.87}$$

or

$$0 = -\rho D_{AB} \frac{dY_A}{dx} + -\rho D_{AB} \frac{dY_B}{dx} \tag{8.88}$$

In general,

$$\sum \dot{m}''_i = 0 \tag{8.89}$$

Equation 8.89 is simply a consequence of the conservation of mass. In the next section, we'll see how D_{AB} is related to temperature and pressure.

8.4.2 Understanding Diffusion from Kinetic Theory of Gases

From the kinetic theory of gases, the following relationships can be established [2, 6, 7] and their derivation is beyond the scope of this book.

If we make certain assumptions [2]

1. Consider a stationary (no bulk flow) plane layer of binary gas mixture consisting of rigid, non-attracting molecules.
2. Molecular masses of gases "a" and "b" are identical.
3. A concentration gradient exists in x-direction, and is sufficiently small that over small enough distances the gradient can be assumed to be linear.

$$\bar{v} = \text{mean speed of species a molecules} - \sqrt{\frac{8k_B T}{\pi m_a}} \tag{8.90}$$

$$Z''_A = \text{Wall collision frequency of a molecules per unit area} = \frac{1}{4} \frac{n_a}{V} \bar{v} \tag{8.91}$$

$$\lambda = \text{Mean free path} = \frac{1}{\sqrt{2}\pi \left(\frac{n_{total}}{V}\right)\sigma^2} \tag{8.92}$$

and

$$a = \text{average perpendicular distance from plane of last}$$
$$\text{collision to plane where next collision occurs} = \frac{2}{3}\lambda \tag{8.93}$$

where k_B is the Boltzmann's constant, ma is the mass of one molecule of gas "a", $\frac{n_a}{V}$ molecular density of gas "a", $\frac{n_{total}}{V}$ molecular density of gas mixture, and σ diameter of molecules "a" and "b".

The next flux of molecule a along the x axis is

$$\dot{m}_A^{''} = \dot{m}_{A\,x-a}^{''} - \dot{m}_{A,x+a}^{''} \tag{8.94}$$

Or in terms of $Z_A^{''}$

$$\dot{m}_A^{''} = \left(Z_A^{''} \right)_{x-a} m_A - \left(Z_A^{''} \right)_{x+a} m_A \tag{8.95}$$

And the density of the binary mixture is

$$\rho = \frac{m_{total}}{v_{total}} \tag{8.96}$$

Recalling the definition of $Z_A^{''}$ is

$$Z_A^{''} = \frac{1}{4} \frac{n_A}{V} \overline{v} \tag{8.97}$$

And substituting Eq. 8.97 into Eq. 8.96 results in

$$Z_A^{''} = \frac{1}{4} \frac{n_A m_A}{m_{total}} \rho \overline{v} \tag{8.98}$$

Or in terms of Y_A

$$Z_A^{''} = \frac{1}{4} Y_A \rho \overline{v} \tag{8.99}$$

And substituting Eq. 8.99 into Eq. 8.95 leads to

$$\dot{m}_A^{''} = \frac{1}{4} \rho \overline{v} \left[Y_{A,x-a} - Y_{A,x+a} \right] \tag{8.100}$$

where $\dfrac{dY}{dx}$ is defined as

$$\frac{dY_A}{dx} = \frac{Y_{A,x-a} - Y_{A,x+a}}{2a} \tag{8.101}$$

or

$$\frac{dY_A}{dx} = \frac{Y_{A,x-a} - Y_{A,x+a}}{\dfrac{4\lambda}{3}} \tag{8.102}$$

Substitution of Eq. 8.102 into Eq. 8.100 for $Y_{A,x-a} - Y_{A,x+a}$, we see that

$$\dot{m}''_A = -\rho \frac{\overline{v}\lambda}{3} \frac{dY_A}{dx} \qquad (8.103)$$

And going back to the definition of binary mass diffusion without bulk flow (Eq. 8.85), we see that

$$D_{AB} = \frac{\overline{v}\lambda}{3} \qquad (8.104)$$

Substitution of Eq. 8.90 for $\overline{v}$ and Eq. 8.92 for λ into Eq. 8.104 results in

$$D_{AB} = \frac{1}{3}\sqrt{\frac{8k_B T}{\pi m_A}} \frac{1}{\sqrt{2}\pi \left(\dfrac{n_{Total}}{V}\right)\sigma^2} \qquad (8.105)$$

Where from ideal gas behavior

$$\rho = \frac{n_{Total}}{V} = \frac{P}{k_B T} \qquad (8.106)$$

Substitution of Eq. 8.106 into Eq. 8.105 leads to

$$D_{AB} = \frac{1}{3}\sqrt{\frac{8k_B T}{\pi m_A}} \frac{1}{\sqrt{2}\pi \dfrac{P}{k_B T}\sigma^2} = \frac{2}{3}\sqrt{\frac{k_B^3 T}{\pi^3 m_A}} \frac{T}{P\sigma^2} \qquad (8.107)$$

And thus

$$D_{AB} \propto \frac{T^{3/2}}{P} \qquad (8.108)$$

8.5 More General Conservation Principles for Premixed Laminar Flames

The developed conservation principles in this section will be utilized in Chap. 12 for planar, one-dimensional systems.

8.5.1 Conservation of Mass

Change in mass with time for the control volume is

$$\frac{dm_{sys}}{dt} = \dot{m}_x - \dot{m}_{x+\Delta x} \tag{8.109}$$

or

$$\frac{d}{dt}\left[\rho A\Delta x\right] = \left[\rho v_x A\right]_x - \left[\rho v_x A\right]_{x+\Delta x} \tag{8.110}$$

Dividing both sides by $A\Delta x$ and taking the limit as $\Delta x \to 0$ result in

$$\frac{\partial \rho}{\partial t} = -\frac{\partial}{\partial x}\left(\rho v_x\right), \frac{\partial \rho}{\partial t} + \frac{\partial}{\partial x}\left(\rho v_x\right) = 0 \tag{8.111}$$

Or for steady state

$$\frac{\partial}{\partial x}\left(\rho v_x\right) = 0 \tag{8.112}$$

Equation 8.112 is appropriate for one dimensional, planar system.

The conservation of species mass with binary diffusion is given for species "A" and it is

$$\dot{m}_A'' = \frac{d}{dx}\left(\dot{m}''Y_A\right) - \frac{d}{dx}\left[\rho D_{AB}\frac{dY_A}{dx}\right] \tag{8.113}$$

8.5.2 Conservation of Momentum

From Newton's second law,

$$\sum F = \left(\dot{m}v\right)_{out} - \left(\dot{m}v\right)_{in} \tag{8.114}$$

For the one-dimensional, planar system

$$\left[PA\right]_x - \left[PA\right]_{x+\Delta x} = \dot{m}\left[v_{x,x+\Delta x} - v_{x,x}\right] \tag{8.115}$$

And dividing both sides by Δx and taking the limit as $\Delta x \to 0$ results in

$$-\frac{dP}{dx} = \dot{m}'' \frac{dv_x}{dx} \tag{8.116}$$

or

$$-\frac{dP}{dx} = \rho v_x \frac{dv_x}{dx} \tag{8.117}$$

8.5.3   Conservation of Energy

Given the diagram (Fig. 8.3).
  The conservation of energy for this steady system is

$$\left[Q_x'' - Q_{x+\Delta x}'' \right] A - \dot{W}_{sys} = \dot{m}'' A \left[\left(h + \frac{v^2}{2} + gz \right)_{x+\Delta x} - \left(h + \frac{v^2}{2} + gz \right)_x \right] \tag{8.118}$$

And assuming $\Delta PE = 0$ and $\dot{W}_{sys} = 0$ results in

$$-\left[Q_{x+\Delta x}'' - Q_x'' \right] = \dot{m}'' \left[\left(h + \frac{v^2}{2} \right)_{x+\Delta x} - \left(h + \frac{v^2}{2} \right)_x \right] \tag{8.119}$$

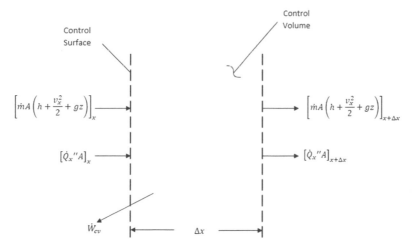

Fig. 8.3 Conservation of Energy (Premixed Deflagrations) [5]

Dividing both sides by Δx and taking the limit as $\Delta x \to 0$ results in

$$-\frac{d\dot{Q}_x''}{dx} = \dot{m}'' \left[\frac{dh}{dx} + v_x \frac{dv_x}{dx} \right] \tag{8.120}$$

There are two contributions to the heat flux, $\dot{Q}_x''$

$$\dot{Q}'' = -k\nabla T + \sum \dot{m}_{i,diff}'' h_i \tag{8.121}$$

For onedimension, planar systems Eq. 8.121 takes the simpler form

$$\dot{Q}_x'' = -k\frac{dT}{dx} + \sum \rho Y_i \left(v_{ix} - v_x \right) h_i \tag{8.122}$$

Or in terms of the bulk and species mass fluxes

$$\dot{Q}_x'' = -k\frac{dT}{dx} + \sum \rho v_{ix} Y_i h_i - \rho v_x \sum Y_i h_i = -k\frac{dT}{dx} + \sum \dot{m}_i'' h_i - \dot{m}'' h \tag{8.123}$$

Noticing that $\dot{m}_i'' = \rho v_{ix}$, $\rho v_x = \dot{m}''$ and $\sum Y_i h_i = h$.
Substituting these three expressions into Eq. 8.123 into Eq. 8.119 results in

$$\frac{d}{dx}\left(\sum h_i \dot{m}_i''\right) + \frac{d}{dx}\left(-k\frac{dT}{dx}\right) + \dot{m}'' v_x \frac{dv_x}{dx} = 0 \tag{8.124}$$

where the first term can be farther expanded as

$$\frac{d}{dx}\left(\sum h_i \dot{m}_i''\right) = \sum \dot{m}_i'' \frac{dh_i}{dx} + \sum h_i \frac{d\dot{m}_i''}{dx} \tag{8.125}$$

And substitution of Eq. 8.125 into Eq. 8.124 results in

$$\sum \dot{m}_i'' \frac{dh_i}{dx} + \frac{d}{dx}\left(-k\frac{dT}{dx}\right) + \dot{m}'' v_x \frac{dv_x}{dx} = -\sum h_i \frac{d\dot{m}_i''}{dx} = -\sum h_i \dot{m}_i''' \tag{8.126}$$

Another form of Eq. 8.126, which is more useful for numerical calculations, is

$$\dot{m}'' C_p \frac{dT}{dx} + \frac{d}{dx}\left(-k\frac{dT}{dx}\right) + \sum \rho Y_i v_i C_{p,i} \frac{dT}{dx} = -\sum h_i \dot{m}_i''' \tag{8.127}$$

where v_i is a species diffusion velocity, and two other forms of conservation of energy that have proven useful are when the system is planar, one-dimensional

$$\dot{m}''C_p\frac{dT}{dx}+\frac{d}{dx}\left(-\rho D\frac{d\int C_p T}{dx}\right)=-\sum h^0_{f,i}\dot{m}'''_i \qquad (8.128)$$

And Eq. 8.128 can be understood as a system where the following mechanisms are important

$$\begin{bmatrix}\text{Specific Rate of Sensible}\\\text{enthalpy transport by}\\\text{convection}\end{bmatrix}+\begin{bmatrix}\text{Specific Rate of Sensible}\\\text{enthalpy transport by}\\\text{diffusion}\end{bmatrix}$$
$$=\begin{bmatrix}\text{Specific Rate of Sensible}\\\text{enthalpy production}\\\text{by chemical reaction}\end{bmatrix} \qquad (8.129)$$

And in an axisymmetric form

$$\frac{1}{r}\frac{\partial\left(r\rho v_r\int c_p dT\right)}{\partial x}+\frac{1}{r}\frac{\partial\left(r\rho v_x\int c_p dT\right)}{\partial r}-\frac{1}{r}\frac{\partial}{\partial r}\left(r\rho D\frac{\partial\int c_p dT}{\partial r}\right)=-\sum h^0_{f,i}\dot{m}'''_i \qquad (8.130)$$

Where now there are two independent variables {x, r} and Eq. 8.130 has an additional term and can be understood as a system where the following mechanisms are important

$$\begin{bmatrix}\text{Specific Rate of Sensible}\\\text{enthalpy transport by}\\\text{axial convection}\end{bmatrix}+\begin{bmatrix}\text{Specific Rate of Sensible}\\\text{enthalpy transport by}\\\text{radial convection}\end{bmatrix}$$
$$+\begin{bmatrix}\text{Specific Rate of Sensible}\\\text{enthalpy transport by}\\\text{radial diffusion}\end{bmatrix}=\begin{bmatrix}\text{Specific Rate of Sensible}\\\text{enthalpy production}\\\text{by chemical reaction}\end{bmatrix} \qquad (8.131)$$

8.6   More General Conservation Principles for a Non-premixed Laminar Flame

The following conservation principles for a non-premixed deflagration system are given without derivation; for more details please refer to [5], which provides a more thorough treatment. Please note these equations are for steady, axial symmetric systems.

These developed conservation principles will be utilized in Chap. 12.

8.6.1 Conservation of Mass

Conservation of mass for a steady, axial symmetric system is

$$\frac{1}{r}\frac{\partial}{\partial r}(r\rho v_r) + \frac{\partial}{\partial x}(\rho v_x) = 0 \tag{8.132}$$

In this coordinate system, there are now two independent variables {r, x} and so two terms.

8.6.2 Conservation of Momentum

For conservation of momentum of non-premixed combustible gases

$$\frac{1}{r}\frac{\partial(r\rho v_x v_x)}{\partial x} + \frac{1}{r}\frac{\partial(\rho v_x v_r)}{\partial r} - \frac{1}{r}\frac{\partial\left(r\mu\dfrac{\partial v_x}{\partial r}\right)}{\partial r} = (\rho_\infty - \rho)g \tag{8.133}$$

Equation 8.133 allows for buoyancy effects.

8.6.3 Conservation of Energy

For conservation of energy of non-premixed combustible gases, which is a slightly different form of Eq. 8.130 and given as

$$\frac{1}{r}\frac{\partial(r\rho v_r Y_i)}{\partial x} + \frac{\partial(r\rho v_x Y_i)}{\partial r} - \frac{1}{r}\frac{\partial}{\partial r}\left(r\rho D\frac{\partial Y_i}{\partial r}\right) = \dot{m}_i''' \tag{8.134}$$

where

$$Y_{\mathrm{Pr}} = 1 - Y_F - Y_{ox} \tag{8.135}$$

8.7 Problems

Problem 8.1 Prove the following relationship

$$\rho^2 c^2 = \frac{P + \dfrac{\partial u}{\partial v}\bigg|_{P,v}}{\dfrac{\partial u}{\partial P}\bigg|_{v,\lambda}}$$

Problem 8.2 Prove the following relationship.

$$\rho c^2 \sigma = -\frac{\dfrac{\partial e}{\partial \lambda}_{P,\lambda}}{\dfrac{\partial e}{\partial P}_{v,\lambda.}}$$

where

$$\frac{\partial P}{\partial e}_{v,\lambda} \frac{\partial e}{\partial \lambda}_{P,v} \frac{\partial \lambda}{\partial P}_{v,e} = -1$$

Problem 8.3 Delineating Compressible and Incompressible Flows

Starting from the conservation of momentum for a steady flow with no heat transfer, no shear stress, nor area change (isentropic flow)

$$u\frac{du}{dx} + \frac{1}{\rho}\frac{dP}{dx} = 0$$

And assuming an ideal, perfect gas show for a density change $\left(\dfrac{d\rho}{\rho}\right)$ of 10% that the corresponding Mach number is 0.3, which is the threshold between treating a moving fluid as incompressible versus compressible. Please note an incompressible fluid field is one where the density is the same throughout the fluid field and where $\left(\dfrac{d\rho}{\rho}\right)$ is significant, then the density is changing with location.

References

1. Powers, J. (2020). *Lecture Notes for Intermediate Fluid Mechanics*. Notre Dame University. Accessed 05/22/2020.
2. Powers, J. (2020). *Lecture Notes for Fundamentals of Combustion*. Notre Dame University. Accessed 05/22/2020.
3. Munson, Young and Okiishi. (2002). *Fundamentals of fluid mechanics* (4th ed.). John Wiley and Sons.
4. http://www2.eng.cam.ac.uk/~mpj1001/learnfluidmechanics.org/LFM_L6.html#:~:text=Pipe%20Flow%20Networks,Blank%20handout&text=The%20pressure%20at%20a%20point,is%20the%20'dynamic%20pressure.
5. Turns, S. R. (2012). *An introduction to combustion (concepts and applications)* (3rd ed.). Mc-Graw Hill.
6. Glassman and Yetter. (2008). *Combustion* (4th ed.). Academic Press.
7. Kuo, K. K. (2005). *Principles of combustion* (2nd ed.). John Wiley and Sons.

Chapter 9
Critical Flow

9.1 Preview

The system under consideration is a pressurized vessel in which the discharge point is referred to as the "choke point," and the center of the tank is denoted the "stagnation point," which is depicted in Fig. 9.1. The control volume could equally be a system where there is a sudden contraction, see Fig. 9.2.

The equation for the mass rate exiting (mass efflux) the system is

$$\frac{dm(sys)}{dt} = -A^* \rho^* u^* \tag{9.1}$$

where m is the mass of fluid within the system at time t, A^* is the choke cross-sectional area, ρ^* is the density at the choke point, and u^* is the local speed of sound at the choke point.

The crux of the problem is to develop a model for $"\rho^* u^*"$ in terms of the properties within the tank, such as T_0 and P_0, where T_0 is the stagnation temperature and P_0 is the stagnation pressure.

Fundamentals of Fluid Mechanics by Munson, Young and Okiishi [2] does an excellent job of explaining static and stagnant conditions in terms of pressures.

The development of a model for single-phase flow for $\rho^* u^*$ involved six assumptions/conditions that are

1. The reversibility of the system
2. An adiabatic system
3. A model for specific heat

Electronic Supplementary Material: The online version of this chapter (https://doi.org/10.1007/978-3-030-87387-5_9) contains supplementary material, which is available to authorized users.

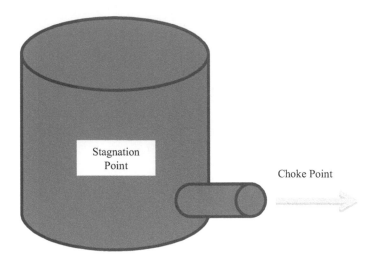

Fig. 9.1 Pressurized vessel

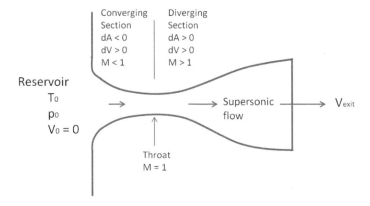

Fig. 9.2 Supersonic nozzle [1]

4. The equation of state
5. The stagnation condition
6. Choked flow

This chapter will discuss critical flow associated with a single phase (single component) and go on to discuss critical flow in a two phase (two-component system). The concept of critical flow for a single-phase fluid can be understood in the following manner.

Given the following equation for first law

$$0 = \Delta h + \Delta KE \tag{9.2}$$

Where the system has no work, no heat transfer, nor changes in potential energy, which has the differential version

$$0 = dh + udu \tag{9.3}$$

And given the following equation for second law for an isentropic flow

$$Tds = dh - \frac{dP}{\rho} = 0 \tag{9.4}$$

Equating the two forms of "*dh*" results in

$$-udu = \frac{dP}{\rho} \tag{9.5}$$

or

$$\frac{du}{u} = -\frac{dP}{\rho u^2} \tag{9.6}$$

And defining the mass flux (*G*) as

$$G = \rho u \tag{9.7}$$

or

$$\ln(G) = \ln(\rho) + \ln(u) \tag{9.8}$$

Taking the derivative of both sides results in

$$\frac{dG}{G} = \frac{d\rho}{\rho} + \frac{du}{u} \tag{9.9}$$

And substituting Eq. 9.6 into Eq. 9.9 results in

$$\frac{dG}{G} = -\frac{dP}{\rho u^2} + \frac{dP}{\rho}\left(\frac{\partial\rho}{\partial P}\right)_s \tag{9.10}$$

or

$$\frac{dG}{G} = \rho dP\left[-\frac{1}{G^2} + \frac{1}{\rho^2}\left(\frac{\partial\rho}{\partial P}\right)_s\right] \tag{9.11}$$

And
G is a maxima when

$$\left(\frac{dG}{dP}\right)_s = 0 \tag{9.12}$$

Such that

$$-\frac{1}{G^2} + \frac{1}{\rho^2}\left(\frac{\partial \rho}{\partial P}\right)_s = 0 \tag{9.13}$$

or

$$G_{max} = \rho\sqrt{\left(\frac{\partial P}{\partial \rho}\right)_s} = \sqrt{\left(-\frac{\partial P}{\partial v}\right)_s} \tag{9.14}$$

Additionally, the speed of sound for a single-phase, single-component substance is

$$a^2 = \left(\frac{\partial P}{\partial \rho}\right)_s \tag{9.15}$$

Such that we see that $\frac{G_{max}}{\rho}$ is equal to the local speed of sound!

We'll see in the next section that this critical flow occurs at the choke point, which is essentially where the area of the conduit is minimal within the fluid flow or a location before the conduit suddenly expands.

We see in Fig. 9.2 that the throat is the choke point and the Mach Number (M) is 1 consistent with the above discussion.

In Sect. 9.2, we will explain Fig. 9.2 and discuss choked flow, which is also termed critical flow.

In Sect. 9.3, we will assume the gas is both ideal and thermodynamically perfect, e.g., specific heat is independent of temperature. With the six above assumptions/ conditions and an ideal, perfect gas, we get two algebraic expressions from the energy and entropy balances.

In Sect. 9.4, we will assume the gas is a van der Waal gas and that the specific heat is independent of temperature and pressure. Again, from the entropy and energy balances we get two algebraic expressions; how differently the two equations of state are in terms of critical flow will be explored through an example.

In Sect. 9.5, liquid/gas flows will be discussed. It is likely that a tank of gas at a high pressure and ambient temperature would "flash." Flashing is not what you'd hope. Flashing is when a gas suddenly experiences a great drop of pressure and begins to form a liquid phase as it expands. It's likely any critical flow from a pressure vessel will experience two-phase flow and so this section discusses liquid/ gas flows.

In Sect. 9.6, speed of sound for either a two-component or liquid/gas flow will be discussed.

In Sect. 9.7, critical flow for a two-phase, one-component system is discussed. Two classes of models will be explored: homogenous models assume average properties between the liquid and gas phase and develop a model based on either the momentum equation or energy equation; non-homogenous models assume an annular flow where a gas core is surrounded by a liquid sheath. Naturally, these two separate flows are traveling at different speeds and a slip velocity exists between the phases. Other non-equilibrium conditions make this model framework a challenge.

9.2 Effect of Area Changes on Gas Dynamic States

From conservation of mass [3–5]

$$\rho u A = \text{constant} \tag{9.16}$$

and

$$\ln(\rho) + \ln(u) + \ln(A) = \text{constant} \tag{9.17}$$

Taking the derivative of both sides

$$\frac{d\rho}{\rho} + \frac{du}{u} + \frac{dA}{A} = 0 \tag{9.18}$$

Substituting Eq. 9.6 for du results in

$$\frac{d\rho}{\rho} + \frac{-dP}{\rho u^2} + \frac{dA}{A} = 0 \tag{9.19}$$

And solving for dP/ρ gives us

$$\frac{dP}{\rho} = u^2 \left(\frac{d\rho}{\rho} + \frac{dA}{A} \right) \tag{9.20}$$

And by definition the speed of sound is

$$a^2 = \left(\frac{\partial P}{\partial \rho} \right)_s \approx \frac{dP}{d\rho} \tag{9.21}$$

And

$$a^2 d\rho \approx dP \tag{9.22}$$

Substituting Eq. 9.22 into Eq. 9.20 results in

$$\frac{d\rho}{\rho} = \frac{u^2}{a^2} \left(\frac{d\rho}{\rho} + \frac{dA}{A} \right) = Ma^2 \left(\frac{d\rho}{\rho} + \frac{dA}{A} \right) \tag{9.23}$$

And solving Eq. 9.23 for $\dfrac{dp}{\rho}$ gives us

$$\frac{d\rho}{\rho} = \left(\frac{Ma^2}{1-Ma^2}\right)\frac{dA}{A} \tag{9.24}$$

Substitute Eq. 9.24 into the differential form of conservation of mass gives as

$$\left(\frac{Ma^2}{1-Ma^2}\right)\frac{dA}{A} + \frac{dA}{A} + \frac{du}{u} = 0 \tag{9.25}$$

And solving for $\dfrac{du}{u}$ results in

$$\frac{du}{u} = -\frac{1}{1-Ma^2}\frac{dA}{A} \tag{9.26}$$

Also, $\dfrac{du}{u}$ is equal to

$$\frac{du}{u} = \frac{-dP}{\rho v^2} \tag{9.27}$$

Such that

$$\frac{dP}{P} = \frac{\rho u^2}{P}\frac{1}{1-Ma^2}\frac{dA}{A} = \gamma Ma^2\left(\frac{1}{1-Ma^2}\right)\frac{dA}{A} \tag{9.28}$$

A summary table is given below

We'll now use Table 9.1 to see how pressure, density, and velocity change through a converging and then diverging engineering device. The results of this analysis are given below.

Table 9.1 Area effects on gas dynamics

$$\frac{dP}{P} = \gamma Ma^2\left(\frac{1}{1-Ma^2}\right)\frac{dA}{A}$$

$$\frac{d\rho}{\rho} = \left(\frac{Ma^2}{1-Ma^2}\right)\frac{dA}{A} \quad (9.27)$$

$$\frac{du}{u} = -\frac{1}{1-Ma^2}\frac{dA}{A} \quad (9.28)$$

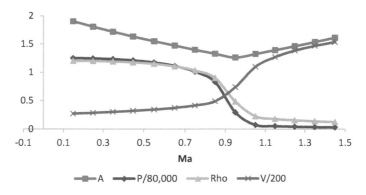

Fig. 9.3 Converging/diverging engineering device

We see in Fig. 9.3 that pressure is continually decreasing which is consistent with a fluid system; we also see the density decrease in a similar fashion with the pressure, which is consistent with ideal gas behavior. Further, we see the area is decreasing for Mach <1 and then increasing for Mach >1 and that the velocity continually increases through the device and drastically increases around Mach 1.

The portion of the device with minimal area can be the point of a choking [3–6]. Choking is a condition where the pressure differential between stagnation pressure and downstream pressure does not dictate the mass flux. It can be shown for $\frac{P_0}{P*} > 2$, where P_0 is the stagnation pressure and $P*$ is the pressure at the choke point, that the Mach number at the choke point is always 1 (for pure substance) and is consistent with Fig. 9.3; the critical flux through a choke point is the maximal flux allowed [3, 6–8] through the system.

9.3 Ideal Gas

In this section, a critical flow model is developed for an ideal, perfect gas.

Entropy Balance
From the first four assumptions/conditions and the entropy balance, we get to the following relationship

$$ds = \frac{c_v dT}{T} + \frac{P dv}{T} \tag{9.29}$$

or

$$s_2 - s_1 = 0 = \frac{c_p}{\gamma} \int_{T_0}^{T^*} \frac{dT}{T} + \int_{v_0}^{v^*} \frac{R_u T}{vT} dv \tag{9.30}$$

where γ is defined as

$$\gamma = \frac{c_p}{c_v} \tag{9.31}$$

and

$$\frac{c_P}{\gamma} \ln\left(\frac{T^*}{T_0}\right) + R_u \ln\left(\frac{v^*}{v_0}\right) = 0 \tag{9.32}$$

For an ideal gas

$$c_p - c_v = R_u \tag{9.33}$$

Substituting Eq. 9.33 into Eq. 9.32 results in

$$\frac{c_P}{\gamma} \ln\left(\frac{T^*}{T_0}\right) + \left(c_p - c_v\right) \ln\left(\frac{v^*}{v_0}\right) = 0 \tag{9.34}$$

And dividing by c_p gives

$$\frac{1}{\gamma} \ln\left(\frac{T^*}{T_0}\right) + \left(1 - \frac{1}{\gamma}\right) \ln\left(\frac{v^*}{v_0}\right) = 0 \tag{9.35}$$

or

$$\ln\left(\frac{T^*}{T_0}\right) + (\gamma - 1) \ln\left(\frac{v^*}{v_0}\right) = 0 \tag{9.36}$$

and

$$\frac{T^*}{T_0} = \left(\frac{v_0}{v^*}\right)^{\gamma - 1} \tag{9.37}$$

Energy Balance
From the last four assumptions/conditions and the energy balance, we get the following relationship

$$h_0 = h^* + \frac{1}{2} c^2 \tag{9.38}$$

where h_0 is the enthalpy at the stagnation point, h^* is the enthalpy at the discharge (choke) point, and c is the local speed of sound [4]. In a converging area, the chocked flow will be sonic when the following condition [4] is met

$$\frac{P^*}{P_0} = \left(\frac{2}{\gamma+1}\right)^{\gamma/\gamma-1}$$

(9.39)

Which as a rule of thumb, essentially states that as long as the stagnation pressure is at least twice of the chocked pressure, then the discharge velocity is equal to the local speed of sound.

We saw in Chap. 7 that c for an ideal, perfect gas is equal to

$$c = \sqrt{\gamma RT}$$

(9.40)

Substituting Eq. 9.40 into 9.38 for a perfect gas results in

$$c_p T_0 = c_p T^* + \frac{1}{2}\gamma RT^*$$

(9.41)

and

$$\frac{T_0}{T^*} = 1 + \frac{1}{2c_p}\gamma R$$

(9.42)

Further,

$$\frac{T_0}{T^*} = 1 + \frac{1}{2}\frac{1}{c_v}\left(c_p - c_v\right)$$

(9.43)

and

$$\frac{T_0}{T^*} = 1 + \frac{1}{2}(\gamma-1) = \frac{1}{2}(\gamma+1)$$

(9.44)

Again the mass rate of discharge is

$$\dot{m}_{sys} = -A^* u^* \rho^*$$

(9.45)

Given above with the definition for the local speed of sound (Eq. 9.40) for an ideal gas and

$$u^* = \sqrt{\gamma R_g T^*}$$

(9.46)

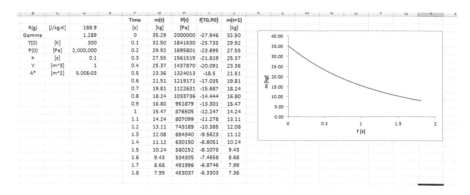

			Time	m(t)	P(t)	f[T0,P0]	m(t+1)
R(g)	[J/kg-K]	188.9	[s]	[kg]	[Pa]		[kg]
Gamma		1.289	0	35.29	2000000	-27.946	32.50
T(0)	[K]	300	0.1	32.50	1841630	-25.733	29.92
P(0)	[Pa]	2,000,000	0.2	29.92	1695801	-23.695	27.55
h	[s]	0.1	0.3	27.55	1561519	-21.819	25.37
V	[m^3]	1	0.4	25.37	1437870	-20.091	23.36
A*	[m^2]	5.00E-03	0.5	23.36	1324013	-18.5	21.51
			0.6	21.51	1219171	-17.035	19.81
			0.7	19.81	1122631	-15.687	18.24
			0.8	18.24	1033736	-14.444	16.80
			0.9	16.80	951879	-13.301	15.47
			1	15.47	876505	-12.247	14.24
			1.1	14.24	807099	-11.278	13.11
			1.2	13.11	743189	-10.385	12.08
			1.3	12.08	684340	-9.5623	11.12
			1.4	11.12	630150	-8.8051	10.24
			1.5	10.24	580252	-8.1078	9.43
			1.6	9.43	534305	-7.4658	8.68
			1.7	8.68	491996	-6.8746	7.99
			1.8	7.99	453037	-6.3303	7.36

Fig. 9.4 Example 9.1 Results

Additionally, we know from Eqs. 9.37 and 9.44 that

$$\frac{\rho^*}{\rho_0} = \left(\frac{T^*}{T_0}\right)^{1/{\gamma-1}} = \left(\frac{2}{\gamma+1}\right)^{1/{\gamma-1}} \tag{9.47}$$

Substitution of Eqs. 9.46 and 9.47 (plus the ideal gas law) into 9.45 results in

$$\dot{m}_{sys} = -A^* \sqrt{\frac{2R\gamma T_0}{\gamma+1}} \frac{P_0}{RT_0}\left(\frac{2}{\gamma+1}\right)^{1/{\gamma-1}} \tag{9.48}$$

Example 9.1 Pressure vessel for ideal gas
Determine the amount of carbon dioxide within a pressure vessel after 1.5 seconds when the initial stagnation pressure is 2 MPa, temperature is 300 K, and the opening is 5e-3 m². The volume of the vessel is 1 m³.

Note that CO_2 at this temperature and initial pressure is not ideal! In fact, using a Peng-Robinson equation of state [7] it can be shown that the Z factor is 0.89, which tells us this gas is not acting ideal.

From the graph above, the answer is about 10.24 kg (Fig. 9.4).

9.4   Van der Waal Gas

In order to incorporate real gas effects, the following approach was adopted and includes the use of the van der Waal's equation of state

$$P = \frac{RT}{v-b} - \frac{a}{v^2} \tag{9.49}$$

where $\{a, b\}$ have been previously defined.

Because the equation of state is more mathematically involved, the entropy and energy balance equation will be more involved.

Energy Balance

From the last four assumptions, an *energy balance* is given as

$$h_0 = h^* + \frac{1}{2}c^2 \tag{9.50}$$

where h_0 is the enthalpy at the stagnation point, h^* is the enthalpy at the discharge (choke) point, and c is the local speed of sound [9] for a van der Waal gas.

What we'll do is define "du," solve this equation and then define "dh" in terms of "du."

$$de = \left(\frac{\partial e}{\partial T}\right)_v dT + \left(\frac{\partial e}{\partial v}\right)_T dv \tag{9.51}$$

where

$$\left(\frac{\partial e}{\partial T}\right)_v = c_v \tag{9.52}$$

and

$$\left(\frac{\partial e}{\partial v}\right)_T = T\left(\frac{\partial P}{\partial T}\right)_v - P = T\frac{R}{v-b} - \left\{\frac{R}{v-b} - \frac{a}{v^2}\right\} = \frac{a}{v^2} \tag{9.53}$$

So that

$$de = c_v dT + \frac{a}{v^2} dv \tag{9.54}$$

And integrating

$$\Delta e = c_v \Delta T + \frac{a}{v_1} - \frac{a}{v_2} \tag{9.55}$$

and

$$\Delta h = h_2 - h_1 = \Delta u + P_2 v_2 - P_1 v_1 = c_v \Delta T + \frac{a}{v_1} - \frac{a}{v_2} + \frac{RT_2 v_2}{v_2 - b} - \frac{a}{v_2} - \frac{RT_1 v_1}{v_1 - b} + \frac{a}{v_1} \tag{9.56}$$

Additionally, the speed of sound for a van der Waal gas is [9]

$$c_{vw}^2 = \left(1+\frac{R}{c_v}\right)\frac{v^2RT}{(v-b)^2} - 2\frac{a}{v} \tag{9.57}$$

And substituting Eqs. 9.56 and 9.57 into 9.50 results in

$$c_v\Delta T + \frac{a}{v_1} - \frac{a}{v_2} + \frac{RT_2v_2}{v_2-b} - \frac{a}{v_2} - \frac{RT_1v_1}{v_1-b} + \frac{a}{v_1} + \frac{1}{2}\left(1+\frac{R}{c_v}\right)\frac{v_2^2RT_2}{(v_2-b)^2} - \frac{a}{v_2} = 0 \tag{9.58}$$

Entropy Balance

From the first four assumptions, an entropy balance is given as

$$Tds = de + Pdv \tag{9.59}$$

and

$$de = c_v dT + \frac{a}{v^2}dv \tag{9.60}$$

Substituting Eq. 9.60 into Eq. 9.59 results in

$$Tds = c_v dT + \left(P + \frac{a}{v^2}\right)dv = c_v dT + \frac{RT}{v-b}dv \tag{9.61}$$

and

$$ds = c_v\frac{dT}{T} + \frac{R}{v-b}dv$$

For isentropic flow

$$c_v\ln\left(\frac{T_2}{T_1}\right) + R\ln\left(\frac{v_2-b}{v_1-b}\right) = 0 \tag{9.62}$$

or

$$\ln\left(\frac{T_2}{T_1}\right) + (\gamma-1)\ln\left(\frac{v_2-b}{v_1-b}\right) = 0 \tag{9.63}$$

And finally

$$\frac{T_2}{T_1} = \left(\frac{v_1 - b}{v_2 - b}\right)^{\gamma-1} \tag{9.64}$$

In Sects. 9.4 and 9.5, relationships were developed between T_0 and T^* and v_0 and v^*. These developed relationships along with the appropriate speed of sound model and conservation of mass provided models for critical flow.

9.5   Liquid/Gas Flows

When we begin to ask questions about quantities such as critical flow in a liquid/gas system, we need to understand there are many types of flow that can be encountered (see Fig. 9.5) and these flows are dependent on a combination of the liquid and gas flows (see Fig. 9.6). Many of the critical flow models for two-phase systems have either assumed homogenous flow or annular flow. Each of these flows is discussed below.

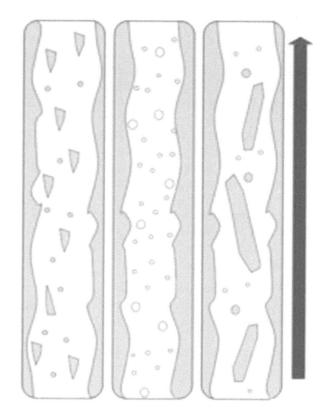

Fig. 9.5 Common vertical flow regimes – from left to right: churn flow, annual flow, and wispy annual flow [10]

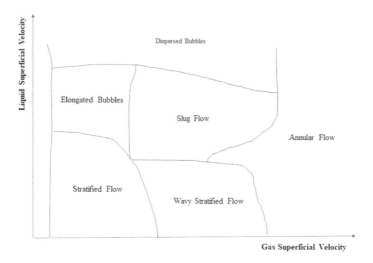

Fig. 9.6 Gas/liquid flow regimes

In *homogenous flows*, the two regimes (liquid flow and gas flow) are assumed to have some average associated with all properties such as density, pressure, or flow velocity; there is equilibrium between the phases. In *annular flow*, and assuming the conduit is a pipe, the gas flows in an inner core with a higher velocity and the liquid flows as an outer sheath with a lower velocity. The reason for the differing velocities is that one pressure is applied to the whole system, but that the effective density of the gas phase is much less than the effective density of the liquid phase and so the gas phase moves at a higher velocity relative to the liquid phase. A slip velocity is created and there is no equilibrium between the phases.

In this section we'll provide some necessary nomenclature to distinguish properties of the liquid phase with properties of the gas phase; we'll then define the conservation principles for a two-phase system.

In Sect. 9.6, we'll explore speed of sound for a liquid/gas flow where it is the case that the gas is supersonic and the liquid is subsonic.

In Sect. 9.7, we'll define critical flow for a liquid/gas flow and see that there are three considerations associated with critical flow as opposed to one in one phase, critical flow. We will then discuss a seminal paper [11] that delineates the two main forms of analytical models for two-phase, critical flow and we'll end with a brief discussion of the Omega method.

Two-Phase Gas Flow Nomenclature

W, mass rate (where g denotes the gas phase and l denotes the liquid phase) and the total mass rate is

$$W = W_g + W_l = \rho_g u_g A_g + \rho_l u_l A_l \tag{9.65}$$

x, quality

$$x = \frac{W_g}{W_g + W_l} \tag{9.66}$$

Q, volumetric flow rate and the total volumetric flow rate is

$$Q = Q_g + Q_l \tag{9.67}$$

And Q_g and Q_l can be defined as

$$Q_g = \frac{W_g}{A_g} \tag{9.68}$$

and

$$Q_l = \frac{W_l}{A_l} \tag{9.69}$$

where α volumetric fraction of gas and defined as

$$\alpha = \frac{Q_g}{Q_l + Q_g} \tag{9.70}$$

G, mass flux (where A is the cross-sectional area orthogonal to the direction of flow)

$$G = \frac{W}{A} = \frac{W_g}{A} + \frac{W_l}{A} = \frac{\rho_g u_g A_g}{A} + \frac{\rho_l u_l A_l}{A} = \alpha \rho_g u_g + (1 - \alpha) \rho_l u_l \tag{9.71}$$

j, volumetric flux and the total volumetric flux is

$$j = \frac{Q}{A} \tag{9.72}$$

Mean density is defined as

$$\bar{\rho} = \alpha \rho_g + (1 - \alpha) \rho_l \tag{9.73}$$

where j_g and j_l are defined as

$$j_g = \frac{Q_g}{A} \tag{9.74}$$

and

$$j_l = \frac{Q_l}{A} \tag{9.75}$$

And the total volumetric flux is

$$j = j_g + j_l = \frac{Q_g + Q_l}{A} = \frac{A_g}{A} u_g + \frac{A_l}{A} u_l = \alpha u_g + (1 - \alpha) u_l \tag{9.76}$$

where

$$u_l = \frac{j_l}{1 - \alpha} \tag{9.77}$$

and

$$u_g = \frac{j_g}{\alpha} \tag{9.78}$$

A summary table is given below (Table 9.2).

Example 9.2
Prove the following relationship

$$\frac{\frac{1-x}{x}}{\frac{1-\alpha}{\alpha}} = \frac{Q_l}{Q_g} \tag{9.79}$$

where

$$\frac{1-x}{x} = \frac{\rho_l u_l A_l}{\rho_g u_g A_g}, \frac{1-\alpha}{\alpha} = \frac{u_l A_l}{u_g A_g}$$

Table 9.2 Two-phase nomenclature

Volumetric Flow Rate $Q_i = \rho_i u_i$	Volumetric Flux $j_i = \frac{Q_i}{A}$	Volumetric Flux (gas) $j_g = \alpha u_g$	Volumetric Flux (liquid) $j_l = (1 - \alpha) u_l$
Mass Flow Rate $W_i = \rho_i u_i A_i$	Mass Flux $G_i = \frac{W_i}{A}$	Mass Flux (gas) $G_g = \alpha \rho_g u_g$	Mass Flux (liquid) $G_l = (1 - \alpha)\rho_l u_l$

and

$$\frac{\dfrac{1-x}{x}}{\dfrac{1-\alpha}{\alpha}} = \frac{\rho_l}{\rho_g} = \frac{Q_l}{Q_g} = \frac{\dfrac{W_l}{A_l}}{\dfrac{W_g}{A_g}}$$

Further,

$$\frac{\rho_l}{\rho_g} = \frac{\dfrac{W_l}{A_l}}{\dfrac{W_g}{A_g}} = \frac{\dfrac{W_l}{A(1-\alpha)}}{\dfrac{W_g}{A\alpha}} = \frac{W_l}{W_g}\frac{\alpha}{1-\alpha}$$

Finally,

$$\frac{\rho_l}{\rho_g} = \frac{W_l}{W_g}\left[\frac{\alpha}{1-\alpha}\right] = \frac{W_l}{W_g}\left[\frac{u_g A_g}{u_l A_l}\right]$$

or

$$\frac{\rho_l}{\rho_g}\frac{u_l A_l}{u_g A_g} = \frac{W_l}{W_g}$$

Conservation Principles for Two-Phase Flow

The conservation principles for a two-phase flow are given below without derivation [3].

Conservation of Momentum

$$\sum dF_z = d\left[\dot{m}_g u_g + \dot{m}_l u_l\right] = dF_{zp} + dF_{zG} + dF_{zw} = 0 \tag{9.80}$$

or

$$\sum dF_z = -A_z dP - \left[\alpha\rho_g + (1-\alpha)\rho_l\right]z dx - \left[\tau_{w,l} + \tau_{w,g}\right]P dX \tag{9.81}$$

where A_z is the cross-sectional area in the "z" direction, $\tau_{w,l}$ is the shear stress associated with the liquid, and $\tau_{w,g}$ is the shear stress associated with the gas.

And dividing by "$A_z dZ$" and solving for dP/dZ results in

$$-\frac{dP}{dZ} = -\left(\frac{dP}{dZ}\right)_g - \left(\frac{dP}{dZ}\right)_f - \left(\frac{dP}{dZ}\right)_{acc} \tag{9.82}$$

or

$$-\frac{dP}{dz} = -\frac{P}{A}\tau_w - \frac{W}{A}\frac{dv}{dz} - \rho_m g\cos\left(\theta\right) \tag{9.83}$$

Conservation of Mass

$$W = \rho_m vA = \text{constant} \tag{9.84}$$

Conservation of Energy

$$\frac{dq}{dz} - \frac{dw}{dz} = W\frac{d}{dz}\left\{h + \frac{u^2}{2} + gZ\right\} \tag{9.85}$$

where q is the heat transfer, w is the work, and the three quantities in brackets are specific enthalpy, specific kinetic energy, and specific potential energy.

9.6   Speed of Sound in a Two-Phase Flow

The first portion of this section develops the speed of sound for a mixture of components [12]; it is assumed that the flow is homogeneous. The effort in this first part doesn't address some of the concerns inherent in two-phase flow where interfacial processes become relevant and the two phases are no longer at equilibrium. It will be shown in this first section that the speed of sound of the mixture is a weighted average of the speed of sound of each component and that the weighting is based on the component density.

The derived relationship for a liquid/gas flow is

$$\frac{1}{c^2} = \frac{\alpha}{\gamma P}\left[\alpha\rho_g + \left(1 - \alpha\rho_l\right)\right] \tag{9.86}$$

In the second portion of this section, a correlation is developed between the speed of sound of the mixture and a variable, ω [13]. The relationship derived is

$$c^* = \frac{c}{\sqrt{Pv}} = \frac{1}{\sqrt{\omega}} \tag{9.87}$$

Speed of Sound for a Two-Phase Flow, Weighted Average of Densities
Consider a very small control volume (with volume equal to 1) where one phase is dispersed (phase A) and the other phase is continuous (phase B). The pressure associated with the dispersed phase will include a surface tension [14] and is given as

$$P_A = P_B + 2\frac{S}{R} \tag{9.88}$$

where P_B is the pressure for phase B, S is the surface tension, and R is the radius of the dispersed phase particles. Eq. 9.88 is Laplace's law for a spherical particle.
    The mass of phase A is

$$m_A = \rho_A \alpha_A + \delta m \tag{9.89}$$

And the mass of phase B is

$$m_B = \rho_B \alpha_B - \delta m \tag{9.90}$$

Where mass is transferring from phase B to A and the amount is δm.
Given the mass above for phase A the volume for phase A is

$$V_A = \frac{\rho_A \alpha_A + \delta m}{\rho_A + \left(\dfrac{\partial \rho_A}{\partial P_A}\right)_{Q_A} \delta P_A} \tag{9.91}$$

And the volume of phase B is

$$V_B = \frac{\rho_B \alpha_B - \delta m}{\rho_B + \left(\dfrac{\partial \rho_B}{\partial P_B}\right)_{Q_B} \delta P_A} \tag{9.92}$$

where Q_A and Q_B are unspecified thermodynamic paths and

$$\frac{1}{C_A^2} = \left(\frac{\partial \rho_A}{\partial P_A}\right)_{Q_A} \tag{9.93}$$

and

$$\frac{1}{C_B^2} = \left(\frac{\partial \rho_B}{\partial P_B}\right)_{Q_B} \tag{9.94}$$

Thus,

$$\delta V = V_A + V_B - 1 \tag{9.95}$$

And the definition of speed of sound is

$$c^{-2} = -\rho \left(\frac{\delta V}{\delta P_B} \right)_{\delta P_B \to 0} \tag{9.96}$$

The following derivation is determined from substituting Eqs. 9.91 to 9.95 into Eq. 9.96

$$c^{-2} = \rho \left[\frac{\alpha_A}{\rho_A} \left(\frac{\partial \rho_A}{\partial P_A} \right)_{QA} \frac{\delta P_A}{\delta P_B} + \frac{\alpha_A}{\rho_B} \left(\frac{\partial \rho_A}{\partial P_A} \right)_{QB} - \frac{\rho_B - \rho_A}{\rho_A \rho_B} \frac{\delta m}{\delta P_B} \right] \tag{9.97}$$

The ratio $\dfrac{\delta P_B}{\delta P_A}$ needs to be determined where no dispersed particles are created, nor destroyed where the new radius of dispersed particles is $R + \delta R$ and

$$\delta P_A = P_A' - P_A = \left(P_B' - P_B \right) + 2S \left[\frac{R}{R(R + \delta R)} - \frac{(R + \delta R)}{R(R + \delta R)} \right] = \delta P_B - 2 \frac{S \delta R}{R^2} \tag{9.98}$$

Thus,

$$\frac{\delta P_B}{\delta P_A} = \frac{1 - \dfrac{2S}{3\alpha_A \rho_A R} \dfrac{\delta m}{\delta P_B}}{1 - \dfrac{2S}{3\rho_A R} \left(\dfrac{\partial \rho_A}{\partial P_A} \right)_{QA}} \tag{9.99}$$

Substituting Eq. 9.99 into 9.97 and utilizing the definitions for speed of sound (Eqs. 9.93 and 9.94), the following relationship is derived

$$\frac{1}{\rho_m c^2} = \frac{\alpha_a}{\rho_B} \frac{1}{c_B^2} + \frac{\left\{ \dfrac{\alpha_A}{\rho_A} \dfrac{1}{c_A^2} - \dfrac{\delta m}{\delta P_B} \left[\dfrac{1}{\rho_A} - \dfrac{1}{\rho_B} + \dfrac{2S}{3\alpha_A \rho_B R} \dfrac{1}{c_A^2} \right] \right\}}{\left[1 - \dfrac{2S}{3\rho_A R} \dfrac{1}{c_A^2} \right]} \tag{9.100}$$

Homogenous flow assumes no mass transfer and typically the second term in the denominator associated with surface tension is much smaller than 1 and thus

$$\frac{1}{c^2} = \left[\alpha_A \rho_A + \alpha_B \rho_B\right]\left[\frac{\alpha_B}{\rho_A c_B^2} + \frac{\alpha_A}{\rho_B c_A^2}\right] \tag{9.101}$$

Equation 9.101 is valid for two-phase or two-component flow. Another form of the equation exists for two-phase flow where the density of one phase is much greater than the density of another phase, which is typical in a one-component, gas/liquid flow. This other form is

$$\frac{1}{c^2} = \left[\alpha \rho_g + (1-\alpha)\rho_l\right]\left[\frac{\alpha}{\gamma P} + \frac{1-\alpha}{\rho_l c_l^2}\right] \tag{9.102}$$

And when $\dfrac{P}{\rho_L c_l^2} \ll 1$ a simpler form is given as

$$\frac{1}{c^2} = \frac{\alpha}{kP}\left[\alpha \rho_g + (1-\alpha)\rho_l\right] \tag{9.103}$$

Example 9.3 Speed of Sound for Water Mixture using Eq. 9.103
Using Eq. 9.103 and saturated water at 100 °C and alpha equal to 0.95, determine the speed of sound. Given below is the solution, which is 54.2 m/s (Fig. 9.7).

Speed of Sound for a Two-Phase Flow, Omega Method [15]
By definition,

$$c^2 = v^2\left[-\left(\frac{\partial v}{\partial P}\right)_s^{-1}\right] \tag{9.104}$$

Where the following thermodynamic relationships will be utilized to develop a homogeneous, two-phase sonic velocity (Clausius-Clapeyron relationships)

Alpha		0.950	
Rho(g)	[kg/m^3]	0.598	
Rho(l)	[kg/m^3]	957.854	
k		1.327	
1/C^2	1/[m/s]^2	3.403E-04	
C	[m/s]	54.205	
C*		1.182	
P/rho(l)*c^2		3.60E-02	

$$\frac{1}{c^2} = \left[\alpha\rho_g + (1-\alpha)\rho_l\right]\left[\frac{\alpha}{kP} + \frac{1-\alpha}{\rho_l c_l^2}\right]$$

And when

a simple form is given as $\qquad \dfrac{P}{\rho_L c_l^2} \ll 1$

$$\frac{1}{c^2} = \frac{\alpha}{kP}\left[\alpha\rho_g + (1-\alpha)\rho_l\right]$$

Fig. 9.7 Solutions for Example 9.3

$$\frac{dP}{dT} = \frac{s_{fg}}{v_{fg}}, s_{fg} = \frac{h_{fg}}{T}, \frac{ds_f}{dT} \approx \frac{c_{pf}}{T} \tag{9.105}$$

Starting from the definition for specific volume of a mixture of gas and liquid

$$v_{mix} = v_l + x\left(v_g - v_l\right) \tag{9.106}$$

And taking the derivation of Eq. 9.106 with respect to P results in

$$\left(\frac{\partial v_{mix}}{\partial P}\right)_s = \frac{\partial v_l}{\partial P} + x\frac{\partial}{\partial P}\left[v_g - v_l\right] + \left[v_g - v_l\right]\frac{\partial x}{\partial P}$$
$$= x\frac{\partial v_g}{\partial P} + v_{fg}\frac{\partial x}{\partial P} + (1-x)\frac{\partial v_l}{\partial P} \tag{9.107}$$

Where we'll assume the liquid phase is an incompressible fluid and so $\frac{\partial v_l}{\partial P} \approx 0$.
Given below are derivation for $\frac{\partial v_g}{\partial P}$ and $\frac{\partial x}{\partial P}$.
Assuming v_g acts ideal

$$v_g = \frac{RT}{P} \tag{9.108}$$

And taking the derivative of both sides with respect to P results in

$$\frac{dv_g}{dP} = R\frac{d}{dP}\left(\frac{T}{P}\right) = R\left[\frac{1}{P}\frac{dT}{dP} - \frac{1}{P^2}T\right] \tag{9.109}$$

and

$$\frac{R}{P} = \frac{v_g}{T} \text{ and } -\frac{1}{P}\left(\frac{RT}{P}\right) = -\frac{v_g}{P} \tag{9.110}$$

and

$$\frac{dP}{dT} = \frac{s_{fg}}{v_{fg}} = \frac{h_{fg}}{Tv_{fg}} \tag{9.111}$$

Substituting Eqs. 9.110 and 9.111 into 9.109 results in

$$\frac{dv_g}{dP} = -\frac{v_g}{P} + \frac{v_g v_{fg}}{h_{fg}} \tag{9.112}$$

and

$$\left(\frac{dx}{dP}\right)_s = \frac{\left(\dfrac{dx}{dT}\right)_s}{\left(\dfrac{dP}{dT}\right)}$$

(9.113)

And the Clausius-Clapeyron relationships is

$$\frac{dP}{dT} = \frac{h_{fg}}{Tv_{fg}}$$

(9.114)

and

$$s_{mix} = s_f + xs_{fg}$$

(9.115)

And taking the derivative of both sides with respect to T for an isentropic process results in

$$0 = \frac{ds_{mix}}{dT} = \frac{ds_f}{dT} + x\frac{ds_{fg}}{dT} + s_{fg}\frac{dx}{dT}$$

(9.116)

or

$$s_{fg}\frac{dx}{dT} = \frac{h_{fg}}{T}\frac{dx}{dT} = -\frac{ds_f}{dT} - x\frac{ds_{fg}}{dT} = -\frac{c_{pf}}{T} - x\frac{d}{dT}\left[\frac{h_{fg}}{T}\right]$$

(9.117)

where h_{fg} is constant with respect to T and so

$$\frac{dx}{dT} = -\left\{\frac{c_{pf}}{T} + x\frac{d}{dT}\left[\frac{h_{fg}}{T^2}\right]\right\}\frac{T}{h_{fg}}$$

(9.118)

and

$$\frac{dx}{dT} = -\frac{c_{pf}}{h_{fg}} + \frac{x}{T}$$

(9.119)

and

$$\left(\frac{dx}{dP}\right)_s = \frac{\left(\dfrac{dx}{dT}\right)_s}{\left(\dfrac{dP}{dT}\right)} = \frac{-\dfrac{c_{pf}}{h_{fg}} + \dfrac{x}{T}}{\dfrac{h_{fg}}{Tv_{fg}}} = -\frac{c_{pf}v_{fg}T}{h_{fg}^2} + \frac{xv_{fg}}{h_{fg}} = \frac{v_{fg}}{h_{fg}}\left[x - \frac{c_{pf}T}{h_{fg}}\right]$$

(9.120)

Substitution of Eqs. 9.112 and 9.120 into Eq. 9.107 results in

$$\left(\frac{\partial v_{mix}}{\partial P}\right)_s = x\frac{\partial v_g}{\partial P} + v_{fg}\frac{\partial x}{\partial P} = -\frac{v_g x}{P} + \frac{v_g v_{fg} x}{h_{fg}} + v_{fg}\left\{\frac{v_{fg}}{h_{fg}}\left[x - \frac{c_{pf}T}{h_{fg}}\right]\right\} \quad (9.121)$$

or

$$\left(\frac{\partial v}{\partial P}\right)_s = \frac{-xv_g}{P}\left[1 - \frac{Pv_{fg}}{h_{fg}}\left(1 + \frac{v_{fg}}{v_g}\right)\right] - c_{pf}T\left(\frac{v_{fg}}{h_{fg}}\right)^2 \quad (9.122)$$

And a dimensionless version of the above equation along with the following observation, $v_{fg} \approx v_g$ in terms of c is

$$\frac{c}{\sqrt{Pv}} = \left\{\frac{xv_g}{v}\left[1 - 2\frac{Pv_{fg}}{h_{fg}}\right] + \frac{c_{pf}TP}{v}\left(\frac{v_{fg}}{h_{fg}}\right)^2\right\}^{-\frac{1}{2}} \quad (9.123)$$

And setting ω equal to

$$\omega = \alpha\left[1 - 2\frac{Pv_{fg}}{h_{fg}}\right] + \frac{c_{pf}TP}{v}\left(\frac{v_{fg}}{h_{fg}}\right)^2 \quad (9.124)$$

results in

$$c^* = \frac{c}{\sqrt{Pv}} = \frac{1}{\sqrt{\omega}} \quad (9.125)$$

Example 9.4 Speed of Sounds Using Omega Method
Using Leung's method as given in Eqs. 9.124 and 9.125, determine the speed of sound of saturated water at 100 °C and quality equal to 50%. Given below is the solution and we can see that the speed of sound for this saturated water mixture is 298 m/s (Fig. 9.8).

Fig. 9.8 Solution for
Example 9.4

Subtance		Water
T(0)	[C]	100
T(0)	[K]	373
P(0)	[kPa]	101.3
x(0)		50%
Gamma		1.327
v(l,0)	[m^3/kg]	0.001044
v(g,0)	[m^3/kg]	1.6729
v(fg0)	[m^3/kg]	1.671856
v(0)	[m^3/kg]	0.836972
c(f0)	[kJ/kg-K]	4.18
h(fg0)	[kJ/kg]	2257.03
Alpha		0.999

Omega(1)		0.85
Omega(2)		0.10
Omega		0.95
c/sqrt(P*nu)		1.02
c	[m/s]	298

$$\omega = \alpha \left[1 - 2\frac{Pv_{fg}}{h_{fg}} \right] + \frac{c_{pf}TP}{v} \left(\frac{v_{fg}}{h_{fg}} \right)^2$$

$$c^* = \frac{c}{\sqrt{Pv}} = \frac{1}{\sqrt{\omega}}$$

9.7 Critical Flow for a Two-Phase Flow System

The first critical flow model for two-phase systems that will be developed [3] incorporates slip that addresses differing velocities between phases but will largely ignore other forms of non-equilibrium behavior, which will be discussed further below.

From the conservation of energy and assuming no heat transfer, no work, no appreciable kinetic energy at the stagnation point, nor changes in potential energy then

$$h_0 = \frac{u_*^2}{2} + h_*$$

(9.126)

where now $u(*)$ is not necessary the local speed of sound, but a mixture of two phases and another form of Eq. 9.126 is

$$2\Delta h = x u_g^2 + (1-x) u_l^2$$

(9.127)

And it can be show that G is equal to

$$G = \rho_l u_l \frac{1-\alpha}{1-x} = Q_l \frac{1-\alpha}{1-x}$$

(9.128)

or

$$G = \rho_g u_g \frac{\alpha}{x} = Q_g \frac{\alpha}{x}$$

(9.129)

From Eq. 9.129,

$$\alpha = \frac{xG}{Q_g}$$

(9.130)

And Substituting Eq. 9.130 into 9.128 results in

$$G = Q_l \frac{1 - \dfrac{xG}{Q_g}}{1-x}$$

(9.131)

and

$$G(1-x) + \frac{xQ_l}{Q_g} G = G\left[(1-x) + \frac{xQ_l}{Q_g}\right] = Q_l$$

(9.132)

And solving for G

$$G = \frac{Q_l}{(1-x) + x\dfrac{Q_l}{Q_g}} = \frac{1}{\dfrac{x}{Q_g} + \dfrac{1-x}{Q_l}} = \frac{1}{\dfrac{x}{\rho_g u_g} + \dfrac{1-x}{\rho_l u_l}}$$

(9.133)

And defining slip velocity (k) as

$$k = \frac{u_g}{u_l} \tag{9.134}$$

It is the case that

$$\frac{2\Delta h}{G^2} = \left[\frac{x}{\rho_g u_g} + \frac{1-x}{\rho_l u_l} \right]^2 \left[xu_g^2 + (1-x)u_l^2 \right] \tag{9.135}$$

And substituting in for u_g the term ku_l results in

$$\frac{2\Delta h}{G^2} = \left[\frac{x}{\rho_g k u_l} + \frac{1-x}{\rho_l u_l} \right]^2 \left[xk^2 u_l^2 + (1-x)u_l^2 \right] \tag{9.136}$$

and

$$\frac{2\Delta h}{G^2} = \left[\frac{x}{\rho_g k} + \frac{1-x}{\rho_l} \right]^2 \left[xk^2 + (1-x) \right] \tag{9.137}$$

For a given x and dh, G can be maximized by taking the derivative of both sides with respect to k.

$$\frac{d}{dk}\left\{ \frac{2\Delta h}{G^2} \right\} = \frac{d}{dk}\left\{ \left[\frac{x}{\rho_g k} + \frac{1-x}{\rho_l} \right]^2 \left[xk^2 + (1-x) \right] \right\} = \frac{d}{dk} AB = A'B + AB' \tag{9.138}$$

where

$$0 = \frac{d}{dk}\left\{ \frac{2\Delta h}{G^2} \right\} = 2kx\left[\frac{x}{k\rho_g} + \frac{1-x}{\rho_l} \right]^2 - 2\left[xk^2 + (1-x) \right]\frac{x}{k^2 \rho_g}\left[\frac{x}{k\rho_g} + \frac{1-x}{\rho_l} \right] \tag{9.139}$$

and

$$k\left[\frac{x}{k\rho_g} + \frac{1-x}{\rho_l} \right]^2 - \left[xk^2 + (1-x) \right]\frac{1}{k^2 \rho_g} = 0 \tag{9.140}$$

or

$$\frac{x}{\rho_g} + k\frac{1-x}{\rho_l} - \frac{x}{\rho_g} - \frac{1-x}{k^2\rho_g} = 0 \tag{9.141}$$

Further

$$k\frac{1-x}{\rho_l} = \frac{1-x}{k^2\rho_g} \tag{9.142}$$

and

$$k^3 = \frac{\rho_l}{\rho_g} = \frac{v_g}{v_l} \tag{9.143}$$

Substituting Eq. 9.143 into Eq. 9.137 results in

$$G_c^2 = \frac{2\Delta h}{\left[\dfrac{1-x}{\rho_l^{.667}} + \dfrac{x}{\rho_g^{.667}}\right]^3} \tag{9.144}$$

Equation 9.144 tends to work well when $x > 10\%$, but below this quality the fluid begins to act homogenous and as x approaches 0 or 100%, the slip goes to 1. For very low quality, another model of two-phase critical flow [16] is required and discussed below.

Example 9.5 Critical Flow using Eq. 9.144
Determine the critical flow for saturated water with a quality of 10% and a pressure of 150 kPa and using Eq. 9.144.

Given below is saturation data for water from 25 to 250 kPa. Using this data and Eq. 9.144, a curve is developed for x equal to 10%. Additionally, curves are provided for $x = 5\%$ and $x = 90\%$.

The answer to the example is 510 kg/m^2-s (Fig. 9.9).

We've seen for one phase, one component flow that

$$\left(\frac{\partial G_c}{\partial P}\right)_s = 0 \tag{9.145}$$

But for two-phase (two-component) flow

$$\left(\frac{\partial G_c}{\partial P}\right)_{Q_A} = 0 \tag{9.146}$$

where the thermodynamic path Q_A is no longer specified to be isentropic

Substance	Water
x	10%
b	0.667
Rg	0.4615

$$G_{max} = \frac{\sqrt{2\Delta h}}{\left[\dfrac{1-x}{\rho_l^{.667}} + \dfrac{x}{\rho_g^{.667}}\right]^{1.5}}$$

$$\ln(P) = -\frac{\Delta h_{sat}}{R_g}\frac{1}{T} + C$$

P [kPa]	T [C]	nu(l) [m^3/kg]	nu(g) [m^3/kg]	h(l) [kJ/kg]	h(g) [kJ/kg]	nu(mix) [m^3/kg]	h(mix) [kJ/kg]	dh [kJ/kg]	Gmax [kg/m^2-s]	Gmax(5%) [kg/m^2-s]	Gmax(90%) [kg/m^2-s]
25	64.97	0.00102	6.20424	271.9	2618.19	0.62134	506.529	234.629	106	203	12
50	81.33	0.00103	3.24034	340.47	2645.87	0.32496	571.01	230.54	197	369	23
75	91.77	0.001037	2.21711	384.36	2662.96	0.22264	612.22	227.86	281	518	34
100	99.62	0.001043	1.694	417.44	2675.46	0.17034	643.242	225.802	361	654	44
125	105.99	0.001048	1.3749	444.3	2685.35	0.13843	668.405	224.105	437	781	54
150	111.37	0.001053	1.15933	467.08	2693.54	0.11688	689.726	222.646	510	901	64
175	116.06	0.001057	1.00363	486.97	2700.53	0.10131	708.326	221.356	580	1013	74
200	120.23	0.001061	0.88573	504.68	2706.63	0.08953	724.875	220.195	648	1121	83
225	124	0.001064	0.79325	520.69	2712.04	0.08028	739.825	219.135	714	1223	93
250	127.43	0.001067	0.71871	535.34	2716.89	0.07283	753.495	218.155	779	1321	102

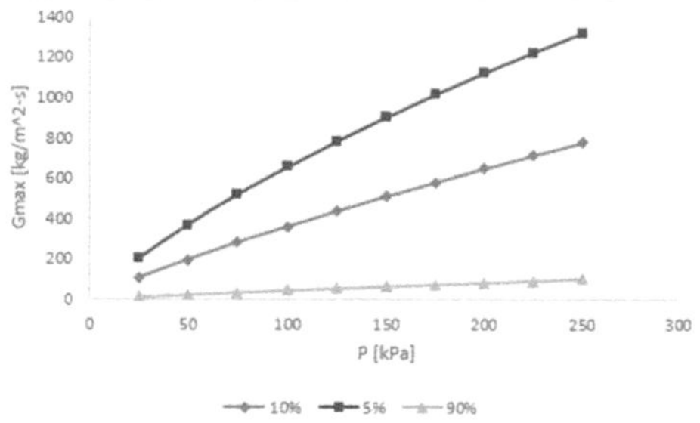

Fig. 9.9 Solution for Example 9.5

And now with a two-phase system

$$v_{mix} = v_l + x\left(v_g - v_l\right) \tag{9.147}$$

And taking the derivation of Eq. 9.147 with respect to P results in

$$\left(\frac{\partial v_{mix}}{\partial P}\right)_s = \frac{\partial v_l}{\partial P} + x\frac{\partial}{\partial P}\left[v_g - v_l\right] + \left[v_g - v_l\right]\frac{\partial x}{\partial P}$$

$$= x\frac{\partial v_g}{\partial P} + v_{fg}\frac{\partial x}{\partial P} + \left(1-x\right)\frac{\partial v_l}{\partial P} \tag{9.148}$$

And if we assume homogenous conditions where both phases are in equilibrium, then we can still assume an isentropic path, such that substituting Eq. 9.148 into Eq. 9.145 results in

$$G_c^2 = -\left(\frac{\partial P}{\partial v}\right)_s = \frac{-1}{\left[x\left(\dfrac{\partial v_g}{\partial P}\right)_s + \left(v_g - v_l\right)\left(\dfrac{\partial x}{\partial P}\right)_s + \left(1-x\right)\left(\dfrac{\partial v_l}{\partial P}\right)_s\right]} \tag{9.149}$$

And assuming an incompressible liquid

$$\left(\frac{\partial v_l}{\partial P}\right)_s \approx 0 \qquad (9.150)$$

And typically

$$v_g \gg v_l \qquad (9.151)$$

Substituting Eqs. 9.150 and 9.151 into 9.149 results in

$$G_c^2 = -\left(\frac{\partial P}{\partial v}\right)_s = \frac{-1}{\left[x\left(\frac{\partial v_g}{\partial P}\right)_s + v_g\left(\frac{\partial x}{\partial P}\right)_s\right]} \qquad (9.152)$$

Equation 9.152 is known as the *homogenous equilibrium model* and has been shown to inadequately predict critical flow because crucial interphase processes have been ignored. One discrepancy is that the densities between the liquid phase and the gas phase are very different and under the same applied pressure would experience very different accelerations and velocities.

Another common model for two-phase flow is the *homogenous frozen model* that assumes the properties through the choke point don't change and specifically x. In Henry and Fauske [11], the underlying assumptions associated with both forms of the two-phase flow model are discussed.

We saw in the previous section that a liquid portion of a flow and a gas portion of a flow are at two very different Mach numbers. Put another way, we will see that phases are not at the same speed and momentum and energy may not be in equilibrium. All these issues will be discussed further when a general critical flow model is derived. This developed model addresses three interfacial transport processes [17]

1. Interfacial heat transfer. The heat transfer rate between the gas phase and the surrounding liquid and solid phase.
2. Interfacial momentum transfer. This transfer determines how fast each phase is accelerating.
3. Interfacial mass transfer. This transfer determines the rate of evaporation or condensation.

A more general definition of critical flow for two-phase (two-component systems) is

$$\left(\frac{\partial G_c}{\partial P}\right)_{H0} = 0 \qquad (9.153)$$

where H_0 denotes a constant stagnation enthalpy [16, 11].

Definition of Critical Flow for a Two-Phase System

Another form of the momentum equation where the effect of gravity and friction are ignored is

$$\rho u \frac{du}{dz} = -\frac{dP}{dz} \tag{9.154}$$

And conservation of mass is

$$\frac{d\rho}{\sigma} + \frac{du}{u} + \frac{dA}{A} = 0 \tag{9.155}$$

And substituting Eq. 9.30 into 9.155 results in

$$\frac{dA}{A} = \frac{dP}{\rho}\left(\frac{1}{u^2} - \frac{d\rho}{dP}\right) \tag{9.156}$$

Where again the critical flow is

$$G_c = \rho \sqrt{\left(\frac{dP}{d\rho}\right)_{H0}} \tag{9.157}$$

Where now the path is constant with regard to stagnation enthalpy. As was just noted above, often $u_g \neq u_l$ and so

$$G\frac{d}{dz}\left[xu_g + (1-x)u_l\right] = -\frac{dP}{dz} \tag{9.158}$$

And by definition

$$\left(\frac{\partial G}{\partial P}\right)_{H0} - 0 \text{ at the critical point} \tag{9.159}$$

Combining Eqs. 9.158 and 9.159 results in

$$-1 = G\frac{\partial}{\partial P}\left[xu_g + (1-x)u_l\right] \tag{9.160}$$

Where the slip velocity is again defined as

$$u_g = ku_l \tag{9.161}$$

Mass flow rate for each phase is defined below.

Where the liquid phase mass flow rate is

$$(1-x)G = (1-\alpha)\frac{u_l}{v_l} \tag{9.162}$$

And the gas phase mass flow rate is

$$xG = \alpha\frac{u_g}{v_g} = \alpha\frac{ku_l}{v_g} \tag{9.163}$$

If we eliminate "G" from Eqs. 9.162 and 9.163 then a relationship between α and x is established

$$\alpha = \frac{xv_g}{k(1-x)v_l + xv_g} \tag{9.164}$$

Combining the previous two equations gives an equation for G solely in terms of u_l

$$G = \frac{k}{k(1-x)v_l + xv_g}u_l \tag{9.165}$$

Eliminating "u_l" from Eqs. 9.165 and 9.163 results in

$$G_c = \frac{1}{\dfrac{\partial}{\partial P}\left[\dfrac{\left[k(1-x)v_l + xv_g\right]\left[xk + (1-x)\right]}{k}\right]_{HO}} = \frac{1}{B_t} \tag{9.166}$$

where B is

$$B_t = \frac{\partial}{\partial P}\left[\frac{xk^2(1-x)v_l + k(1-x)^2 v_l + x^2 kv_g + x(1-x)v_g}{k}\right]_{HO} \tag{9.167}$$

or

$$B_t = \frac{\partial}{\partial P}\left[xk(1-x)v_l + (1-x)^2 v_l + x^2 v_g + \frac{x(1-x)v_g}{k}\right]_{HO}$$

And the following derivatives will have to be determined either experimentally or theoretically

$$\left(\frac{\partial x}{\partial P}\right)_t, \left(\frac{\partial k}{\partial P}\right)_t, \left(\frac{\partial v_l}{\partial P}\right)_t, \left(\frac{\partial v_g}{\partial P}\right)_t \tag{9.168}$$

Henry and Fauske [11, 16, 18] have shown that

$$G_c^2 = \left\{ \frac{d}{dP} \left[\frac{xk + (1-x)}{k} \left[(1-x)v_l + xv_g \right] \right]^{-1} \right\}_{HO}$$

(9.169)

Expanding Eq. 9.169 results in

$$G_c^2 = \left\{ k\left[1 + x(k-1)\right]x\frac{dv_g}{dP} + \left[v_g\left\{1 + 2x(k-1) + kv_l 2(k-1)\right\}\frac{dx}{dP} + k\left[1 + x(k-2) - x^2(k-1)\right]\frac{dv_l}{dP} + x(1-x)\left(kv_l - \frac{v_g}{k}\right)\right]\frac{dk}{dP} \right\}$$

(9.170)

Again, using an order of magnitude arguments Eq. 9.170 can be simplified

$$k \sim 0(1) \ll \sqrt{\frac{v_g}{v_l}} \sim O(10) \text{ and } v_g \gg v_l$$

(9.171)

and

$$\left(\frac{\partial v_l}{\partial P}\right)_{HO} \approx 0$$

(9.172)

And thus Eq. 9.170 is simplified to

$$G_c^2 = \frac{k}{\left\{\left[1 + x(k-1)\right]x\left(\frac{\partial v_g}{\partial P}\right)_{HO} + v_g\left[1 + 2x(k-1)\right]\left(\frac{\partial x}{\partial P}\right)_{HO} - x(1-x)\frac{v_g}{k}\left(\frac{\partial k}{\partial P}\right)_{HO}\right\}}$$

(9.173)

Given below are determinations through theory and experiments values or formulas for the derivatives appearing in Eq. 9.173 [16, 11].

$$\left(\frac{dk}{dP}\right)_{HO} = \frac{\left(\frac{dk}{dz}\right)_t}{\left(\frac{dP}{dz}\right)_t}$$

(9.174)

Below are graphs for k versus z and P versus z where it is seen in Fig. 9.10 that k is a minima at $z = z_t = 1.6''$ (where z_t is the throat of the nozzle) with the first derivative equal to zero and from Fig. 9.11 P is an inflection point at $z = z_t$ where the first derivative is not equal to zero.

Fig. 9.10 k versus z [11]
for various qualities

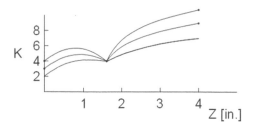

Fig. 9.11 Pressure versus
z [11] for various qualities

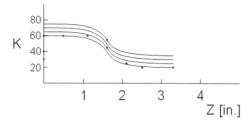

Evaluating Eq. 9.174 in light of these two figures it is seen that

$$\left(\frac{dk}{dP}\right)_{HO} = 0 \tag{9.175}$$

The differential involving v_g has been determined to be a polytropic process [11, 16, 18] given below as

$$\left(\frac{dv_g}{dP}\right)_{HO} = N\left(\frac{v_g}{nP}\right)_s \tag{9.176}$$

where N is dependent on the geometry of the conduit and provides an estimate for the fact that the actual path "H_0" is between isentropic and isenthalpic where a portion of the thermodynamic energy is converted to kinetic energy.

And n equals

$$n = \frac{(1-x)\dfrac{c_f}{c_{pg}} + 1}{(1-x)\dfrac{c_f}{c_{pg}} + \dfrac{1}{\gamma}} \tag{9.177}$$

Further [16]

$$\left(\frac{\partial x}{\partial P}\right)_{QA} = N\left(\frac{\partial x}{\partial P}\right)_s \tag{9.178}$$

And from Fauske [18] a model for $\left(\dfrac{\partial x}{\partial P}\right)_s$ is

$$\left(\frac{\partial x}{\partial P}\right)_s = -\frac{1}{h_{fg}}\left[\frac{dh_f}{dP} + x\frac{dh_{fg}}{dP}\right] \approx -\frac{1}{h_{fg}}\left[\frac{\Delta h_f}{\Delta P} + x\frac{\Delta h_{fg}}{\Delta P}\right] \tag{9.179}$$

Substituting Eqs. 9.175 through 9.179 into Eq. 9.173 results in

$$G_c^2 = \frac{k}{\left\{\left[1+x(k-1)\right]xN\dfrac{v_g}{nP} + Nv_g\left[1+2x(k-1)\right]\left[-\dfrac{1}{h_{fg}}\left[\dfrac{\Delta h_f}{\Delta P} + x\dfrac{\Delta h_{fg}}{\Delta P}\right]\right]\right\}} \tag{9.180}$$

And it has been found [18, 16] for $x < 5\%$ that $k = 1$ such that

$$G_c^2 = \frac{1}{\left\{xN\dfrac{v_g}{nP} + Nv_g\left[-\dfrac{1}{h_{fg}}\left[\dfrac{\Delta h_f}{\Delta P} + x\dfrac{\Delta h_{fg}}{\Delta P}\right]\right]\right\}} \tag{9.181}$$

Equation 9.181 is similar to a model, Eq. 3.25, given in Henry [16].

Example 9.6 Compare Three Critical Flow Models
In this example, three models are compared for critical flow. These models include the model at the beginning of this section given from Wallis [3], the model just developed, and a model attributed to Henry and Fauske [11].

Given below is the analysis for Eqs. 9.140 and 9.181 for water at a pressure of 600 psia as Fig. 9.12 for quality ranging from 15% to 80%. Figure 9.13 provides a

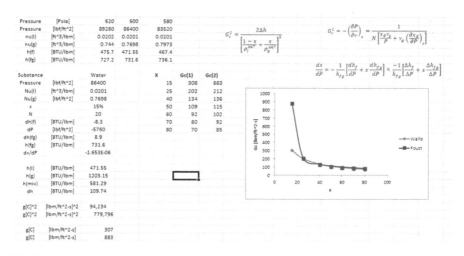

Fig. 9.12 Solution for Example 9.6

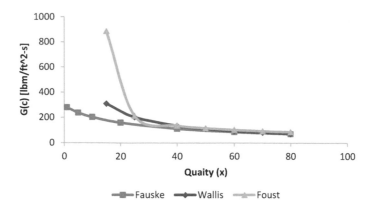

Fig. 9.13 Comparison of three critical flow models for two-phase systems

comparison between the two discussed models and a model developed by Henry and Fauske [11]. It is seen as quality decreases that Eq. 9.181 approaches the mass flow flux of liquid as computed from Bernoulli's equation, which is 1807 lbm/ft.^2-s and based on $u_l = \sqrt{2 \dfrac{\Delta P}{\rho}}$.

Critical Flow Utilizing the Omega Method
Another method exists that is much more empirical and much more straightfor-ward – the ω method. We utilized this method earlier to look at determinations of speed of sound in two-phase flows.

The critical flow of a homogeneous two-phase flow can be expressed in terms of the omega correlation [17, 19, 13]. The equation is given as

$$G_c^2 = \eta^2 \frac{P_0}{\omega v_0} \tag{9.182}$$

where ω takes a slightly different form from Eq. 9.124 where the first terms differs slightly.

$$\omega = \alpha_0 + \frac{c_{f0} T_0 P_0}{v_0} \left(\frac{v_{fg0}}{h_{fg0}} \right)^2 \tag{9.183}$$

And η, which is the critical pressure ratio $\left(\dfrac{P_0}{P_C} \right)$, has been determined to be

$$\eta^2 + \left(\omega^2 - 2\omega \right)\left(1 - \eta \right)^2 + 2\omega^2 \ln\left(\eta \right) + 2\omega^2 \left(1 - \eta \right) = 0 \tag{9.184}$$

where omega can be determined either graphically or through empirical relationships [14] such as
When

$$\omega \geq 4$$

$$\frac{G}{\sqrt{\frac{P_0}{v_0}}} = \frac{0.6055 + 0.1356 * \ln(\omega) - 0.0131 * (\ln(w))^2}{\sqrt{\omega}} \tag{9.185}$$

Or when

$$\omega < 4$$

$$\frac{G}{\sqrt{\frac{P_0}{v_0}}} = \frac{0.66}{\omega^{0.39}} \tag{9.186}$$

Please note the following

$$\omega < 1 \rightarrow Non - Flashing\ Flow, \omega = \alpha_0 \tag{9.187}$$

$$\omega > 1 \rightarrow Flashing\ Flow, \omega = \alpha_0 + \frac{c_{f0}T_0 P_0}{v_0}\left(\frac{v_{fg0}}{h_{fg0}}\right)^2$$

Example 9.7 Critical Flow Using Omega Method
Determine the critical flow [kg/m^2-s] for saturated water at 100 °C and quality equal to 50% utilizing the Omega method discussed above.
   It is seen from below that the calculated omega is 1.10 (Flashing) and so Eq. 9.186 is utilized. For the given stagnation pressure (101.3 kPa) and specific volume of the mixture (0.84 m^3/kg), the critical flow is determined to be 185 kg/m^2-s and about 20% of the value for $G(L)$, which is the critical flow for liquid water at the same temperature and pressure (Fig. 9.14).

9.8 Problems

Problem 9.1 Ideal, Perfect Critical Flow
Determine mass rate versus time for methane with an initial pressure of 5 MPa and temperature of 300 K. The pressure vessel is composed of a cylinder with two hemispherical end-caps where the diameter is 36″ and the length of the cylinder is 60″. The opening has a diameter of 1/8″.

Fig. 9.14 Solution for
Example 9.7

Subtance		Water
T(0)	[C]	100
T(0)	[K]	373
P(0)	[kPa]	101.3
x(0)		50%
Gamma		1.327
v(l,0)	[m^3/kg]	0.001044
v(g,0)	[m^3/kg]	1.6729
v(fg0)	[m^3/kg]	1.671856
v(0)	[m^3/kg]	0.836972
c(f0)	[kJ/kg-K]	4.18
h(fg0)	[kJ/kg]	2257.03
Alpha		0.999
Omega(1)		1.00
Omega(2)		0.10
Omega		1.10
G(L)/sqrt[P(0)/nu(0)]		3.11
G/sqrt[P(0)/nu(0)]	For Omega < 4.0	0.64
G/sqrt[P(0)/nu(0)]	For Omega => 4.0	0.59
G/sqrt[P(0)/nu(0)]		0.64
G/G(L)		0.20
G	[kg/m^2-s]	185
G(L)	[kg/m^2-s]	905

Problem 9.2 Two-Phase Flows
Prove the following

$$G = Q_l \frac{1-\alpha}{1-x}$$

Problem 9.3 Speed of Sound for Two-Phase System
Determine the speed of sound for saturated methane at 95 K where the quality is 30%.

Problem 9.4 Speed of Sound for a Two-Component System
Determine the speed of sound for water and air at STP where α for air is 50%.

Problem 9.5 Two-Phase Critical Flow
Using the two-phase model for critical flow given from Wallis [3] determine the critical flow rate [lbm/ft^2-s] for saturated carbon dioxide at 50 °F with a quality of 50%.

Problem 9.6 Two-Phase Critical Flow
Using the two-phase model for critical flow given from Henry and Fauske [11] determine the critical flow rate [lbm/ft^2-s] for carbon dioxide at 50 °F with a quality of 50%.

Appendix 9.1: Critical Flow, Ideal Gas (website)

An Excel Spreadsheet was developed to illustrate a particular concept and is given on the companion website.

Appendix 9.2: Speed of Sound in a Two-Phase Flow (website)

An Excel Spreadsheet was developed to illustrate a particular concept and is given on the companion website.

Appendix 9.3: Omega Method (website)

An Excel Spreadsheet was developed to illustrate a particular concept and is given on the companion website.

References

1. https://en.wikipedia.org/wiki/Isentropic_nozzle_flow. Accessed 6/12/2021.
2. Munson, B. R., Young, D. F., & Okiishi, T. H. (2002). *Fundamentals of fluid mechanics* (4th ed.). John Wiley and Sons.
3. Wallis, G. B. (1969). *One-dimensional two-phase flow*. McGraw Hill.
4. White, F. M. (1999). *Fluid mechanics* (4th ed.). Mc-Graw Hill.
5. Zucker, R. D., & Biblarz, O. (2002). *Fundamentals of gas dynamics* (2nd ed.). John Wiley and Sons.
6. Powers, J. M. (2012). *Lecture notes for introduction to thermodynamics*. Notre Dame University.

7. Peng, D. Y., & Robinson, D. B. (1976). A new two-constant equation of state. *Industrial Engineering Chemistry, Fundamental, 15*(1), 59–64.
8. Petrovic, M. M., & Stevanovic, V. D. (2016). "Two-component two-phase critical flow"; FME Transactions, 44, pp. 109–114.
9. Donaldson, C. D. duP. (1948). *The importance of imperfect-gas effects and variation of heath capacities on the isentropic flow of gases*; NACA RM No. L8J14.
10. https://en.wikipedia.org/wiki/Multiphase_flow. Accessed 6/12/2021.
11. Henry, R. E. & Fauske H. K. (1971). The Two-Phase Critical Flow of One-Component Mixtures in Nozzles, Orifices and Short Tubes. *Journal of Heat Transfer*, May issue, pp. 179–187.
12. Brennen, C. E. (2005). *Fundamentals of multiphase flows*. Cambridge University Press.
13. Leung, J. C. (1996). On the application of the method of Landau and Lifshitz to sonic velocities in homogenous two-phase mixtures. *Transactions of the ASME, 118*, 186–188.
14. Hiemenz, P. C. (1986). *Principles of colloid and surface chemistry* (2nd ed.). Marcel Dekker.
15. Leung, J. C. (1986). A generalized correlation for one-component homogeneous equilibrium flashing choked flow. *AIChE Journal, 32*(10), 1743–1746.
16. Henry, R. E. (1968). *A study of one- and two-component, two-phase critical flows at low qualities;* Ph.D. Thesis, University of Notre Dame.
17. Hsu, Y. Y. (1972). *Review of the critical flow rate, propagation of pressure pulse, and sonic velocity in two phase media*; NASA Technical Note, NASA TN D-6814, June.
18. Fauske, H. K. (1962). *Contribution to the theory of two-phase, one-component critical flow*; Argonne National Laboratory, ANL-6633.
19. Leung, J. C. (1990). Two phase flow discharge in nozzles and pipes – A unified approach. *Journal of Loss Prevention Process Industry, 3*, 27–32.

Part III
Fundamentals of Combustion

Chapter 10
Physically Based Combustion

10.1 Preview

To this point, we have gathered a foundation in thermodynamics and applied thermodynamics to the area of gas dynamics to understand speed of sound, normal shocks, and critical flow. Another important application of thermodynamics is combustion.

A combustion system is simply a wave which involves a chemical reaction that sustains the wave while oxidizer (usually air) and fuel are available; when the combustion is a detonation, the wave is traveling greater than Mach 1 and is a shock wave. When the combustion is deflagration, the wave is traveling much less than Mach 1. This chapter will mainly address detonation; deflagration will be addressed in Chap. 12.

Various properties (states) of a detonation or deflagration are given as Table 10.1 where the subscript "1" denotes unburned gas and "2" denotes burnt gas.

One representation of a detonation system is known as the Zeldovic, von Neumann, and Doren, which are commonly known as ZND models [1–4] and given as Fig. 10.1; in Chap. 13, we'll discuss ZND models and call them dynamic detonation models. In this figure the shock wave and attached reaction zone are moving from right to left and the reaction zone includes an induction zone.

This chapter asks the following question "for a given initial states (1) what are the final states at either points (2) or (3)"? and assumes the heat release is instantaneous. The ZND model deals with the fact that the heat is released as a result of a chemical reaction and allows for characterization of the states within the induction/reaction zone (2- > 3).

Much of our experimental knowledge for combustion comes from shock tubes (see Fig. 10.2), which are essentially tubes divided by a diaphragm and have a

Electronic Supplementary Material: The online version of this chapter (https://doi.org/10.1007/978-3-030-87387-5_10) contains supplementary material, which is available to authorized users.

Table 10.1 Properties of combustions [1]

	Detonation	Deflagration
$\text{Mach}_1 = u_1/c_1$	5–10	0.0001–0.03
u_2/u	0.4–0.7 (Deceleration)	4–6 (Acceleration)
P_2/P_1	13–55 (Compression)	.98 (Slight Expansion)
T_2/T_1	8–21 (Heat Addition)	4–16 (Heat Addition)
ρ_2/ρ_1	1.7–2.6	0.06–0.25

Fig. 10.1 ZND model [5]

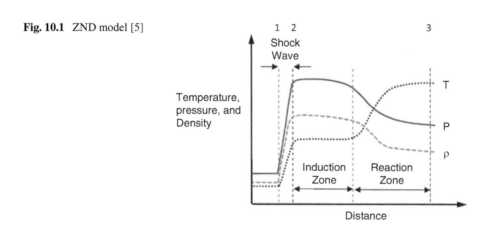

Fig. 10.2 Shock tube with states versus time [6]

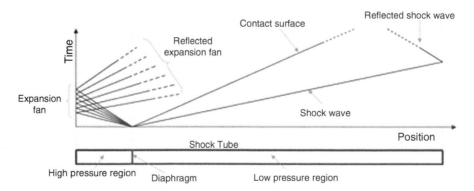

region of higher pressure (*Driver Section*) and a region of lower, ambient, pressure (*Driven Section*). If the differential pressure is great enough, the thin-skinned diaphragm "bursts" and a shock wave travels down the tube and obeys Eqs. 10.1, 10.2, and 10.3 where q is zero for a normal shock or has some positive value for a combustion system.

Conservation Equations
Conservation of mass:

$$\rho_1 u_1 = \rho_2 u_2 = \dot{m} \tag{10.1}$$

Conservation of energy:

$$h_1 + \frac{1}{2} u_1^2 + q = h_2 + \frac{1}{2} u_2^2 \tag{10.2}$$

Conservation of momentum:

$$P_1 + \rho_1 u_1^2 = P_2 + \rho_2 u_2^2 \tag{10.3}$$

When the gas is an air/fuel mixture and ignites in some fashion as the diagram bursts, the system around the shock and attached reaction zone can be treated as one dimensional and friction and turbulence ignored. A graphical solution for Eqs. 10.1, 10.2, and 10.3 is known as the Rayleigh-Hugoniot system (see Fig. 10.3) and transforms the equations into a line (Rayleigh line) and curve (Hugoniot curve); the point where the line and curve intersect is an endpoint of the system in terms of the point between the compressive shock and beginning of the reaction zone ($\lambda = 0\%$) or the end of the reaction zone ($\lambda = 100\%$) where λ is the percent reaction complete. For detonation systems, it is often assumed that the gas at the end of the reaction zone

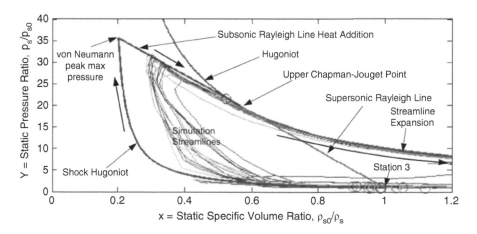

Fig. 10.3 Rankine/Hugoniot system [4]

travels at Mach 1 and this point is known as the Chapman-Jouget (CJ) point; states associated with the "CJ" point can be determined by assuming that the Rayleigh line is tangent to the Hugoniot curve. The determined velocity, Ma(cj), for a detonation system has been shown to be within a few percent of experimentally determined velocities; unfortunately, this is not the case for the "CJ" point associated with a deflagration [1, 7, 8] where important transport processes have been ignored.

Figure 10.3 represents the states associated with a rotational detonation engine [4] where the pressure rises along the "shock Hugoniot" to the "von Neumann peak" ($\lambda = 0\%$) and then the reaction occurs along the "subsonic Rayleigh line" and the reaction is complete at the "upper Chapman-Jouget point" where $\lambda = 100\%$.

The chapter goes on to discuss a RH system for partially complete combustions ($\lambda < 100\%$) and extends the RH framework to a system where issues of boundary layers are considered, which addresses some of the issues friction. It has been observed [1, 3, 4] that gas velocities within the reaction zone are sub-mach and that issues of friction become important, which is the purpose of Fay's system and will be discussed in Sect. 10.5.

The chapter ends will an alternate method to determine the Mach number of the shock wave with or without friction; this section parallels the discussion of Norman shocks in Chap. 7 and includes equations for ratios of the states.

10.2   Standard Rankine-Hugoniot Theory

Staring from Eqs. 10.1 to 10.3, this section will develop the Rankine line and Hugoniot curve that graphically depicts the conservation principles.

10.2.1   Deriving the Rankine Line

From Eq. 10.2,

$$P_2 - P_1 = \rho_1 u_1^2 - \rho_2 u_2^2 \tag{10.4}$$

From Eq. 10.1,

$$\frac{\rho_1}{\rho_2} u_1 = u_2, \left(\frac{\rho_1}{\rho_2} u_1 \right)^2 = u_2^2 \tag{10.5}$$

Substituting Eq. 10.5 into Eq. 10.4 results in

$$P_2 - P_1 = \rho_1 u_1^2 - \rho_2 \left[\frac{\rho_1}{\rho_2} u_1 \right]^2 = \rho_1 u_1^2 \left[1 - \frac{\rho_1}{\rho_2} \right] \tag{10.6}$$

And dividing both sides by P_1 and substituting $y = \dfrac{P_2}{P_1}, x = \dfrac{P_1}{\rho_2}$ results in

$$y - 1 = \frac{u_1^2}{P_1 v_1}\left[1 - x\right] \tag{10.7}$$

which is the celebrated Rankine line. Another form of Eq. 10.7 is

$$y - 1 = \gamma_1 Ma_1^2 \left[1 - x\right] \tag{10.8}$$

10.2.2   Mass Flux

$$u_1 = \frac{\dot{m}}{\rho_1}, u_2 = \frac{\dot{m}}{\rho_2} \tag{10.9}$$

Substituting Eq. 10.9 into Eq. 10.4 results in

$$P_2 - P_1 = \frac{\dot{m}^2}{\rho_1} - \frac{\dot{m}^2}{\rho_2} \tag{10.10}$$

and

$$P_2 - P_1 = \dot{m}^2\left[\frac{1}{\rho_1} - \frac{1}{\rho_2}\right] \tag{10.11}$$

Further

$$\dot{m}^2 = \frac{P_2 - P_1}{\dfrac{1}{\rho_1} - \dfrac{1}{\rho_2}} \tag{10.12}$$

10.2.3   Derivation for $-\Delta KE$

And $-\Delta KE$ is defined as

$$-\Delta KE = \frac{1}{2}\left[u_1^2 - u_2^2\right] = \frac{1}{2}\left[\frac{\dot{m}^2}{\rho_1^2} - \frac{\dot{m}^2}{\rho_2^2}\right] = \frac{\dot{m}^2}{2}\left[\frac{1}{\rho_1^2} - \frac{1}{\rho_2^2}\right] \tag{10.13}$$

And substituting Eq. 10.12 into Eq. 10.13 results in

$$-\Delta KE = \frac{1}{2}\frac{P_2 - P_1}{\dfrac{1}{\rho_1} - \dfrac{1}{\rho_2}}\left[\frac{1}{\rho_1^2} - \frac{1}{\rho_2^2}\right] \tag{10.14}$$

and

$$\frac{\left[\dfrac{1}{\rho_1^2} - \dfrac{1}{\rho_2^2}\right]}{\dfrac{1}{\rho_1} - \dfrac{1}{\rho_2}} = \frac{\rho_1\rho_2}{\rho_2 - \rho_1}\left[\frac{1}{\rho_1^2} - \frac{1}{\rho_2^2}\right] = \frac{1}{\rho_2 - \rho_1}\left[\frac{\rho_2}{\rho_1} - \frac{\rho_1}{\rho_2}\right]$$

$$= \frac{1}{\rho_2 - \rho_1}\left[\frac{\rho_2^2 - \rho_1^2}{\rho_1\rho_2}\right] = \frac{\rho_1 + \rho_2}{\rho_1\rho_2} = \left(\frac{1}{\rho_1} + \frac{1}{\rho_2}\right) \tag{10.15}$$

Therefore, substituting Eq. 10.15 into Eq. 10.14 results in

$$-\Delta KE = \frac{1}{2}\left(P_2 - P_1\right)\left(\frac{1}{\rho_1} + \frac{1}{\rho_2}\right) \tag{10.16}$$

10.2.4   *Deriving the Hugoniot Curve*

The energy balance for the system, Eq. 10.2, is

$$-\Delta KE = h_2 - h_1 - q = \frac{1}{2}\left(P_2 - P_1\right)\left(\frac{1}{\rho_1} + \frac{1}{\rho_2}\right) \tag{10.17}$$

and

$$C_P^2 T_2 - C_P^1 T_1 - q = \frac{1}{2}\left(P_2 - P_1\right)\left(\frac{1}{\rho_1} + \frac{1}{\rho_2}\right) \tag{10.18}$$

and

$$RT_i = \frac{P_i}{\rho_i}, C_P^i = \frac{R\gamma_i}{\gamma_i - 1} \tag{10.19}$$

Substituting Eq. 10.19 into Eq. 10.18 gives

$$\frac{P_2}{\rho_2}\frac{\gamma_2}{\gamma_2-1}-\frac{P_1}{\rho_1}\frac{\gamma_1}{\gamma_1-1}-q=\frac{1}{2}\left(P_2-P_1\right)\left(\frac{1}{\rho_1}+\frac{1}{\rho_2}\right) \tag{10.20}$$

or

$$2\frac{P_2}{\rho_2}\frac{\gamma_2}{\gamma_2-1}-2\frac{P_1}{\rho_1}\frac{\gamma_1}{\gamma_1-1}-2q=\left(P_2-P_1\right)\left(\frac{1}{\rho_1}+\frac{1}{\rho_2}\right) \tag{10.21}$$

and

$$-2q=\left(P_2-P_1\right)\left(\upsilon_1+v_2\right)-2\frac{P_2}{\rho_2}\frac{\gamma_2}{\gamma_2-1}+2\frac{P_1}{\rho_1}\frac{\gamma_1}{\gamma_1-1} \tag{10.22}$$

$$-2q=\left[P_2\upsilon_1+P_2\upsilon_2-P_1\upsilon_1-P_1\upsilon_2\right]-2\frac{P_2}{\rho_2}\frac{\gamma_2}{\gamma_2-1}+2\frac{P_1}{\rho_1}\frac{\gamma_1}{\gamma_1-1} \tag{10.23}$$

And diving both sides of Eq. 10.23 by $P_1\upsilon_1$ results in

$$\left[y+xy-1-x\right]-2\frac{\gamma_2}{\gamma_2-1}xy+2\frac{\gamma_1}{\gamma_1-1}=-\frac{2q}{P_1\upsilon_1}=-2q' \tag{10.24}$$

and

$$y\left[1+x-\left(\frac{2\gamma_2}{\gamma_2-1}\right)x\right]=-2q'+1+x-2\frac{\gamma_1}{\gamma_1-1} \tag{10.25}$$

Solving for y

$$y=\frac{2q'-x-1+\dfrac{2\gamma_1}{\gamma_1-1}}{\left(\dfrac{2\gamma_2}{\gamma_2-1}\right)x-x-1} \tag{10.26}$$

Equation 10.26 can be simplified using the following relationships

$$\frac{2\gamma_1}{\gamma_1-1}-1=\frac{2\gamma_1}{\gamma_1-1}-\frac{\gamma_1-1}{\gamma_1-1}=\frac{\gamma_1+1}{\gamma_1-1}=a \tag{10.27}$$

and

$$\left(\frac{2\gamma_2}{\gamma_2-1}\right)x-x-1=\left(\frac{2\gamma_2}{\gamma_2-1}-1\right)x-1=\frac{\gamma_2+1}{\gamma_2-1}x-1=bx-1 \tag{10.28}$$

Substituting Eqs. 10.27 and 10.28 into Eq. 10.26 results in

$$y=\frac{2q'-x+a}{bx-1} \tag{10.29}$$

which is the celebrated Hugoniot curve.

We will now apply all this theory to a practical problem, but first more theory!

10.2.5   Delineating Combustion Regions

Given Rankine lines and a Hugoniot curve, we want to understand the various regions of the graph given below [1]. Looking at Fig. 10.4, the initial point is "A" and possible endpoints include "U" and "L", which are Chapman-Jouget points where the Rankine line and Hugoniot curve are tangent.

Using the developed equation for mass flux

$$\dot{m}=\sqrt{\frac{P_2-P_1}{\dfrac{1}{\rho_1}-\dfrac{1}{\rho_2}}} \tag{10.30}$$

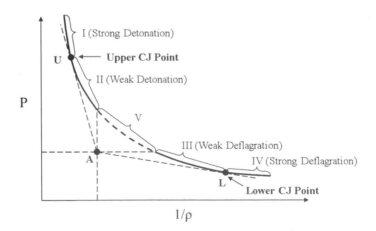

Fig. 10.4 Jouget's rules [1]

and

$$u_2 - u_1 = \dot{m}\left[v_2 - v_1\right] \tag{10.31}$$

If we square both sides of Eq. 10.30, we see that this is the Rankine line and dividing by $P_1\rho_1$ results in

$$\frac{\dot{m}^2}{P_1\rho_1} = \frac{y-1}{1-x} = \gamma_1 Ma_1^2 \tag{10.32}$$

We'll see in the next section that

$$Ma_1^2 \approx \frac{2\left(\gamma_2^2 - 1\right)q'}{\gamma_1} \tag{10.33}$$

Substituting Eq. 10.33 into Eq. 10.32 results in

$$\frac{\dot{m}^2}{P_1\rho_1} = \frac{y-1}{1-x} = \gamma_1 Ma_1^2 = 2\left(\gamma_2^2 - 1\right)q' = 2\left(\gamma_2^2 - 1\right)\frac{q}{P_1 v_1} \tag{10.34}$$

Using Fig. 10.4 and Eqs. 10.30, 10.31, and 10.34 observations will be made about the following quantities

1. Relationship between P_2 and P_1 for a particular region
2. Relationship between v_2 and v_1 for a particular region
3. Relationship between u_2 and u_1 for a particular region
4. Relationship between slope of Rankine line, Ma_1, and q for a particular region

Analysis is given as Tables 10.2, 10.3, and 10.4.

The only portions of the graph observed experimentally (there are rare exceptions) are the Detonation CJ point and the weak deflagration region (Region III) [3]; the weak deflagration region is understood through a knowledge of laminar flame speed of premixed gases [1, 5, 7] and will not be discussed in this chapter. This chapter will focus solely on the detonation CJ point and properties of this point such as $\{x_{cj}, Ma_{cj}\}$.

Table 10.2 Jouget rules, Part I

Region	P_2 versus P_1	v_2 versus v_1	Designation	Comments
I	$P_2 > P_1$	$v_2 < v_1$	Strong Detonation	$u_1 > u_2$
II	$P_2 > P_1$	$v_2 < v_1$	Weak Detonation	$u_1 > u_2$
V	$P_2 > P_1$	$v_2 > v_1$	Impossible	Quantity under radical would be negative!
III	$P_2 < P_1$	$v_2 > v_1$	Weak Deflagration	$u_1 < u_2$
IV	$P_2 < P_1$	$v_2 > v_1$	Strong Deflagration	$u_1 < u_2$

Table 10.3 Jouget rules, Part II [1]

Region	Slope of Rankine Line	Ma_1	q
I	Steep	$Ma_1 > 1$	High
II	Steepest	$Ma_1 > 1$	Highest
V	Impossible		
III	Shallowest	$Ma_1 < 1$	Lowest
IV	Shallow	$Ma_1 < 1$	Low

Table 10.4 Jouget rules, Part III [1]

Region	Observed Experimentally	Comments		
I	CJ Detonation Point Only	$u_1 > u_2$	Flow supersonic in front, subsonic behind
II	No	$u_1 > u_2$	Flow supersonic in front, supersonic behind
V	Impossible		
III	Yes	$u_1 < u_2$	Flow subsonic in front, subsonic behind
IV	No	$u_1 < u_2$	Flow subsonic in front, supersonic behind

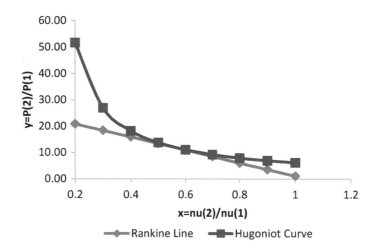

Fig. 10.5 Solution to Example 10.1

Example 10.1 RH Theory

Given ambient conditions of 1 atmosphere, 25 C, γ_1 is 1.4, γ_2 is 1.2, and the heat released is 2400 kJ/kg. Develop the Rankine line and Hugoniot for this system and graph. Assume $Ma(cj)$ is 4.20. We see from the graphs that $x(cj)$ is between 0.5 and 0.6 and $y(cj)$ is between 10 and 15, which provides an estimate for $x(cj)$ and $y(cj)$. Knowing ρ_1 and P_1 we can determine ρ_2 and P_2. We will take this problem further in the next section, but first we need to understand the CJ condition (Fig. 10.5).

10.3   Chapman-Jouget (CJ) Point for Standard RH System

When the Rankine line and Hugoniot curve are tangent to each other, the endpoint of the combustion is known as the Chapman-Jouget condition and there are two possible "CJ" points; one is associated with detonation and the other is associated with deflagration.

This section will derive the states (X_{cj}, Ma_{cj}) associated with a detonation at the CJ point and also the point where the reaction is complete ($\lambda = 100\%$).

At the CJ point

$$y_{\text{Rankine Line}} = y_{\text{Hugoniot Curve}} \tag{10.35}$$

and

$$\frac{d}{dx} y_{\text{Rankine Line}} = \frac{d}{dx} y_{\text{Hugoniot Curve}} \tag{10.36}$$

Equation 10.35 will be utilized to determine X_{cj} and Eq. 10.36 will be utilized to determine Ma_{cj}.

10.3.1   Derivation for X(cj)

$$1 + \gamma_1 Ma_1^2 \left(1 - x\right) = \frac{a - x + 2q'}{bx - 1} \tag{10.37}$$

$$bx - 1 + \gamma_1 Ma_1^2 \left(1 - x\right)\left(bx - 1\right) = a - x + 2q' \tag{10.38}$$

or

$$bx - 1 + \gamma_1 Ma_1^2 bx - b\gamma_1 Ma_1^2 x^2 + \gamma_1 Ma_1^2 x - \gamma_1 Ma_1^2 - a - 2q' + x = 0 \tag{10.39}$$

which is of the form

$$Ax^2 + Bx + C = 0 \tag{10.40}$$

where

$$A = -b\gamma_1 Ma_1^2, B = b + \gamma_1 Ma_1^2 b + \gamma_1 Ma_1^2 + 1, C = -1 - \gamma_1 Ma_1^2 - a - 2q' \tag{10.41}$$

The CJ point, by definition, is a point of tangency and Eq. 10.39 has a unique solution. Therefore,

$$x_{cj} = -\frac{B}{2A} = \frac{1}{2\gamma_1 Ma_1^2} + \frac{1}{2} + \frac{1}{2b} + \frac{1}{2b\gamma_1 Ma_1^2} \approx \frac{1}{2}\left(1+\frac{1}{b}\right) = \frac{1}{2b}(b+1) \tag{10.42}$$

Recalling that b is defined as

$$b = \frac{\gamma_2 + 1}{\gamma_2 - 1} \tag{10.43}$$

And $1 \le \gamma_2 \le 2 \rightarrow x_{cj} \in \left[\dfrac{1}{2}, \dfrac{2}{3}\right]$ and for air/fuel mixtures is more likely between 0.5 and 0.6.

Now solve for Ma_{cj}

10.3.2   Derivation for Ma(cj)

Given

$$y_{RL} = \left(1 + \gamma_1 Ma_1^2\right) - \gamma_1 Ma_1^2 x \tag{10.44}$$

and

$$y_{HC} = \frac{a - x + 2q'}{bx - 1} \tag{10.45}$$

and

$$y'_{RL} = -\gamma_1 Ma_1^2 \tag{10.46}$$

and

$$y'_{HC} = \frac{1 - ab - 2bq'}{(bx - 1)^2} \tag{10.47}$$

Set Eqs. 10.46 and 10.47 equal to each other and solve for Ma_1.

$$-\gamma_1 Ma_1^2 = \frac{1 - ab - 2bq'}{(bx - 1)^2}, \gamma_1 Ma_1^2 = \frac{ab + 2bq' - 1}{(bx - 1)^2} \tag{10.48}$$

and

$$x_{cj} = \frac{1}{2}\left(1 + \frac{1}{b}\right), bx = \frac{1}{2}(b+1), bx - 1 = \frac{1}{2}(b-1) \tag{10.49}$$

Further,

$$\left(bx-1\right)^2 = \frac{1}{4}\left(b-1\right)^2 \tag{10.50}$$

where b is defined as

$$b = \frac{\gamma_2+1}{\gamma_2-1}, b-1 = \frac{\gamma_2+1}{\gamma_1-1} - \frac{\gamma_2-1}{\gamma_2-1} = \frac{2}{\gamma_2-1} \tag{10.51}$$

Therefore,

$$\frac{1}{4}\left(b-1\right)^2 = \frac{1}{\left(\gamma_2-1\right)^2} \tag{10.52}$$

Substituting Eq. 10.52 into Eq. 10.48 results in

$$\gamma_1 Ma_1^2 = \frac{\gamma_2+1}{\gamma_2-1}\left(\gamma_2-1\right)^2\left[a+2q'\right] - \left(\gamma_2-1\right)^2 \tag{10.53}$$

and

$$Ma_1^2 \approx \frac{2\left(\gamma_2^2-1\right)q'}{\gamma_1} \tag{10.54}$$

Example 10.2 RH Theory and CJ Point

Complete Example 10.1 by determining Ma_{cj}, x_{cj}, ν_2, T_2, and P_2. (Fig. 10.6)

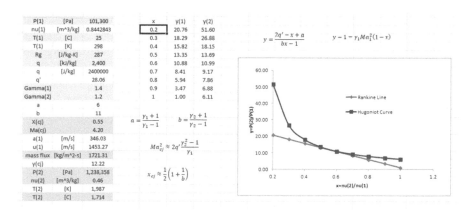

Fig. 10.6 Solution for Example 10.2

10.4   Partially Complete Reactions

When a reaction does not go to completion ($\lambda < 100 \%$), a series of Hugoniot Curves are graphed for each λ and when this series of curves intersects the Rankine Line, and then the endpoint of the system is determined.

Let's start by formally defining λ.

$$\lambda = \frac{[F]_0 - [F]_f}{[F]_0} * 100\% \tag{10.55}$$

where $[F]_0$ is the initial concentration of fuel and $[F]_f$ is the final concentration of fuel.

The only change to the RH theory is to multiple the q term within the Hugoniot curve by λ. An example using partial Hugoniot curves is given in Chap. 13.

10.5   Fay's System and RH Theory

This section will focus on algebraically deriving the equations for the Rankine line and Hugoniot curve starting from the given conservation principles for Fay's system. These equations will be obtained in terms of initial and final conditions of detonations, as well as Mach numbers. The different equations for conservations will take into account the mass divergence considerations as devised by Fay.

The theory for combustion presented in Sects. 10.2 and 10.3 assumes one-dimensional flow that ignores the effects of friction and turbulence and the theory has been shown over 100 years to work surprisingly well. There are exceptions though. It was seen that when a shock tube was utilized with a smaller inside diameter the predicted $Ma(cj)$ and actual $Ma(cj)$ were significantly different. In Fay [3], "It was argued that the effects of the wall were confined to a thin layer of fluid closed to the wall, and could only influence the major portion of the flow through changes in pressure propagated through the subsonic reaction zone. The effect of this boundary layer is to cause the streamlines in the reaction zone to diverge and thereby reduce the propagation velocity." Given as Eqs. 10.56, 10.57, and 10.58 are the conservation equations for a system that addresses this area divergence.

The derivation for the Rankine line and Hugoniot curve in this section is solely the work of Imane Ennadi [2] who derived these equations as part of her senior honor's thesis at the University of Saint Thomas.

Equations of Conservation
Conservation of mass:

$$\rho_1 u_1 = \rho_2 u_2 \left[1 + \xi \right] = \dot{m} \tag{10.56}$$

Conservation of energy:

$$h_1 + \frac{1}{2}u_1^2 = h_2 + \frac{1}{2}u_2^2 \tag{10.57}$$

Conservation of momentum:

$$P_1 + \rho_1 u_1^2 = P_2\left[1+\xi\left(1-\epsilon\right)\right] + \rho_2 u_2^2\left[1+\xi\right] \tag{10.58}$$

where $\xi = \dfrac{A_2}{A_1} - 1$ and $1 < \epsilon \leq 2$.

10.5.1 Rankine Line

Using Eq. 10.58, we can rearrange it to get:

$$P_2\left[1+\xi\left(1-\epsilon\right)\right] - P_1 = \rho_1 u_1^2 - \rho_2 u_2^2\left[1+\xi\right] \tag{10.59}$$

Using Eq. 10.56, we get:

$$\frac{\rho_1}{\rho_2\left[1+\xi\right]}u_1 = u_2 \tag{10.60}$$

$$\left(\frac{\rho_1}{\rho_2\left[1+\xi\right]}u_1\right)^2 = u_2^2 \tag{10.61}$$

Thus, we can plug in Eq. 10.61 into Eq. 10.59

$$P_2\left[1+\xi\left(1-\epsilon\right)\right] - P_1 = \rho_1 u_1^2 - \rho_2\left[1+\xi\right]\left(\frac{\rho_1}{\rho_2\left[1+\xi\right]}u_1\right)^2 \tag{10.62}$$

$$P_2\left[1+\xi\left(1-\epsilon\right)\right] - P_1 = \rho_1 u_1^2 - \frac{\rho_1^2}{\rho_2\left[1+\xi\right]}u_1^2 \tag{10.63}$$

$$P_2\left[1+\xi\left(1-\epsilon\right)\right] - P_1 = \rho_1 u_1^2\left(1 - \frac{\rho_1}{\rho_2\left[1+\xi\right]}\right) \tag{10.64}$$

$$P_2\left[1+\xi\left(1-\epsilon\right)\right] - P_1 = \rho_1 u_1^2\left(\frac{\rho_2\left[1+\xi\right] - \rho_1}{\rho_2\left[1+\xi\right]}\right) \tag{10.65}$$

We define Y and X to be:

$$Y = \frac{P_2}{P_1} \tag{10.66}$$

$$X = \frac{\rho_1}{\rho_2} \tag{10.67}$$

To introduce these two variables, we divide Eq. 10.65 by variable P_1, obtaining:

$$\frac{P_2\left[1+\xi(1-\epsilon)\right]}{P_1} - \frac{P_1}{P_1} = \frac{\rho_1}{P_1}u_1^2\left(\frac{\rho_2[1+\xi]}{\rho_2[1+\xi]} - \frac{\rho_1}{\rho_2[1+\xi]}\right) \tag{10.68}$$

Replacing the variables, we obtain the equation:

$$Y\left[1+\xi(1-\epsilon)\right] - 1 = \frac{\rho_1}{P_1}u_1^2\left(1 - \frac{X}{[1+\xi]}\right) \tag{10.69}$$

Determining this equation in term of Mach numbers, we obtain:

$$Y\left[1+\xi(1-\epsilon)\right] - 1 = Y_1\text{Mach}_1^2\left(1 - \frac{X}{[1+\xi]}\right) \tag{10.70}$$

This equation is the equation of the *Rayleigh line*.

10.5.2   Mass Flux

Using Eq. 10.56, we can derive two equations:

$$u_1 = \frac{\dot{m}}{\rho_1} \tag{10.71}$$

$$u_2 = \frac{\dot{m}}{\rho_2[1+\xi]}$$

Using Eq. 10.71, we can replace the variables u_1 and u_2 in Eq. 10.58:

$$P_2\left[1+\xi\left(1-\epsilon\right)\right]-P_1 = \frac{\dot{m}^2}{\rho_1} - \frac{\dot{m}^2}{\rho_2\left[1+\xi\right]} \qquad (10.72)$$

$$P_2\left[1+\xi\left(1-\epsilon\right)\right]-P_1 = \dot{m}^2\left(\frac{1}{\rho_1} - \frac{1}{\rho_2\left[1+\xi\right]}\right) \qquad (10.73)$$

$$\frac{P_2\left[1+\xi\left(1-\epsilon\right)\right]-P_1}{\dfrac{1}{\rho_1} - \dfrac{1}{\rho_2\left[1+\xi\right]}} = \dot{m}^2 \qquad (10.74)$$

Following, we get the equation:

$$\frac{P_2\left[1+\xi\left(1-\epsilon\right)\right]-P_1}{\dfrac{1}{\rho_2\left[1+\xi\right]} - \dfrac{1}{\rho_1}} = -\dot{m}^2 \qquad (10.75)$$

10.5.3 Deriving Hugoniot Curve

We first start by rewriting Eq. 10.57:

$$-\Delta KE = \frac{1}{2}\left(u_1^{\,2} - u_2^{\,2}\right) \qquad (10.76)$$

Using Eqs. 10.71, we obtain:

$$-\Delta KE = \frac{1}{2}\left(\frac{\dot{m}^2}{\rho_1^{\,2}} - \frac{\dot{m}^2}{\rho_2^{\,2}\left[1+\xi\right]^2}\right) \qquad (10.77)$$

$$\Delta KE = \frac{1}{2}\dot{m}^2\left(\frac{1}{\rho_1^{\,2}} - \frac{1}{\rho_2^{\,2}\left[1+\xi\right]^2}\right) \qquad (10.78)$$

Using Eq. 10.74, we obtain:

$$\Delta KE = \frac{1}{2}\left(\frac{P_2\left[1+\xi\left(1-\epsilon\right)\right]-P_1}{\dfrac{1}{\rho_1} - \dfrac{1}{\rho_2\left[1+\xi\right]}}\right)\left(\frac{1}{\rho_1^{\,2}} - \frac{1}{\rho_2^{\,2}\left[1+\xi\right]^2}\right) \qquad (10.79)$$

where:

$$\left(\cfrac{1}{\cfrac{1}{\rho_1}-\cfrac{1}{\rho_2[1+\xi]}}\right)\left(\cfrac{1}{\rho_1^{\,2}}-\cfrac{1}{\rho_2^{\,2}[1+\xi]^2}\right)=\left(\cfrac{\rho_2\rho_1[1+\xi]}{\rho_2[1+\xi]-\rho_1}\right)\left(\cfrac{1}{\rho_1^{\,2}}-\cfrac{1}{\rho_2^{\,2}[1+\xi]^2}\right) \tag{10.80}$$

$$\left(\cfrac{\rho_2\rho_1[1+\xi]}{\rho_2[1+\xi]-\rho_1}\right)\left(\cfrac{1}{\rho_1^{\,2}}-\cfrac{1}{\rho_2^{\,2}[1+\xi]^2}\right)=\left(\cfrac{\rho_2[1+\xi]+\rho_1}{\rho_2[1+\xi]\rho_1}\right)=\left(\cfrac{[1+\xi]+\cfrac{\rho_1}{\rho_2}}{\rho_2[1+\xi]\cfrac{\rho_1}{\rho_2}}\right) \tag{10.81}$$

Using Eq. 10.67, we obtain:

$$\left(\cfrac{[1+\xi]+\cfrac{\rho_1}{\rho_2}}{\rho_2[1+\xi]\cfrac{\rho_1}{\rho_2}}\right)=\left(\cfrac{[1+\xi]+X}{\rho_2[1+\xi]X}\right) \tag{10.82}$$

Thus, Eq. 10.81 becomes:

$$\Delta KE=\frac{1}{2}\left(\cfrac{[1+\xi]+X}{\rho_2[1+\xi]X}\right)\left(P_2\left[1+\xi(1-\epsilon)\right]-P_1\right) \tag{10.83}$$

where, following the ideal gas law, we know that:

$$\Delta KE=q+C_{p2}T_2-C_{p1}T_1 \tag{10.84}$$

where:

$$R_uT_2=\frac{P_2}{\rho_2} \tag{10.85}$$

$$R_uT_1=\frac{P_1}{\rho_1}$$

And, following the ideal gas law:

$$C_p = R_u \frac{\Upsilon}{\Upsilon - 1} \tag{10.86}$$

Thus, we obtain:

$$\Delta KE = q + \left(\frac{\Upsilon_2}{\Upsilon_2 - 1} \right) \frac{P_2}{\rho_2} - \left(\frac{\Upsilon_1}{\Upsilon_1 - 1} \right) \frac{P_1}{\rho_1} \tag{10.87}$$

$$\Delta KE - \left(\frac{\Upsilon_2}{\Upsilon_2 - 1} \right) \frac{P_2}{\rho_2} + \left(\frac{\Upsilon_1}{\Upsilon_1 - 1} \right) \frac{P_1}{\rho_1} = q \tag{10.88}$$

Using Eq. 10.83, we obtain:

$$\frac{1}{2} \left(\frac{[1+\xi]+X}{\rho_2[1+\xi]X} \right) \left(P_2 \left[1 + \xi(1-\epsilon) \right] - P_1 \right) - \left(\frac{\Upsilon_2}{\Upsilon_2 - 1} \right) \frac{P_2}{\rho_2} + \left(\frac{\Upsilon_1}{\Upsilon_1 - 1} \right) \frac{P_1}{\rho_1} = q \tag{10.89}$$

$$\left(\frac{[1+\xi]+X}{\rho_2[1+\xi]X} \right) \left(P_2 \left[1 + \xi(1-\epsilon) \right] - P_1 \right) - \left(\frac{2\Upsilon_2}{\Upsilon_2 - 1} \right) \frac{P_2}{\rho_2} + \left(\frac{2\Upsilon_1}{\Upsilon_1 - 1} \right) \frac{P_1}{\rho_1} = 2q \tag{10.90}$$

where we know that:

$$v_1 = \frac{1}{\rho_1} \tag{10.91}$$

$$v_2 = \frac{1}{\rho_2} \tag{10.92}$$

Thus, we get:

$$u_2 \left(\frac{[1+\xi]+X}{[1+\xi]X} \right) \left(P_2 \left[1 + \xi(1-\epsilon) \right] - P_1 \right) - \left(\frac{2\Upsilon_2}{\Upsilon_2 - 1} \right) P_2 v_2 + \left(\frac{2\Upsilon_1}{\Upsilon_1 - 1} \right) P_1 v_1 = 2q \tag{10.92}$$

$$\frac{v_2}{v_1} \left(\frac{[1+\xi]+X}{[1+\xi]X} \right) \left(\frac{P_2 \left[1 + \xi(1-\epsilon) \right]}{P_1} - 1 \right) - \left(\frac{2\Upsilon_2}{\Upsilon_2 - 1} \right) \frac{P_2 v_2}{P_1 v_1} + \left(\frac{2\Upsilon_1}{\Upsilon_1 - 1} \right) = \frac{2q}{P_1 v_1} \tag{10.93}$$

Using density instead of specific volume, we get:

$$\frac{\rho_1}{\rho_2}\left(\frac{[1+\xi]+X}{[1+\xi]X}\right)\left(\frac{P_2\left[1+\xi(1-\epsilon)\right]}{P_1}-1\right)-\left(\frac{2\Upsilon_2}{\Upsilon_2-1}\right)\frac{P_2\rho_1}{P_1\rho_2}+\left(\frac{2\Upsilon_1}{\Upsilon_1-1}\right)=\frac{2q}{P_1v_1} \quad (10.94)$$

Using Eqs. 10.66 and 10.67, we obtain:

$$X\left(\frac{[1+\xi]+X}{[1+\xi]X}\right)\left(Y\left[1+\xi(1-\epsilon)\right]-1\right)-\left(\frac{2\Upsilon_2}{\Upsilon_2-1}\right)YX+\left(\frac{2\Upsilon_1}{\Upsilon_1-1}\right)=\frac{2q}{P_1v_1} \quad (10.95)$$

$$\left(1+\frac{X}{[1+\xi]}\right)\left(Y\left[1+\xi(1-\epsilon)\right]-1\right)-\left(\frac{2\Upsilon_2}{\Upsilon_2-1}\right)YX+\left(\frac{2\Upsilon_1}{\Upsilon_1-1}\right)=\frac{2q}{P_1v_1} \quad (10.96)$$

$$Y\left[1+\xi(1-\epsilon)\right]+\frac{X\left[1+\xi(1-\epsilon)\right]}{[1+\xi]}-\left(\frac{2\Upsilon_2}{\Upsilon_2-1}\right)X=\frac{2q}{P_1v_1}-\frac{2\Upsilon_1}{\Upsilon_1-1}+1-\frac{X}{[1+\xi]} \quad (10.97)$$

As such, we obtain the equation for the Hugoniot curve to be:

$$Y=\frac{1-\dfrac{2\Upsilon_1}{\Upsilon_1-1}+\dfrac{2q}{P_1v_1}-\dfrac{X}{[1+\xi]}}{\left[1+\xi(1-\epsilon)\right]\left(1+\dfrac{X}{[1+\xi]}\right)-\left(\dfrac{2\Upsilon_2}{\Upsilon_2-1}\right)X} \quad (10.98)$$

And can be simplified to

$$Y=\frac{\dfrac{2q}{P_1v_1}-\dfrac{X}{[1+\xi]}-\dfrac{\gamma_1+1}{\gamma_1-1}}{\left[1+\xi(1-\epsilon)\right]\left(1+\dfrac{X}{[1+\xi]}\right)-\left(\dfrac{2\Upsilon_2}{\Upsilon_2-1}\right)X}=\frac{2q'-X'-a}{\left[1+\xi(1-\epsilon)\right](1+X')-bX} \quad (10.99)$$

10.6   Determination of States to Include Ma(cj)

An alternative approach exists for determining the *Ma(cj)* for a system with or without considerations of friction, which is the issue that Fay [3] addressed; this alternate approach will also provide equations for ratios of the states. This approach [9–12] parallels the method we utilized for determining the equations for ratio of states associated with a normal shock, Chap. 7.

10.6.1 Determination of States without Thermodynamic Changes (Coleman)

In this section, equations are derived for the following non-dimensionalized states $\left\{ \dfrac{T_2}{T_1}, \dfrac{P_2}{P_1}, \dfrac{\rho_2}{\rho_1}, \dfrac{Ma_2^2}{Ma_1^2} \right\}$. Please note in this section thermodynamics properties such as $\{c_P, \gamma\}$ will be considered constant, but will be allowed to varying in Sect. 10.6.2.

Derivation for $\left\{ \dfrac{u_2}{u_1} \right\}$ and $\left\{ \dfrac{\rho_1}{\rho_2} \right\}$

From Eqs. 10.1 to 10.3 and other considerations for an ideal, perfect gas it can be shown (see Appendix 10.1) that

$$1 - \frac{u_2}{u_1} = 1 - \frac{\rho_1}{\rho_2} = \frac{Ma_1^2 - 1}{Ma_1^2(\gamma + 1)}\left[1 + \sqrt{1 - \frac{2(\gamma + 1)Ma_1^2}{\left(Ma_1^2 - 1\right)^2}\frac{q}{C_P T}} \right] = \frac{Ma_1^2 - 1}{Ma_1^2(\gamma + 1)}F \tag{10.100}$$

Coleman [9] has shown that F characterizes the type of detonation system where

$$F = 1 \rightarrow \text{Chapman} - \text{Jouget Detonation}$$

$$1 \le F \le 2 \rightarrow \text{Strong Detonation}$$

$$F = 2 \rightarrow \text{Adiabatic Shock}$$

Let

$$A = \frac{Ma_1^2 - 1}{(\gamma + 1)}F \tag{10.101}$$

And Substituting Eq. 10.101 into Eq. 10.100 results in

$$1 - \frac{u_2}{u_1} = 1 - \frac{\rho_1}{\rho_2} = \frac{A}{Ma_1^2} \tag{10.102}$$

Derivation for $\left\{ \dfrac{P_2}{P_1} \right\}$
From Eq. 10.2,

$$\frac{P_1}{P_1} + \frac{\rho_1}{P_1}u_1^2 = \frac{P_2}{P_1} + \frac{\rho_2}{P_1}u_2^2 \tag{10.103}$$

and

$$\frac{P_2}{P_1} = 1 + \frac{\rho_1}{P_1}u_1^2 - \frac{\rho_2}{P_1}u_2^2 \tag{10.104}$$

or

$$\frac{P_2}{P_1} = 1 + \gamma \frac{u_1^2}{a_1^2} - \frac{P_2}{P_1} u_2^2 = 1 + \gamma Ma_1^2 - \frac{P_2}{P_1} u_2^2 \qquad (10.105)$$

where u_2 is also equal to

$$u_2 = \frac{\rho_1}{\rho_2} u_1 \qquad (10.106)$$

Substituting Eq. 10.106 into Eq. 10.105 equals

$$\frac{P_2}{P_1} = 1 + \gamma Ma_1^2 - \frac{P_2}{P_1}\left[\left(\frac{\rho_1}{\rho_2}\right)^2 u_1^2\right] = 1 + \gamma Ma_1^2 - \frac{\rho_1^2}{P_1} \frac{1}{\rho_2} u_1^2 \frac{\gamma}{\gamma} = 1 + \gamma Ma_1^2\left[1 - \frac{\rho_1}{\rho_2}\right] \qquad (10.107)$$

And substituting Eq. 10.102 into Eq. 10.107 results in

$$\frac{P_2}{P_1} = 1 + \gamma A \qquad (10.108)$$

Derivation for $\left\{\dfrac{Ma_2^2}{Ma_1^2}\right\}$

The ratio of Mach numbers squared is derived as a function of the quantities given above. By definition

$$\frac{Ma_2^2}{Ma_1^2} = \frac{\dfrac{u_2^2}{a_2^2}}{\dfrac{u_1^2}{a_1^2}} = \frac{\gamma \dfrac{u_2^2}{P_2}}{\gamma \dfrac{u_1^2}{P_1}}{\dfrac{\rho_2}{\rho_1}} = \frac{u_2^2}{u_1^2}\frac{P_1}{P_2}\frac{\rho_2}{\rho_1} = \frac{\left[1 - \dfrac{A}{Ma_1^2}\right]^2}{\left[1 - \dfrac{A}{Ma_1^2}\right][1 + \gamma A]} = \frac{\left[1 - \dfrac{A}{Ma_1^2}\right]}{1 + \gamma A} \qquad (10.109)$$

10.6.2   Determination of States with Thermodynamic Changes (Adamson)

In this section, thermodynamic properties across the shock are allowed to vary, but essentially the equations keep the same form [12]. The only difference is in how "F" and "A" are defined.

Again, starting from Eqs. 10.1, 10.2, and 10.3 one could get to

$$\frac{P_2}{P_1} - 1 = \gamma_1 A \tag{10.110}$$

$$\frac{\rho_1}{\rho_2} = 1 - \frac{A}{Ma_1^2} = \frac{u_2}{u_1} \tag{10.111}$$

where

$$F = 1 + \sqrt{1 - 2\frac{\gamma_2^2 - 1}{\gamma_1 - 1}\frac{Ma_1^2}{\left[Ma_1^2 - \frac{\gamma_2}{\gamma_1}\right]^2}\left\{\frac{q}{C_P T} - \frac{\gamma_2 - \gamma_1}{\gamma_1(\gamma_2 - 1)}\right\}} \tag{10.112}$$

and

$$A = \frac{F}{\gamma_2 + 1}\left[Ma_1^2 - \frac{\gamma_2}{\gamma_1}\right] \tag{10.113}$$

We'll derive relationships for $\left\{\frac{T_2}{T_1}\right\}$ and $\left\{\frac{Ma_2^2}{Ma_1^2}\right\}$ using the relationships given above.

Derivation for $\left\{\frac{T_2}{T_1}\right\}$

Given

$$\frac{P}{\rho} = RT \tag{10.114}$$

Therefore,

$$\frac{\dfrac{P_2}{\rho_2}}{\dfrac{P_1}{\rho_1}} = \frac{T_2}{T_1} = \frac{P_2}{P_1}\frac{\rho_1}{\rho_2} = (1 + \gamma_1 A)\left(1 - \frac{A}{Ma_1^2}\right) \tag{10.115}$$

Derivation for $\left\{\dfrac{Ma_2^2}{Ma_1^2}\right\}$

The ratio of Mach numbers squared can be defined as

$$\frac{Ma_2^2}{Ma_1^2} = \frac{\dfrac{u_2^2}{a_2^2}}{\dfrac{u_1^2}{a_1^2}} = \frac{\dfrac{u_2^2}{\gamma\dfrac{P_2}{\rho_2}}}{\dfrac{u_1^2}{\gamma\dfrac{P_1}{\rho_1}}} = \frac{u_2^2}{u_1^2}\frac{P_1}{P_2}\frac{\rho_2}{\rho_1} = \frac{\left[1 - \dfrac{A}{Ma_1^2}\right]^2}{\left[1 - \dfrac{A}{Ma_1^2}\right][1 + \gamma_1 A]} = \frac{\left[1 - \dfrac{A}{Ma_1^2}\right]}{1 + \gamma_1 A} \tag{10.116}$$

10.6.3   Ma(cj) for a Detonation System with and without Area Divergence

Without Area Divergence

The determination of the Mach number associated with the CJ detonation wave is determined from Eq. 10.112, which is the F associated with Adamson's system [12].

$$F = 1 + \sqrt{1 - 2\frac{\gamma_2^2 - 1}{\gamma_1 - 1}\frac{Ma_1^2}{\left[Ma_1^2 - \frac{\gamma_2}{\gamma_1} \right]^2}\left[\frac{q}{C_P T_1} - \frac{\gamma_1 - \gamma_2}{\gamma_1(\gamma_2 - 1)} \right]} \qquad (10.117)$$

Setting $F = 1$ results in

$$1 = 2\frac{\gamma_2^2 - 1}{\gamma_1 - 1}\frac{Ma_1^2}{\left[Ma_1^2 - \frac{\gamma_2}{\gamma_1} \right]^2}\left[\frac{q}{C_P T_1} - \frac{\gamma_1 - \gamma_2}{\gamma_1(\gamma_2 - 1)} \right] \qquad (10.118)$$

And solving for Ma_1 we get to

$$\frac{\left[Ma_1^2 - \frac{\gamma_2}{\gamma_1} \right]^2}{Ma_1^2} = 2\frac{\gamma_2^2 - 1}{\gamma_1 - 1}\left[\frac{q}{C_P T_1} - \frac{\gamma_1 - \gamma_2}{\gamma_1(\gamma_2 - 1)} \right] \qquad (10.119)$$

With Area Divergence [10]

From the conservation principles associated with Fay's system, one with great patience could get to the following relationship

$$\frac{\left[Ma_1^2 - \frac{\gamma_2}{\gamma_1} \right]^2}{Ma_1^2} + \psi\gamma_2^2\frac{\left[Ma_1^2 + \frac{1}{\gamma_1} \right]^2}{Ma_1^2} = 2\frac{\gamma_2^2 - 1}{\gamma_1 - 1}\left[\frac{q}{C_P T_1} - \frac{\gamma_1 - \gamma_2}{\gamma_1(\gamma_2 - 1)} \right] \qquad (10.120)$$

where ψ is defined as

$$1 + \psi = \left[\frac{1}{1 - \frac{\epsilon\xi}{(1 + \gamma_2)(1 + \epsilon)}} \right]^2 \qquad (10.121)$$

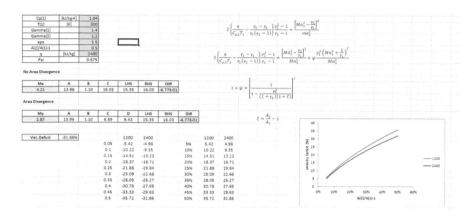

Fig. 10.7 Solution for Example 10.3

Example 10.3 Sensitivity Analysis of Area Divergence Effects on Velocity Deficit

Show how area divergence affects the velocity deficit for two values of q. Velocity deficit is defined as (Fig. 10.7)

$$\Delta V = \frac{\left| Ma_{cj,RH} - Ma_{cj,Fay} \right|}{Ma_{cj,RH}} \tag{10.122}$$

10.7   Problems

Problem 10.1 Standard RH Problem Around RDE

Given a pressure of 2 atmospheres, 250 K, $\gamma_1 = \gamma_2 = 1.4$ and a heat release of 20 kJ/kg create the Rankine line and Hugoniot curve for this system. Also determine the states at the CJ point.

Problem 10.2 Partially Complete Combustion

Repeat the analysis of Problem 10.1 but assume that $\lambda = 60\%$.

Problem 10.3 Fay's RH Problem

Using the Rankine line and Hugoniot curve for Fay's system with $\epsilon = 1.5$ and $\xi = 0.1$ for Problem 10.1 determine the $Ma(cj)$. This will be an iterative problem where $Ma(cj)$ will be varied until the Rankine line and Hugoniot curve are tangent to each other.

Problem 10.4 Ma(cj) from Alternate Method

Using the tools developed in Sect. 10.6, determine $Ma(cj)$ with and without area divergence for the system described in Problems 10.1 and 10.3.

Appendix 10.1: Derivation for Eq. 10.100

Given

$$\rho_1 u_1 = \rho_2 u_2 \tag{C.1}$$

$$P_1 + \rho_1 u_1^2 = P_2 + \rho_2 u_2^2 \tag{C.2}$$

$$\frac{1}{2}u_1^2 + \frac{a_1^2}{\gamma - 1} + q = \frac{1}{2}u_2^2 + \frac{a_2^2}{\gamma - 1} \tag{C.3}$$

$$a^2 = \gamma RT = \gamma \frac{P}{\rho} \tag{C.4}$$

and

$$C_P = \frac{\gamma R}{\gamma - 1} \tag{C.5}$$

Dividing Eq. C.2 by P_1 results in

$$1 + \frac{\gamma u_1^2}{\gamma \dfrac{P_1}{\rho_1}} = \frac{P_2}{P_1} + \frac{\rho_2}{P_1}u_2^2 \tag{C.6}$$

Substituting in $u_2^2 = \left[\dfrac{\rho_1}{\rho_2}u_1\right]^2$ results in

$$1 + \frac{\gamma u_1^2}{\gamma \dfrac{P_1}{\rho_1}} = \frac{P_2}{P_1} + \frac{\rho_2}{P_1}\left[\frac{\rho_1^2}{\rho_2^2}u_1^2\right] \tag{C.7}$$

and

$$1 + \gamma Ma_1^2 = \frac{\rho_2 T_2}{\rho_1 T_1} + \frac{\rho_2}{P_1}\left[\frac{\rho_1^2}{\rho_2^2}u_1^2\right] \tag{C.8}$$

Further

$$1 + \gamma Ma_1^2 = \frac{\rho_2 a_2^2}{\rho_1 a_1^2} + \frac{\rho_2}{P_1}\left[\frac{\rho_1^2}{\rho_2^2}u_1^2\right] \tag{C.9}$$

And multiplying both sides by $\dfrac{\rho_1}{\rho_2}$ results in

$$\left(1+\gamma Ma_1^2\right)\frac{\rho_1}{\rho_2} = \frac{a_2^2}{a_1^2} + \frac{\rho_1}{\rho_2}\frac{P_2}{P_1}\left[\frac{\rho_1^2}{\rho_2^2}u_1^2\right] \tag{C.10}$$

$$\left(1+\gamma Ma_1^2\right)\frac{u_2}{u_1} = \frac{a_2^2}{a_1^2} + \frac{\rho_1}{\rho_2}\frac{P_2}{P_1}\left[\frac{u_2^2}{u_1^2}u_1^2\right] = \frac{a_2^2}{a_1^2} + \gamma Ma_1^2\left(\frac{u_2^2}{u_1^2}\right) \tag{C.11}$$

Starting from Eqs. C.4 and C.5

$$\frac{1}{2}\frac{u_1^2}{u_1^2} + \frac{1}{(\gamma-1)}\frac{a_1^2}{u_1^2} + \frac{q}{u_1^2} = \frac{1}{2}\left(\frac{u_2}{u_1}\right)^2 + \frac{a_2^2}{(\gamma-1)u_1^2} \tag{C.12}$$

And multiplying both sides by 2

$$1 + \frac{2}{(\gamma-1)}\frac{a_1^2}{u_1^2} + \frac{q}{u_1^2} = \left(\frac{u_2}{u_1}\right)^2 + \frac{2a_2^2}{(\gamma-1)u_1^2} \tag{C.13}$$

Further

$$1 + \frac{2}{(\gamma-1)Ma_1^2} + \frac{q}{u_1^2} = \left(\frac{u_2}{u_1}\right)^2 + \frac{2a_2^2}{(\gamma-1)u_1^2} \tag{C.14}$$

And solving for $\left(\dfrac{u_2}{u_1}\right)^2$

$$\left(\frac{u_2}{u_1}\right)^2 = 1 + \frac{2}{(\gamma-1)Ma_1^2}\left[1 + \frac{q}{C_P T_1} - \left(\frac{a_2}{a_1}\right)^2\right] \tag{C.15}$$

where

$$\frac{2}{(\gamma-1)Ma_1^2}\frac{q}{C_P T_1} = \frac{2q}{(\gamma-1)\dfrac{u_1^2}{\gamma RT_1}C_P T_1} = \frac{2q}{(\gamma-1)\dfrac{u_1^2}{\gamma RT_1}\dfrac{\gamma R}{\gamma-1}T_1} = \frac{2q}{u_1^2} \tag{C.16}$$

Substituting Eq. C.11 into C.15 results in

$$\left(\frac{u_2}{u_1}\right)^2 = 1 + \frac{2}{(\gamma-1)Ma_1^2}\left[1 + \frac{q}{C_P T_1} - \left\{\left(1+\gamma Ma_1^2\right)\frac{u_2}{u_1} - \gamma Ma_1^2\left(\frac{u_2^2}{u_1^2}\right)\right\}\right] \tag{C.17}$$

or

$$\left(\frac{u_2}{u_1}\right)^2 = 1 + \frac{2}{(\gamma-1)Ma_1^2}\left[1 + \frac{q}{C_pT_1} - \left(1+\gamma Ma_1^2\right)\frac{u_2}{u_1} + \gamma Ma_1^2\left(\frac{u_2^2}{u_1^2}\right)\right] \quad (C.18)$$

or

$$\left\{\frac{2\gamma Ma_1^2}{(\gamma-1)Ma_1^2}-1\right\}\left(\frac{u_2}{u_1}\right)^2 - \frac{2\left(1+\gamma Ma_1^2\right)}{(\gamma-1)Ma_1^2}\frac{u_2}{u_1} + \frac{2}{(\gamma-1)Ma_1^2}\left[1+\frac{q}{C_pT_1}\right]+1=0 \quad (C.19)$$

It can be shown that

$$1 - \frac{u_2}{u_1} = 1 - \frac{\rho_1}{\rho_2} = \frac{Ma_1^2-1}{Ma_1^2(\gamma+1)}\left[1+\sqrt{1-\frac{2(\gamma+1)Ma_1^2}{\left(Ma_1^2-1\right)^2}\frac{q}{C_pT}}\right] = \frac{Ma_1^2-1}{Ma_1^2(\gamma+1)}F \quad (C.20)$$

Appendix 10.2: More Exact Solution for CJ Conditions

The following appendix derives more exact solutions for $X(cj)$ and $Ma(cj)$, which does not include changes in thermodynamic properties across the shock.

Another form of the Hugoniot curve is

$$\left(y+\mu^2\right)\left(x-\mu^2\right) = 1-\mu^4 + 2\mu^2\lambda q' \quad (E.1)$$

where

$$\mu^2 = \frac{\gamma-1}{\gamma+1} \quad (E.2)$$

If we substitute the Rankine line into Eq. E.1, then we get

$$\left[1+\gamma Ma_1^2\left(1-x\right)+\mu^2\right]\left[x-\mu^2\right] = 1-\mu^4 + 2\mu^2\lambda q' \quad (E.3)$$

And expanding the terms

$$x+\gamma Ma_1^2\left(1-x\right)x+\mu^2 x-\mu^2-\mu^2\gamma Ma_1^2\left(1-x\right)-\mu^4 = 1-\mu^4 + 2\mu^2\lambda q' \quad (E.4)$$

which is quadratic in x

$$-\gamma Ma_1^2 x^2 + \left[1 + \gamma Ma_1^2 + \mu^2 + \mu^2 \gamma Ma_1^2\right] x - \mu^2 \left[1 + \gamma Ma_1^2 + 2\lambda q'\right] - 1 = 0 \qquad \text{(E.5)}$$

or

$$-\gamma Ma_1^2 x^2 + \left[\left(1 + \mu^2\right)\left(1 + \gamma Ma_1^2\right)\right] x - \mu^2 \left[1 + \gamma Ma_1^2 + 2\lambda q'\right] - 1 = 0 \qquad \text{(E.6)}$$

When $B^2 = 4AC$, a unique solution exists. From this fact, more exact solutions exist for $x(cj)$ and $Ma(cj)$, which are given below.

$X(cj)$ is

$$x(cj) = \frac{\left[\left(1 + \mu^2\right)\left(1 + \gamma Ma_1^2\right)\right]}{2\gamma Ma_1^2} \qquad \text{(E.7)}$$

And $Ma(cj)$ can be determined from

$$\left[\left(1 + \mu^2\right)\left(1 + \gamma Ma_1^2\right)\right]^2 = 4\left[\gamma Ma_1^2\right]\left[\mu^2\left[1 + \gamma Ma_1^2 + 2\lambda q'\right] + 1\right] \qquad \text{(E.8)}$$

Appendix 10.3: Standard Rankine-Hugoniot Worksheet (website)

An Excel Spreadsheet was developed to illustrate a particular concept and is given on the companion website.

Appendix 10.4: Partially Combusted Rankine-Hugoniot Worksheet (website)

An Excel Spreadsheet was developed to illustrate a particular concept and is given on the companion website.

Appendix 10.5: Ma(cj) Worksheet (website)

An Excel Spreadsheet was developed to illustrate a particular concept and is given on the companion website.

References

1. Kuo K (2005). Principles of combustion; , 2nd.
2. Ennadi, I. (2019). *Numerical and graphical solutions to more general Rankine Hugoniot systems*; Senior Honor's Thesis, Department of Mathematics and Computer Science, University of St. Thomas.
3. Fay, J. (1959). Two-dimensional gaseous detonation: Velocity deficit. *The Physics of Fluids, 2*(3), 283–289.
4. Norden, C. (2013). *Thermodynamics of a rotating detonation engine*; University of Connecticut, Dissertation.
5. Rouser, Kurt P., personal Communications, April 4, 2021.
6. Wikipedia. https://en.wikipedia.org/wiki/Shock_tube. Accessed 6/12/2021.
7. Lee, J. H. S. (2008). *The detonation phenomenon*. Cambridge University Press.
8. Turns, S. R. (2012). *An introduction to combustion concepts and applications* (3rd ed.). Mc-Graw Hill.
9. Coleman, L. H. (1970). An analysis of oblique and Normal detonation waves; Ministry of Technology, Aeronautical Research Council, Reports and Memoranda, R&M No. 3638.
10. Dabora, E. K. (1963). *The influence of a compressible boundary on the propagation of gaseous detonations*; Dissertation, University of Michigan.
11. Evans and Ablow (1960). *Theories of detonation*; Standard Research Institute, Received May 16, 1960.
12. Murray, S. B. (1985). *The influence of initial and boundary conditions on gaseous detonation waves*; Dissertation, Mc-Gill University, Suffield Report 411.

Chapter 11
Combustion Chemistry

11.1 Preview

We saw in Chap. 10 how to determine the endpoints of either the shock or combustion utilizing the Rankine-Hugoniot theory and the Chapman-Jouget theory. This theory has been shown over the last century to accurately determine the endpoint states of the reaction zone for a detonation, but doesn't work as well for deflagrations. The theory in Chap. 10 was from a purely physical basis.

In Chaps. 12 (Deflagrations) and 13 (Detonation), we will incorporate chemistry and other considerations to see how states change within the control volume, but first we need to learn some chemistry.

In this chapter, we will learn stoichiometry, which is the idea that for a particular set of chemistry molecules react in certain proportions and produce other molecules in other proportions. When a fuel and oxidizing run stoichiometric, there can be advantages to this situation.

Other areas covered in this chapter are a more general definition of enthalpy, discussion of chemical equilibrium and kinetics, and a discussion of adiabatic temperature for a constant pressure and constant volume process. The discussion of adiabatic temperature will include complete combustion (reactions) and incomplete combustion.

11.2 Stoichiometry

Stoichiometry is a branch of chemistry that asks very practical questions such as the following: if I have 10 kg of gasoline, how many kilograms of air do I need to completely combust the fuel? It's also an area of chemistry that utilizes conservation of

Electronic Supplementary Material: The online version of this chapter (https://doi.org/10.1007/978-3-030-87387-5_11) contains supplementary material, which is available to authorized users.

H. C. Foust III, *Thermodynamics, Gas Dynamics, and Combustion*,
https://doi.org/10.1007/978-3-030-87387-5_11

mass and shows that molecules associated with particular reactions react in certain amounts.

The basic oxidation reaction for hydrocarbons of the form C_nH_{2n+2} is given as

$$C_xH_y + a(O_2 + 3.76N_2) \rightarrow xCO_2 + \frac{y}{2} H_2O + 3.76aN_2 \qquad (11.1)$$

where $a = x + \frac{y}{4}$

In order to completely combust the fuel, C_xH_y, the following ratio of fuel to oxidizer (air), $O_2 + 3.76N_2$, is required

$$(A/F)_{stoich} = \left(\frac{m_{air}}{m_{fuel}}\right)_{stoich} = \frac{4.76a}{1} \frac{MW_{air}}{MW_{fuel}} \qquad (11.2)$$

Combustions do not always run at these stoichiometric conditions, and as such, an equivalence ratio is defined as

$$\Phi = \frac{\left(\dfrac{A}{F}\right)_{Stoich}}{\left(\dfrac{A}{F}\right)} \qquad (11.3)$$

when $\phi < 1 \rightarrow$ fuel-lean and when $\phi > 1 \rightarrow$ fuel rich

It can be seen in Fig. 11.1 there are times to run fuel-lean and other times to run fuel-rich. Also, graphing different quantities of concern against the non-dimensional quantity Φ provides insight on how the chemistry affects performance (see Fig. 11.2) where in Fig. 11.2 complete combustion results in the highest adiabatic temperature.

Example 11.1 Stoichiometric example
Determine the equivalence ratio for a mixture of methane and air where the mass rate of methane is 0.5 kg/s and the mass rate of air is 16 kg/s.

$$a = x + \frac{y}{4} = 1 + 1 = 2$$

$$\left(\frac{A}{F}\right)_{stoich} = \frac{4.76(a)}{1} \frac{MW_{air}}{MW_{fuel}} = \frac{4.76(2)}{1} \frac{29}{16.032} = 17.2$$

$$\left(\frac{A}{F}\right) = \frac{\dot{m}_{air}}{\dot{m}_{fuel}} = 32$$

$$\Phi = \frac{\left(\dfrac{A}{F}\right)_{stoich}}{\left(\dfrac{A}{F}\right)} = \frac{17.2}{32} = 0.5$$

Fig. 11.1 *A/F* ratio versus power [12]

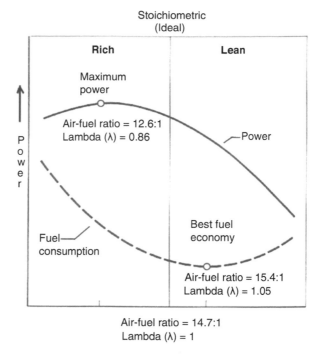

Fig. 11.2 Adiabatic flame temperature versus equivalence ratio (constant volume and air) [13]

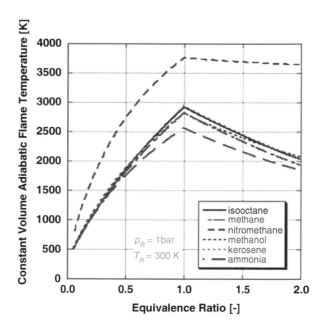

11.3 Enthalpy (Revisited)

To this point, we have ignored the fact that there are different forms of enthalpy and now we'll look at three main forms:

- Sensible enthalpy
- Latent enthalpy
- Enthalpy of formation

Each form is defined below and more details are given in separate sections.

Sensible enthalpy is the amount of energy released to either the system or the surrounds to change the temperature by a given amount and does not involve phase changes, nor changes in chemical composition.

Latent enthalpy is the amount of energy released to either the system or the surrounds to go from one phase into another phase. We dealt with this previously when we discussed Clausius-Clapeyron equation in Chap. 7.

Enthalpy of formation is the amount of energy contained within a molecule's molecular bonds and is typically given at some standard set of conditions (298 K and 1 atmosphere).

A more general energy balance to incorporate these three forms of enthalpy is given as Eq. 11.4.

$$\delta q - \delta w = \Delta h_{heat} + \Delta h_{latent} + h_{formation} + \Delta KE + \Delta PE \qquad (11.4)$$

How to compute the amount of energy released for each type of enthalpy is the subject of the sections below.

11.3.1 Sensible Enthalpy

The type of enthalpy we have considered to this point is either determined from thermodynamic tables or the definition of specific heat, constant pressure (c_p) and given as

$$c_p(T) = \left.\frac{\partial h}{\partial T}\right|_p \qquad (11.5)$$

where the gas is assumed ideal and c_p is solely a function of temperature

11.3.2 Latent Enthalpy

Latent enthalpy can be computed from the thermodynamic tables or the Clausius-Clapeyron equation, which is given as

$$\frac{dP_{sat}}{P_{sat}} = \frac{h_{fg}}{R} \frac{dT_{sat}}{T_{sat}^2} \tag{11.6}$$

where T_{sat} is the saturation temperature, P_{sat} is the saturation pressure, R is the gas constant, and h_{fg} is the amount of energy released to either the system or surroundings based on the phase change.

11.3.3 Enthalpy of Formation

Chemical reaction enthalpy is the type of enthalpy to be considered in this chapter.
   Consider the following reaction

$$C(s) + O_2 \rightarrow CO_2 + \delta q \tag{11.7}$$

Where the heat released is -393.522 kJ/mole and using the energy balance it is seen

$$\delta q = H_2 - H_1 = H_{products} - H_{reactants} = 1 * h_{f,298}(CO_2) - 1$$
$$* \left[h_{f,298}(C(s)) + h_{f,298}(O_2) \right] \tag{11.8}$$

where by definition, "the enthalpy of formation is zero for reference elements [9]" and as such $H_1 = 0$. It needs to be noted that the temperature was the same between the reactants and products and at STP; this is detonated as *standard enthalpy of formation* and has a special designation, h_f^0.
   How do we determine the heat release for a reaction when the reactants and products are at different temperatures?
   The standard enthalpy at temperature T [9] is given as

$$h_i(T) = h_0^f + \Delta h_{s,i}(T) = h_{o,i}^f + \left\{ h(T) - h_{ref} \right\}_i \tag{11.9}$$

   which states that the standard enthalpy at temperature T_{ref} is equal to the enthalpy of formation at reference conditions plus the change in latent energy between temperature T and T_{ref}.
   A generalization of the above can be stated as

$$\delta q = h_{products} - h_{reactants} \tag{11.10}$$

where

$$h_j = \sum_i \chi_i h_i \qquad (11.11)$$

And j is 1 for the reactions and 2 for the products.
An example of these ideas is the following.

Example 11.2
Given a gas stream containing CO, CO$_2$, and N$_2$ at 1 atmosphere and 1800 K where
the molar fraction for CO is 0.1 and CO$_2$ is 0.2 determine the enthalpy of the mixture in terms of moles and kilograms.
  Given

$$\chi_{CO} = 0.1, \chi_{CO_2} = 0.2 \rightarrow \chi_{N_2} = 0.7$$

And

Substance	MW [kg/kmole]	Temperature [K]	h(formation) [kJ/kmole]	h(T) − h(ref) [kJ/kmole]
CO	28.01	298	−110,541	
CO	28.01	1800		49,517
CO$_2$	44.01	298	−393,546	
CO$_2$	44.01	1800		79,399
N$_2$	28.01	298	0	
N$_2$	28.01	1800		48,971

$$h_{mix}^T = \sum_i \chi_i h_i = \chi_{CO} \left\{ h_{f,CO}^0 + \left(h(T) - h_f^0 \right)_{CO} \right\} + \chi_{CO_2} \left\{ h_{f,CO_2}^0 + \left(h(T) - h_f^0 \right)_{CO_2} \right\}$$
$$+ \chi_{N_2} \left\{ h_{f,N_2}^0 + \left(h(T) - h_f^0 \right)_{N_2} \right\}$$

$$h_{mix}^T = \sum_i \chi_i h_i = (0.1)\{-110,541 + 49,517\} + (0.2)\{-393,546 + 79,399\}$$
$$+ (0.7)\{0 + 48,971\} = -34,652 \, kJ/kmole$$

Also,

$$MW_{mix} = \sum_i \chi_i MW_i \qquad (11.12)$$

and

$$h_{mix}\left[\frac{kJ}{kg}\right] = \frac{h_{mix}\left[\dfrac{kJ}{kmole}\right]}{MW_{mix}} = \frac{-34,652 \, kj/kmole}{32.2 \, kg/kmole} = -1110.3 \, kJ/kg \quad (11.13)$$

Accurately knowing the thermodynamic properties before and after the shock for a given set of reactants and products

$$C_xH_y + a(O_2 + 3.76N_2) \rightarrow xCO_2 + \frac{y}{2}H_2 + 3.76aN_2 \qquad (11.14)$$

where $a = x + \dfrac{y}{4}$

can provide a better estimate of the detonation velocity (u_2) and temperature after the shock, T_2.

This analysis involves determining the following thermodynamic properties before and after the shock

- Specific heat, constant pressure (c_p)
- Gas constant (R_g)
- γ

where

$$C_{p,1} = \frac{\sum_{\text{Reactants}} \chi_i C(T)_{p,i}^{\text{Reactants}}}{\sum_{\text{Reactants}} \text{MW}_i} \qquad (11.15)$$

$$C_{p,2} = \frac{\sum_{\text{Products}} \chi_i C(T)_{p,i}^{\text{Products}}}{\sum_{\text{Products}} \text{MW}_i} \qquad (11.16)$$

And for the gas constants

$$R_1 = \frac{R_u}{\sum_{\text{Reactants}} \text{MW}_i} \qquad (11.17)$$

and

$$R_2 = \frac{R_u}{\sum_{\text{Products}} \text{MW}_i} \qquad (11.18)$$

where $\gamma_i = \dfrac{C_{p,i}}{R_i - C_{p,i}}$.

It is noted that $c_{p,2}$ and γ_2 are dependent on T_2 and this process usually involves an iteration methodology.

Knowing the value of T_2 and u_1 allows us to determine if the selected T_2 was accurate and if it is, then the thermodynamic properties calculated and u_1 are accurate. Turns [9] provides equations for T_2 and u_1 in terms of thermodynamic properties before and after the shock

$$T_2 = \frac{2\gamma_2^2}{\gamma_2 + 1}\left[\frac{C_{p,1}}{C_{p,2}}T_2 + \frac{q}{C_{p,2}}\right] = \frac{2\gamma_2^2}{\gamma_2 + 1}A \qquad (11.19)$$

and

$$u_1^2 = 2(\gamma_2 + 1)\gamma_2 R_2\left[\frac{C_{p,1}}{C_{p,2}}T_2 + \frac{q}{C_{p,2}}\right] = 2(\gamma_2 + 1)\gamma_2 R_2 A \qquad (11.20)$$

An example of these ideas is the following.

Example 11.3

A rotational detonation engine at the Air Force Research Laboratory on Wright Patterson Air Force Base is running a mixture of H_2/Air at 226 K and 145 kPa and it is running with an equivalence ratio of 1.0.

This information will be useful in the illustrated example in Chap. 13.

What are the values of the thermodynamic properties, T_2 and u_2?

Given

x		0
y		2
a		0.5
$T(1)$	[K]	226
$T(2)$	[K]	2975

The reactant thermodynamic properties are

Reactants	MW(i)	N(i)	$X(i) = N(i)/N(\text{total})$	$X(i) *$ MW(i)	$Y(i) = X(i) *$ MW(i)/MW(mix)	$c_p(I,T1)$	$X(i) *$ $c_p(I)$
C_xH_y	2	1	0.30	0.59	0.03	**28.61**	8.47
O_2	31.999	0.5	0.15	4.73	0.23	**28.70**	4.24
N_2	28.013	1.88	0.56	15.58	0.75	**28.87**	16.06
		3.38		20.91			28.77

The product thermodynamic properties are

Products	MW(i)	N(i)	$X(i) = N(i)/N(\text{total})$	$X(i) *$ MW(i)	$Y(i) = X(i) *$ MW(i)/MW(mix)	$c_p(I,T2)$	$X(i) *$ $c_p(I)$
CO_2	44.011	0	0.00	0.00	0.00	**0**	0.00
H_2O	18.016	1	0.35	6.26	0.25	**55.779**	19.37
N_2	28.013	1.88	0.65	18.29	0.75	**37.028**	24.17
		2.88		24.54			43.54

The reactant enthalpies are

Reactants	$Y(i)$	$h(i)$	$Y(i) * h(i)$
C_2H_2	0.03	**−1027.549603**	−29.08
O_2	0.23	**0**	0.00
N_2	0.75	**0**	0.00
			−29.08

The product enthalpies are

Products	$Y(i)$	$h(i)$	$Y(i) * h(i)$
CO_2	0.00	**0**	0.00
H_2O	0.25	**−14043.22547**	−3579.53
N_2	0.75	**0**	0.00
			−3579.53

And the computed thermodynamic properties are

$c_p(1)$	[kJ/kg-K]	1.38
$c_p(2)$	[kJ/kg-K]	1.77
Ru	[J/mole-K]	8.315
$R(1)$	[kJ/kg-K]	0.40
$R(2)$	[kJ/kg-K]	0.34
Gamma(1)		1.41
Gamma(2)		1.24

The results for T_2 and u_1 are

q [kJ/kg]	[kJ/kg]	3550.45
A	[K]	2176.593
$v(D)$ [m/s]	[m/s]	2019.03
$T(2)$	[K]	2974.43

Because $T_2(i = 1)$ and $T_2(i = 2)$ are so close, no further iterations are required.

11.4 Chemical Equilibrium

Much of this section comes from [6, 9, 10].

11.4.1   Gibb's Free Energy and Chemical Potential

Given our definition for first law

$$\delta q - \delta w = \Delta h + \Delta PE + \Delta KE \qquad (11.21)$$

And a definition for entropy

$$ds = \frac{\delta q}{T}, \; Tds = \delta q \qquad (11.22)$$

Where we assume no change in kinetic energy, no change in potential energy If we substitute Eq. 11.22 into Eq. 11.21 the result is

$$-\delta w = \Delta h - Tds \qquad (11.23)$$

And Gibb's free energy, which we'll show is a measure of the available energy of the system to do work is

$$\Delta g = \Delta h - Tds \qquad (11.24)$$

or

$$g = u + Pv - sT \qquad (11.25)$$

And taking the derivative of both sides

$$dg = du + Pdv + vdP - sdT - Tds \qquad (11.26)$$

where dg is defined for an equilibrium condition and so $\{P, T\}$ are fixed, which results in

$$dg = du + Pdv - Tds \qquad (11.27)$$

Internal energy (e) for an ideal gas can be determined through a knowledge of entropy and specific volume

$$e = f(s,v) \qquad (11.28)$$

But when different chemical species (molecules) are involved this function is extended

$$e = f(s,v,n_1,n_2 \cdots) \qquad (11.29)$$

And the total differential of both sides is

$$de = \left(\frac{\partial e}{\partial s}\right)_{v,n_1,n_2...} ds + \left(\frac{\partial e}{\partial v}\right)_{s,n_1,n_2...} dv + \left(\frac{\partial e}{\partial n_1}\right)_{s,v,n_2,n_3...} dn_1 + \left(\frac{\partial e}{\partial s}\right)_{s,,n_1,n_3...} dn_2 \cdots (11.30)$$

Or more generally

$$de = \left(\frac{\partial e}{\partial s}\right)_{v,n_1,n_2...} ds + \left(\frac{\partial e}{\partial v}\right)_{s,n_1,n_2...} dv + \sum_{1}^{c} \left(\frac{\partial e}{\partial n_i}\right)_{s,v,n_j} dn_i \qquad (11.31)$$

where the index n_j represents that the last partial derivative is not fixed for n_i. It was shown in Chap. 6 when we derived the Maxwell relationships that

$$\left(\frac{\partial e}{\partial s}\right)_v = T \text{ and } \left(\frac{\partial e}{\partial v}\right)_s = -P \qquad (11.32)$$

These two relationships can be generalized to

$$\left(\frac{\partial e}{\partial s}\right)_{v,n_j} = T \text{ and } \left(\frac{\partial e}{\partial v}\right)_{s,n_j} = -P \qquad (11.33)$$

Substitution of Eq. 11.33 into Eq. 11.31 results in

$$de = Tds - Pv + \sum_{i=1}^{c} \left(\frac{\partial e}{\partial n_i}\right)_{s,v,n_j} \qquad (11.34)$$

Where the last partial derivative is so important it has a name, chemical potential (μ_I) and is defined as

$$\mu_i = \left(\frac{\partial e}{\partial n_i}\right)_{s,v,n_j} \qquad (11.35)$$

Thus, Eq. 11.34 takes the form

$$de = Tds - Pdv + \sum_{i=1}^{c} \mu_i dn_i \qquad (11.36)$$

Substitution of Eq. 11.36 into our definition of dg (Eq. 11.27) results in

$$dg = Pdv - Tds + Tds - Pdv + \sum_{i=1}^{c} \mu_i dn_i \qquad (11.37)$$

or

$$\mu_i = \left(\frac{\partial G}{\partial n_i}\right)_{T,P,n_j} = \left(\frac{\partial [Ng]}{\partial N}\right)_{T,P} = g_i \tag{11.38}$$

A physical interpretation of the chemical potential is as follows. The first law for this system takes the form

$$\delta q - \delta w = du = Tds - Pdv + \sum_{i=1}^{c}\mu_i dn_i \tag{11.39}$$

and

$$\delta q = Tds \tag{11.40}$$

Substituting Eq. 11.40 into Eq. 11.39 results in

$$Tds - \delta w = Tds - Pdv + \sum_{i=1}^{c}\mu_i dn_i \tag{11.41}$$

or

$$-\delta w = -Pdv + \sum_{i=1}^{c}\mu_i dn_i \tag{11.42}$$

Equation 11.42 provides an equation for the available work of the system to include work due to chemical reactions or mass transfer [10].

We'll see that at chemical equilibrium

$$dG = 0 = \sum_{i=1}^{c}\mu_i dn_i \tag{11.43}$$

We now want to extend the model for chemical potential to non-standard (reference) conditions.

For an ideal gas

$$\mu_i = g_i\left(T,P_i\right) \tag{11.44}$$

where μ_i is the chemical potential for species "i" and g_i is the free energy for species "i" for a given temperature and pressure.

Additionally,

$$g_i = \mu_i = h_i - Ts_i \tag{11.45}$$

where

$$\mu_i = h_i - T\left[s_i^0 - R\ln\left(\frac{y_iP}{P_0}\right)\right] \tag{11.46}$$

and

$$\mu_i = h_i - Ts_i^0 + RT\ln\left(\frac{y_iP}{P_0}\right) = \mu_i^0 + RT\ln\left(\frac{y_iP}{P_0}\right) \tag{11.47}$$

11.4.2   Chemical Reactions

Given a chemical reaction

$$aA + bB \leftrightarrow cC + dD \tag{11.48}$$

or

$$cD + cD - aA - bB = 0 = \sum_{i=1}^{4}\nu_iI_i \tag{11.49}$$

where ν_i is the stoichiometric coefficient of the species I_i.

And the amount of moles of I_i at time t can be related to the amount of moles of I_i at time to by the equation

$$N_i(t) = N_i(0) + \nu_i\xi \tag{11.50}$$

where ξ, the molar extent of reaction, is defined as

$$\xi = \frac{N_i(t) - N_i(0)}{\nu_i} \tag{11.51}$$

Another common metric is λ, which is the % reaction complete and defined as

$$\lambda_i = \frac{N_i(t) - N_i(0)}{N_i(0)} \tag{11.52}$$

A relationship between ξ and λ is

$$\xi\nu_i = \lambda_iN_i(0) \tag{11.53}$$

Example 11.4

Consider the following gas phase reaction at 400 K and 2 atm where ideal gas behavior is assumed.

$$A + 2B \leftrightarrow C \qquad (11.54)$$

The mole fractions of the reactive species at equilibrium satisfy the relationship

$$\frac{y_c}{y_A y_B^2} = 1.1124 \qquad (11.55)$$

Starting with equal moles of A and B and no C where molar mass is represented either in terms of coefficients associated with the chemical reaction equation or using the notation $[X]$, determine y_A, y_B, y_C, λ, and $1 - \xi$ at chemical equilibrium.

	$[A]$	$[B]$	$[C]$	Total
Initial	1	1	0	2
Final	$1 - \xi$	$1 - 2\xi$	ξ	$2(1 - \zeta)$

where

$$y_A = \frac{1-\xi}{2(1-\xi)} = \frac{1}{2}, \; y_B = \frac{1-2\xi}{2(1-\xi)}, \; y_C = \frac{\xi}{2(1-\xi)} \qquad (11.56)$$

And Substituting Eq. 11.56 into Eq. 11.55 results in

$$\frac{\xi}{2(1-\xi)} = \frac{1.1124}{2}\left[\frac{1-2\xi}{2(1-\xi)}\right]^2 \qquad (11.57)$$

Equation 11.57 has solutions $\{0.16, 0.84\}$ but when $\xi = 0.84 \rightarrow y_i < 0$, therefore $\xi = 0.16$.

More details of the solution are given as Appendix 11.1 and the solutions are also given in the table below

	zeta	A	B	C	Total	$\lambda[A]$	$\lambda[B]$
Initial	0.00	1.00	1.00	0.00	2.00		
Final	0.16	0.84	0.69	0.16	1.69	15.6%	31.2%

11.4.3   Chemical Reactions and Gibb's Free Energy

Example 11.5

Consider the following ideal gas reaction at 1 bar [10].

$$aA + bB \leftrightarrow cAB \tag{11.58}$$

where $a = b = 1$ moles initially and

$$a = b = 1 - \xi \tag{11.59}$$

	a	b	c	Total
Initial	1	1	0	2
Final	$1 - \xi$	$1 - \xi$	2ξ	2

The total Gibb's free energy for the system is

$$G = \sum n_i g_i = \sum n_i \mu_i = n_A \mu_A + n_b \mu_B + n_{AB} \mu_{AB} \tag{11.60}$$

and

$$\mu_i = g_i^0 + RT \ln\left(\frac{P_i}{1\,\text{Bar}}\right) \tag{11.61}$$

Substitution of Eq. 11.61 into Eq. 11.60 results in

$$G = n_A g_A^0 + n_B g_B^0 + n_{AB} g_{AB}^0 + RT\left\{ n_A \ln\left(P_A\right) + n_B \ln\left(P_B\right) + n_{AB} \ln\left(P_A B\right) \right\} \tag{11.62}$$

and

$$P_i = y_i P \tag{11.63}$$

Substitution of Eq. 11.63 into Eq. 11.62 results in

$$G = n_A g_A^0 + n_B g_B^0 + n_{AB} g_{AB}^0$$
$$+ RT\left\{ n_A \ln\left[y_A(\xi)\right] + n_B \ln\left[y_B(\xi)\right] + n_{AB} \ln\left[y_{AB}(\xi)\right] \right\} + RT\left\{ n_A + n_B + n_{AB} \right\} \ln(P) \tag{11.64}$$

In terms of ξ Eq. 11.64 becomes

$$G = (1 - \xi)\left(g_A^0 + g_B^0\right) + 2\xi g_{AB}^0$$
$$+ RT\left\{ (1-\xi)\ln\left[y_A(\xi)\right] + (1-\xi)\ln\left[y_B(\xi)\right] + 2\xi \ln\left[y_{AB}(\xi)\right] \right\} + RT \ln(P) \tag{11.65}$$

Given a system where

$$T = 1000\,\text{K}, \ P = 1\,\text{Bar}, \ R = 8.3124 \frac{\text{J}}{\text{mole}} - \text{K}, \ g_A^0 = g_B^0 = 0, \text{ and } g_{AB}^0$$
$$= -9.5\,\text{kJ/mole}$$

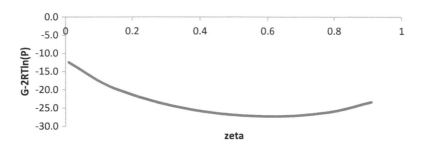

Fig. 11.3 $G - RT\ln(P)$ versus zeta

Equation 11.65 reduces to

$$G - RT\ln(P) = 2\xi g_{AB}^0$$
$$+ RT\left\{(1-\xi)\ln\left[y_A(\xi)\right] + (1-\xi)\ln\left[y_B(\xi)\right] + 2\xi\ln\left[y_{AB}(\xi)\right]\right\} \quad (11.66)$$

The solution for Eq. 11.66 is given in Fig. 11.3 and more details can be found in Appendix 11.1.

Figure 11.3 graphically illustrates that where $dG = 0$ is the chemical equilibrium of the system and ξ is roughly 0.6.

11.4.4 Fugacity

Fugacity is a derived thermodynamic property measured in terms of pressure and provides an experimental means of determining phase and chemical equilibrium. For a pure substance at a given T and P,

$$d\mu = v\,dP - s\,dT \quad (11.67)$$

And when T is constant

$$d\mu = v\,dP \quad (11.68)$$

Assuming ideal gas behavior

$$v = R\frac{T}{P} \quad (11.69)$$

Substitution of Eq. 11.69 into Eq. 11.68 results in

$$d\mu = \frac{RT}{P}\,dP = RT\,d\left\{\ln(P)\right\} \quad (11.70)$$

And knowing for a real gas $v \neq \dfrac{RT}{P}$
Defining fugacity as

$$d\mu = RTd\ln\{f\} \tag{11.71}$$

where as $P \to 0$ implies $\dfrac{f}{P} \to 1$, which is a correction for non-ideal behavior. Integrating Eq. 11.71 results in

$$\mu - \mu^0 = RT\ln\left(\frac{f}{f^0}\right) \tag{11.72}$$

For phase equilibrium between phases α and β,

$$\mu_i^{\alpha} = \mu_i^{\beta} \tag{11.73}$$

Substitution of Eq. 11.72 into Eq. 11.73 results in

$$\mu_i^{o,\alpha} + RT\ln\left(\frac{f_i^{\alpha}}{f_i^{\alpha,0}}\right) = \mu_i^{o,\beta} + RT\ln\left(\frac{f_i^{\beta}}{f_i^{\beta,0}}\right) \tag{11.74}$$

Another form of Eq. 11.74 is

$$\mu_i^{o,\alpha} - \mu_i^{o,\beta} = RT\ln\left(\frac{f_i^{\alpha,0}}{f_i^{\beta,0}}\right) + RT\ln\left(\frac{f_i^{\beta}}{f_i^{\alpha}}\right) \tag{11.75}$$

From Eq. 11.72

$$\mu_i^{o,\alpha} - \mu_i^{o,\beta} = RT\ln\left(\frac{f_i^{\alpha,0}}{f_i^{\beta,0}}\right) \tag{11.76}$$

Therefore,

$$RT\ln\left(\frac{f_i^{\beta}}{f_i^{\alpha}}\right) = 0 \to f_i^{\alpha} = f_i^{\beta} \tag{11.77}$$

Example 11.6
Determine the fugacity of liquid water at 30 °C and at the saturation pressures, 10 Bar and 100 Bar.
   At 20 °C, $P^{sat} = 0.0424$ bar. In the limit as $P \to 0$, $f \to P$, therefore $f^{sat} = P^{sat} = 0.0424$ bar.

From the definition of fugacity

$$d\mu = RTd\ln(f) = vdP \tag{11.78}$$

Integrating from $P(\text{sat})$ to P results in

$$RT\ln\left(\frac{f^L}{f^{\text{sat}}}\right) = v\left(P - P^{\text{sat}}\right) \rightarrow f^L = f^{\text{sat}}\exp\left[\frac{v\left(P - P^{\text{sat}}\right)}{RT}\right] \tag{11.79}$$

where specific volume is

$$v = v_l\left[\frac{\text{m}^3}{\text{kg}}\right]\left(18.02\frac{\text{kg}}{\text{kmole}}\right) = 0.01809\frac{\text{m}^3}{\text{kmole}} \tag{11.80}$$

Such that the fugacity at 10 Bars is

$$f^L = 0.0424 * \exp\left[\frac{0.01809(10 - 0.0424)*1e5}{8314.3*303.15}\right] = 0.0427\,\text{Bars} \tag{11.81}$$

And the fugacity at 100 Bars is

$$f^L = 0.0424 * \exp\left[\frac{0.01809(100 - 0.0424)*1e5}{8314.3*303.15}\right] = 0.0455\,\text{Bars} \tag{11.82}$$

11.4.5   Chemical Equilibrium Constant

Previously, we've seen at chemical equilibrium
$$dG_{T,P} = \Sigma v_i\mu_i d\xi \tag{11.83}$$

or

$$dG_{T,P} = 0 \rightarrow \left(\frac{\partial G}{\partial \xi}\right)_{T,P} = \Sigma v_i\mu_i = 0 \tag{11.84}$$

and

$$d\mu_i = RTd\ln(f_i) \tag{11.85}$$

And integrating Eq. 11.85 results in

$$\mu_i = g_i^0 + RT \ln\left(\frac{f_i}{f_i^0}\right)$$ (11.86)

Substituting of Eq. 11.86 into Eq. 11.84 results in

$$\Sigma v_i g_i^0 + RT \Sigma v_i \ln\left(\frac{f_i}{f_i^0}\right) = 0$$ (11.87)

And defining activity, a_i

$$a_i = \frac{f_i}{f_i^0}$$ (11.88)

Another form of Eq. 11.87 is

$$\ln \prod\left(\frac{f_i}{f_i^0}\right)^{v_i} = \frac{-\Sigma v_i g_i^0}{RT} = -\frac{\Delta g_{rxn}^0}{RT}$$ (11.89)

where Δg_{rxn}^0 is the standard Gibb's free energy change of reaction.
The left-hand side of Eq. 11.89 is the logarithm of the chemical equilibrium constant (K) such that

$$\ln(K) = -\frac{\Delta g_{rxn}^0}{RT}$$ (11.90)

Example 11.7
Calculate the equilibrium constant for the following reaction using the data at 298 K

$$H_2O(g) + CH_3OH(g) \leftrightarrow CO_2(g) + 3H_2(g)$$ (11.91)

where

	$H_2O(g)$	$CH_3OH(g)$	$CO_2(g)$	$H_2(g)$
$\Delta g_{rxn}^0 \left[\dfrac{kJ}{mole}\right]$	−228.57	−161.96	−394.36	0

and

$$\Delta g_{rxn}^0 = \Sigma v_i \Delta g_i^0 = -394.36 + 3(0) - (-228.57) - (-161.96) = -3.83 \, kJ/mole \quad (11.92)$$

Thus

$$K = \exp\left(-\frac{\Delta g^0_{rxn}}{RT}\right) = \exp\left[\frac{3830}{(8.314)(298.15)}\right] = 4.69 \qquad (11.93)$$

Example 11.8

Find the equilibrium concentration of $N_2O(g)$ due to the following chemical reaction that occurs at 25 °C and 1 atmosphere.

$$N_2O_4(g) \leftrightarrow 2NO_2(g) \qquad (11.94)$$

where the following reactions occur
Chemical Reaction 1 (R_1)

$$N_2O_4(g) \leftrightarrow N_2(g) + 2O_2(g), \ \Delta g^0_{rxn} = -23.14\frac{kcal}{mole} \qquad (11.95)$$

and
Chemical Reaction 2 (R_2)

$$0.5N_2(g) + O_2(g) \leftrightarrow NO_2(g), \Delta g^0_{rxn} = 12.24\,kcal/mole \qquad (11.96)$$

The overall reaction is a combination of R_1 and R_2 where

$$\text{Overall} = R_1 + 2R_2 \qquad (11.97)$$

Thus,

$$\Delta g^0_{rxn} = \Delta g^0_{rxn,\,R1} + \Delta g^0_{rxn,R2} = -23.41 + 2(12.24) = 1.11\frac{kcal}{mole}$$
$$= 4644\,\frac{J}{mole} = -RT\ln(K) \qquad (11.98)$$

and

$$K = \exp\left(-\frac{\Delta g^0_{rxn}}{RT}\right) = \exp\left[\frac{-4644}{(8.314)(298.15)}\right] = 0.1563 \qquad (11.99)$$

And the chemical equilibrium constant (K) in terms of activities (fugacity) is

$$K = \frac{a_{NO_2}^2}{a_{M2P4}} = \frac{\left(\dfrac{f_{NO_2}}{f_{NO_2}^0}\right)^2}{\left(\dfrac{f_{N_2O_4}}{f_{N_2O_4}^0}\right)} = \frac{\left(\dfrac{P_{NO_2}}{P_{NO_2}^0}\right)^2}{\left(\dfrac{P_{N_2O_4}}{P_{N_2O_4}^0}\right)} = \frac{\left(\dfrac{y_{NO_2} P}{1\,atm}\right)^2}{\left(\dfrac{y_{N_2O_4} P}{1\,atm}\right)} = \frac{y_{NO_2}^2}{y_{N_2O_4}} = 0.1563 \quad (11.100)$$

and

	$[N_2O_4]$	$[NO_2]$	Total
Initial	1	0	1
Final	$1 - \xi$	2ξ	$1 + \xi$

Substitution of the values for $y(NO_2)$ and $y(N_2O_4)$ from the table into Eq. 11.100 results in

$$\frac{\left(\dfrac{2\xi}{1+\xi}\right)^2}{\left(\dfrac{1-\xi}{1+\xi}\right)} = \frac{4\xi^2}{1-\xi^2} = K = 0.1563 \quad (11.101)$$

With solution $\xi = 0.194$ and a complete solution is

	$[N_2O_4]$	$[NO_2]$	Total	$y[N_2O_4]$	$y[NO_2]$
Initial	1	0	1		
Final	0.81	0.39	1.19	0.68	0.32

11.5   Chemical Kinetics

11.5.1   Reaction Fundamentals

Much of this section comes from [1].

Definitions

Heterogeneous Reactions involves more than one phase and likely occurs at the interfacial boundary

Homogenous Reactions involves a single phase

Reversible Reactions can proceed forward from reactants to products or backwards from products to reactants

Irreversible Reactions proceeds in only one direction

Equilibrium Chemistry composition of species when a reaction is complete

Rate Law

The chemical reaction

$$aA + bB \leftrightarrow cC + dD \tag{11.102}$$

Has the following reaction rate

$$\frac{dc_A}{dt} = -kc_A^{\alpha} c_B^{\beta} \tag{11.103}$$

Where $\{\alpha, \beta\}$ are empirical constants determined from regression analysis of experimental results.

But in this section, we'll assume the following form

$$\frac{dc_A}{dt} = -kc_A^{\alpha} \tag{11.104}$$

Zero Order

A zero-order reaction has the following form

$$\frac{dc}{dt} = -k \tag{11.105}$$

And solution

$$c = c_0 - kt \tag{11.106}$$

First Order

A first-order reaction has the following form

$$\frac{dc}{dt} = -kc \tag{11.107}$$

And solution

$$c = c_0 e^{kt} \tag{11.108}$$

Second Order

A second-order reaction has the following form

$$\frac{dc}{dt} = -kc^2 \tag{11.109}$$

And solution

$$\frac{1}{c} = \frac{1}{c_0} + kt \tag{11.110}$$

Higher Order Reactions

A second-order reaction has the following form

$$\frac{dc}{dt} = -kc^n \tag{11.111}$$

And solution

$$\frac{1}{c^{n-1}} = \frac{1}{c_0^{n-1}} + (n-1)kt \tag{11.112}$$

In order to determine "what is the proper reaction rate for a given chemical reaction?", reaction rate experiments must be conducted. From the data, graphs are made of

1. C versus Time
2. $Ln[c]$ versus Time
3. $1/C$ versus Time

Example 11.9

Given the following reaction rate data (Table 11.1)

Graph the three graphs listed above and based on the coefficient of correlation (R^2), the best fit is the first-order model (Figs. 11.4, 11.5, and 11.6).

11.5.2 Chemical Kinetic Complexity

The number of reactions involved with the combustion of hydrocarbons can be over 100 and associated with each reaction is a reaction rate equation developed from kinetics studies and this whole process involves a huge amount of effort. Many

Table 11.1 Concentration versus Time Data

Time [days]	0	1	3	5	10	15	20
C [mg/L]	12	10.7	9	7.1	4.6	2.5	1.8

Fig. 11.4 Zero-order model

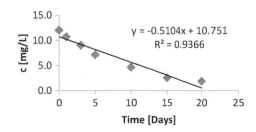

Fig. 11.5 First-order
model

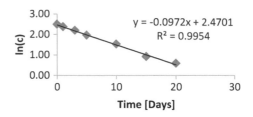

Fig. 11.6 Second-order
model

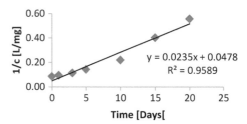

scientists [2, 4, 5, 7] have determined various ways to reduce the chemical kinetics
complexity and one approach is the work of Westbrook and Dryer [10] who have
determined a global reaction rate equation of the form

$$k_{av} = \dot{\omega} = AT^n \exp\left[-\frac{E_a}{RT}\right][\text{Fuel}]^a [\text{Oxidizer}]^b \qquad (11.113)$$

where n is set to zero and $\{A, a, b\}$ are determined through a regression method
that involves fitting the model for the laminar flame speed equation, which is given
below (see Chap. 12 for a full derivation)

$$S_L^2 = \frac{\lambda}{\rho C_p} \frac{\left(T_f - T_i\right)}{\left(T_i - T_0\right)} \frac{\dot{\omega}}{\rho} \qquad (11.114)$$

where λ is the heat convection, ρ is density, c_p is the specific heat, T_f is the final
temperature, T_0 is the initial temperature, T_i is the ignition temperature, and $\dot{\omega}$ is the
reaction rate.

The goal was to fit the laminar flame speed equation (Eq. 11.114) that includes
the global reaction rate (Eq. 11.113) to an experimental curve of laminar flame
speed versus equivalence ratio; an example is given below as Fig. 11.7. Westbrook
and Dryer did this analysis for 19 hydrocarbons [11].

The parameters $\{A, a, b\}$ are determined to ensure that the following laminar
flame speed characteristics match between the model and experimental results:

1. Peak S_L
2. ϕ_L', which is the lean flammability limit

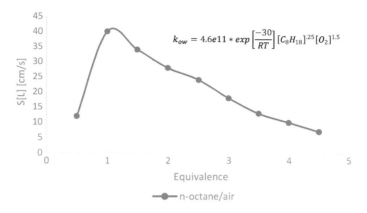

Fig. 11.7 Laminar flame speed versus equivalence [11]

3. ϕ_R', which is the rich flammability limit
4. General shape

Other efforts to reduce the chemical kinetic complexity include the work of Hautman et al. [2] and Sichel et al. [8]. Hautman et al. will be briefly discussed.

Hautman, Dryer, Schug, and Glassman [2] reduced the reactions associated with certain hydrocarbons (aliphatic hydrocarbons) to four reactions

$$C_n H_{2n+2} \rightarrow \frac{n}{2} C_2 H_2 + H_2 \tag{11.115}$$

$$C_2 H_2 \rightarrow 2CO + 2H_2 \tag{11.116}$$

$$CO + \frac{1}{2} O_2 \rightarrow CO_2 \tag{11.117}$$

$$H_2 + \frac{1}{2} O_2 \rightarrow H_2 O \tag{11.118}$$

With four reaction rate equations

$$\frac{d\left[C_n H_{2n+2}\right]}{dt} = -10^x \exp\left[-\frac{E}{RT}\right] \left[C_n H_{2n+2}\right]^a \left[O_2\right]^b \left[C_2 H_2\right]^c \tag{11.119}$$

$$\frac{d\left[C_2 H_2\right]}{dt} = -10^x \exp\left[-\frac{E}{RT}\right] \left[C_2 H_2\right]^a \left[O_2\right]^b \left[C_n H_{2n+2}\right]^c \tag{11.120}$$

Table 11.2 Coefficients associated with Eqs. 11.119, 11.120, 11.121, and 11.122

Reaction	x	E	a	b	c
C(n)H($2n$ + 2)	17.32	49,600	0.5	1.07	0.4
C(2)H(4)	14.7	50,000	0.9	1.18	−0.37
CO	13.52	41,000	0.85	1.42	−0.56
H(2)	14.6	40,000	1	0.25	0.5

$$\frac{d[CO]}{dt} = -10^x \exp\left[-\frac{E}{RT}\right][CO]^a [O_2]^b [H_2O]^c S \tag{11.121}$$

$$\frac{d[C_nH_{2n+2}]}{dt} = -10^x \exp\left[-\frac{E}{RT}\right][H_2]^a [O_2]^b [C_2H_2]^c \tag{11.122}$$

where $S = 7.93 \exp(-2.48\phi_0)$ and ϕ_0 is the initial equivalence ratio.

Where the coefficients are given as Table 11.2

Hautman et al.'s [2] work provides a means to observe the development of the various stages of hydrocarbon oxidation where Eqs. 11.119 and 11.122 provide reaction steps associated with the heat release.

Turns [8] has provided for hydrocarbons of the form C_nH_{2n+2} a three-step process for the combustion chemistry that is given as

1. The fuel molecule is attached by O and H atoms and breaks down, primarily forming alkenes and hydrogen. The hydrogen oxidizes to water, subject to available oxygen.
2. The unsaturated alkenes further oxidize to CO and H_2. Essentially, all of the H_2 is converted to water.
3. The CO burns out via the following reaction $CO + H_2O + \frac{1}{2}O_2 \rightarrow CO_2 + H_2O$. Nearly all of the heat released associated with the overall combustion process occurs in this step.

The following reactions summarize these three steps for methane [2]

Induction Phase

$$CH_4 + \frac{3}{2}O_2 \rightarrow 2H_2O + CO \tag{11.123}$$

and

Reaction Phase

$$H_2O + CO + \frac{1}{2}O_2 \rightarrow CO_2 + H_2O \tag{11.124}$$

Where these two equations represent two distinct phases, which are called the induction phase (Eq. 11.123) and the reaction phase (Eq. 11.124). More will be made of the induction phase (zone) and reaction phase (zone) in Chap. 13, Detonations.

Table 11.3 Reaction rate data

T	k(f)
[K]	[cm³/mole-s]
592	498
603.5	775
627	1810
651.5	4110
656	4740

It also needs to be noted with the incredible power of modern computers at our disposal the full combustion chemistry, which involves dozens of pathways and hundreds of reactions for a particular fuel and oxidizer, are now routinely solved, but deemed beyond the scope of this book.

11.5.3 Temperature Effects on Reaction Rates

Given the Arrhenius reaction rate

$$k = A \exp\left[-\frac{E_a}{R_u T}\right] \tag{11.125}$$

where k is the reaction rate, A is a constant dependent on several factors, E_a is the activation energy, R_u is the universal gas constant, and T is the temperature.

or

$$\ln(k) = \ln(A) - \frac{E_a}{R_u}\frac{1}{T} \tag{11.126}$$

A graph of experimental data in terms of $\ln(k)$ versus $\frac{1}{T}$ provides an estimate for $\{E_a, A\}$.

Example 11.10 Determination of Activation Energy

Given from [3] for the decomposition of nitrogen dioxide is the following experimental reaction rate data (Table 11.3)

A graph of $\ln(k)$ versus $1/T$ provides (see Fig. 11.8) an estimate for $\{E_a, R_u\}$ where the slope of the line is equal to $\frac{-E_a}{R_u}$ and $\ln(A)$ is equal to the y intercept. For this problem,

$$\frac{E_a}{R_u} = 13{,}671, \quad R_u = 1.9872\frac{\text{cal}}{\text{mole}\ \ \text{K}}, \quad E_a = 27.2\frac{\text{kcal}}{\text{mole}}$$

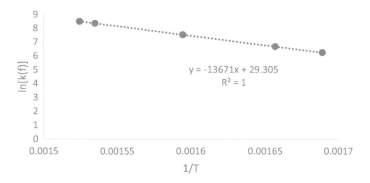

Fig. 11.8 ln(k) versus 1/T

and

$$\ln(A) = 29.305, A = 5.33e12\,\frac{\text{cm}^3}{\text{mole} - \text{s}}$$

11.6   Adiabatic Flame Temperature

Back in Chap. 2, we mentioned that the thermodynamic energy for a closed system is generally in terms of internal energy and for an open system it is generally in terms of enthalpy. This is relevant in this section because a closed system is a *constant volume* system and an open system is a *constant pressure* system.

11.6.1   Complete Reaction

The adiabatic flame temperature for a *constant pressure* system can be determined in the following fashion. The energy balance for this system is as follows

$$h_{\text{reactants}}(T_i, P) = h_{\text{products}}(T_{\text{adiabatic}}, P) + \delta q \qquad (11.127)$$

were for an adiabatic process $\delta q = 0$.

And the adiabatic flame temperature for a *constant volume* process can be determined in the following fashion. The energy balance for this system is as follows

$$u_{\text{reactants}}(T_i, P) = u_{\text{products}}(T_{\text{adiabatic}}, P) + \delta q \qquad (11.128)$$

where for an adiabatic process $\delta q = 0$.

By determining the enthalpy of formation and knowing the specific heat for a given substance and temperature, the value of T_{ad} can be determined. An example is as follows.

Example 11.11 Adiabatic Flame Temperature
Estimate the constant pressure adiabatic flame temperature for complete combustion of the following reaction, which is initially at 298 K and 1 atmosphere

$$CH_4 + 2(O_2 + 3.76N_2) \rightarrow CO_2 + 2H_2O + 7.52N_2 \qquad (11.129)$$

Where the following properties will be utilized (Table 11.4)
Going back to the energy balance for the system, Eq. 11.127, and plugging in the enthalpies and specific heats results in

$$
\begin{aligned}
(1)(-74,831) + 2(0) + 7.52(0) = &1\left[-393,546 + 56.21(T_{ad} - 298)\right] \\
+ 2\left[-241,845 + 43.87(T_{ad} - 298)\right] &+ 7.52\left[0 + 33.71(T_{ad} - 298)\right]
\end{aligned}
\qquad (11.130)
$$

Solving Eq. 11.130 for results in T_{ad} = 2381 K, which is about 100 K high due to not accounting for disassociation; disassociation is the fact that at higher temperatures water breaks down to simpler molecules.

The distinction between constant volume and constant pressure combustion will be explored further in the next chapter.

In the next section, we'll discuss incomplete combustion.

11.6.2   Incomplete Reactions

Much of this section comes from [10].

System I: Complete Combustion in Oxygen
One combustion reaction is the following

$$C(s) + O_2(g) \leftrightarrow CO_2(g) \qquad (11.131)$$

Table 11.4 Enthalpy of formation and specific heat associated with Eq.10.129

Chemical speciation	Enthalpy of formation [kJ/kmole]	Specific heat @ 1200 K [kJ/kmole-K]
CH_4	−74,831	
CO_2	−393.546	56.21
H_2O	−241,845	43.87
N_2	0	33.71
O_2	0	

System II: Incomplete Combustion in Oxygen
But from a knowledge of Gibb's free energy and the understanding that the G is minimized when

$$\frac{\partial G_{T,P}}{\partial \xi} = 0 \tag{11.132}$$

leads us to a more realistic reaction

$$C(s) + O_2 \leftrightarrow xCO_2 + yCO + zO_2 \tag{11.133}$$

where now the unknowns are $\{T_{ad}, x, y, z\}$ versus T_{ad} seen in the last section.

In order to solve for four unknowns, we need four equations and these equations are

Mass balance for carbon

$$x + y = 1 \tag{11.134}$$

Mass balance for oxygen

$$2x + y + 2z = 2 \tag{11.135}$$

Equilibrium constant for this system is

$$k_p = \exp\left[-\frac{\Delta G_{rxn}^0}{RT_{ad}}\right] = \frac{P_{CO} P_{O_2}^{0.5}}{P_{CO_2}} = \frac{x\sqrt{z}}{y}\sqrt{\frac{P}{x+y+z}} \tag{11.136}$$

Please note that the equilibrium constant would vary with temperature where this effect has been ignored in the example. Given that

$$K = e^{-\Delta G_{rxn}^0 / T}$$

is only good for T at standard conditions, i.e., 298.15 K, a better expression, though somewhat still an approximation, is the so-called short-cut van't Hoff equation

$$K = \frac{\Delta G_T^0}{RT} = \frac{\Delta G^o}{RT^o} + \frac{\Delta H^o}{R}\left(\frac{1}{T} - \frac{1}{T^o}\right) \tag{11.137}$$

And finally, the energy balance for the system is

$$-\left[xh_f^0(CO_2) + yh_f^0(CO)\right] = x\left[h(T_{ad}) - h_f^0\right]_{CO_2} + y\left[h(T_{ad}) - h_f^0\right]_{CO} + z\left[h(T_{ad}) - h_f^0\right]_{O_2} \tag{11.138}$$

Equations 11.134, 11.134, 11.135, 11.136, 11.137, and 11.138 are a set of non-linear algebraic equations and can be solved. The solution is found to be

$$x = 0.23 \text{ moles}, \quad y = 0.77 \text{ moles}, \quad z = 0.38 \text{ moles and } T_{ad} = 3537 \text{ K} \qquad (11.139)$$

System III: Incomplete Combustion in Air
If now the oxidizer is air, then the equations become.

$$x + y = 1 \qquad (11.140)$$

$$2x + y + 2z = 2 \qquad (11.141)$$

$$k_p = \exp\left[-\frac{\Delta G_{rxn}^0}{RT_{ad}}\right] = \frac{P_{CO} P_{O_2}^{0.5}}{P_{CO_2}} = \frac{x\sqrt{z}}{y}\sqrt{\frac{P}{x+y+z+79/21}} \qquad (11.142)$$

$$-\left[xh_f^0(CO_2) + yh_f^0(CO)\right] = x\left[h(T_{ad}) - h_f^0\right]_{CO_2} + y\left[h(T_{ad}) - h_f^0\right]_{CO}$$
$$+ z\left[h(T_{ad}) - h_f^0\right]_{O_2} + \frac{79}{21}\left[h(T_{ad}) - h_f^0\right]_{N_2} \qquad (11.143)$$

The above set has solution

$$x = 0.89 \text{ moles}, \quad y = 0.11 \text{ moles}, \quad z = 0.05 \text{ moles and } T_{ad} = 2312 \text{ K} \qquad (11.144)$$

Comparing the solutions for System II and III, it is seen that

1. Adiabatic flame temperature is significantly lowering when using air
2. The reaction is less complete when using pure oxygen

11.7 Problems

Problem 11.1 Stoichiometry Problem
Determine the equivalence ratio for a mixture of propane and air where the mass rate of propane is 0.5 kg/s and the mass rate of air is 16 kg/s.

Problem 11.2 Chemical Equilibrium Problem
Given the following reaction

$$A + 2B \leftrightarrow 2C$$

Where the mole fractions of reactive species at equilibrium satisfy the relationship

$$\frac{y_C^2}{y_A y_B^2} = 1.25$$

And initially $\{y_A, y_B, y_C\} = \{1, 1, 0\}$
Determine the final concentrations for $\{y_A, y_B, y_C\}$

Problem 11.3 Chemical kinetics problems
Given Eqn. 11.115 through Eqn. 11.121 and Table 11.2 and the following initial conditions

$$\dot{m} = 0.97 \frac{\text{moles}}{\text{s}}, \ [C_n H_{2n+2}] = 4.25e - 3 \frac{\text{mole}}{\text{cc}}, \ [O_2] = 1.7e - 6 \frac{\text{mole}}{\text{cc}}$$

Graph the change with time for the quantities

$$[C_n H_{2n+2}], [C_2 H_4], [CO], [H_2]$$

Problem 11.4 Determining q from a chemical reaction
Determine the heat release from Example 11.11 for complete combustion of the fuel/air mixture and a constant pressure process at 300 K and 1 atmosphere.

Problem 11.5 Adiabatic Temperature, Constant Pressure
Determine the adiabatic temperature, constant pressure for the complete combustion of stoichiometric mixture of propane and air at 400 K and 1 atmosphere.

Problem 11.6 Adiabatic Temperature, Constant Pressure
Determine the adiabatic temperature, constant pressure for the complete combustion of stoichiometric mixture of propane and air at 800 K and 1 atmosphere.

Problem 11.7 Adiabatic Temperature, Constant Volume
Determine the adiabatic temperature, constant volume for the complete combustion of stoichiometric mixture of propane and air at 400 K and 1 atmosphere.

Appendix 11.1: Chemical Equilibrium (website)

An Excel Spreadsheet was developed to illustrate a particular concept and is given on the companion website.

Appendix 11.2: Chemical Kinetics (website)

An Excel Spreadsheet was developed to illustrate a particular concept and is given on the companion website.

References

1. Chapra, S. C. (1997). *Surface water quality modeling*. McGraw Hill.
2. Hautman, D. J., Dryer, F. L., Schug, K. P., & Glassman, I. (1981). A multiple step overall kinetic mechanism for the oxidation of hydrocarbons. *Combustion Science and Technology, 25*, 219–235.
3. Kuo, K. K. (2005). *Principles of combustion* (2nd ed.). Wiley.
4. Montgomery, C. J., Cannon, S. M., Mawid, M. A., Sekar B. (2002). *Reduced chemical kinetic mechanisms for JP-8 combustion*. AIAA 2002-0336.
5. Montgomery, C. J., Cremer, M. A., Chen J.-Y., Westbrook, C. K., Maurice, L. Q. (2006). *Reduced chemical kinetic mechanism for hydrocarbon fuels*. Project report.
6. Nguyen, T. K. (2009). *Lecture notes for "chemical engineering thermodynamics II"*. Chemical and Materials Engineering.
7. Powers, J. M., & Paolucci, S. (2005). Accurate spatial resolution estimates for reactive supersonic flow with detailed chemistry. *AIAA Journal, 43*(5), 1088–1099.
8. Sichel, M., Tonello, N. A., Oran, E. S., Jones, D. A. (2002). *A two-step kinetics model for numerical simulation of explosions and detonations in H2-O2 mixtures*. Proceedings Royal Society of London, 458, pp. 49–82.
9. Turns, S. R. (2012). *An introduction to combustion (concepts and applications)* (3rd ed.). Mc-Graw Hill.
10. Wang, H. (2012). *Lecture notes for "combustion chemistry"*. CEFRC Summer School on Combustion, June 25–29.
11. Westbrook, C. K., & Dryer, F. L. (1981). Simplified reaction mechanisms for oxidation of hydrocarbon fuels in flames. *Combustion Science and Technology, 27*, 31–43.
12. https://en.wikipedia.org/wiki/Air%E2%80%93fuel_ratio. Accessed 10/17/2020.
13. https://en.wikipedia.org/wiki/Adiabatic_flame_temperature. Accessed 6/12/2021.

Chapter 12
Deflagration

12.1 Preview

In this chapter, two types of deflagration systems will be discussed:

1. Pre-mixed laminar flames
2. Non-premixed laminar flames

An example of a premixed laminar flame is a Bunsen burner.

Examples of a non-premixed laminar flame are a match burning or a pool of gasoline on fire. Non-premixed laminar flame systems are more numerous in nature than premixed laminar flames.

Turbulence will not be discussed but note an understanding of turbulence is predicated on an understanding of the laminar theory.

As we saw in Chap. 10, the Rankine-Hugoniot (RH) theory doesn't predict well the wave speed for a CJ deflagration and this is due to some fundamental transport processes being ignored where the RH theory ignores all transport processes such as mass transfer, momentum diffusion, and process of the heat transfer as expressed by a particular reaction rate equation. In the RH theory, heat is assumed instantaneously released and the path is ignored.

The first deflagration system to be explored is a laminar, premixed system where the object is to determine the laminar flame speed, which is essentially the speed of the deflagration wave. The second deflagration system to be explored is a laminar, non-premixed system (also known as diffusion flames) and in this portion of the chapter, it will be asked "what is the height of the flame, concentration of fuel and air at certain locations within the flame, and the mass rate of fuel and air consumptions?"

Electronic Supplementary Material: The online version of this chapter (https://doi. org/10.1007/978-3-030-87387-5_12) contains supplementary material, which is available to authorized users.

12.2 Qualitative Differences Between Various Combustion Phenomena

Imagine a very long shock tube (insulated from the environment) where air and a fuel have been premixed. Now imagine that the mixture is ignited, which is the process of increasing the temperature and pressure high enough until a combustion occurs and a wave travels down the tube toward the right. If the left end of the tube is open, the speed of the wave will be between 20 and 200 cm/s [5], which constitutes a constant pressure combustion and the energy balance of the system is

$$h_1 + \lambda q = h_2 \tag{12.1}$$

where "1" is the initial state and "2" is the state associated with the wave, h_i is the enthalpy associated with state "i", and λ is the percent reaction complete and discussed in Chap. 11.

Or for an ideal gas

$$c_p T_1 + \lambda q = c_p T_2^P \tag{12.2}$$

And when $\lambda = 100\%$, then T_2^P is the adiabatic flame temperature, constant pressure and this type of combustion is a deflagration.

Now imagine the same process, but the left end is capped and now the combustion is constant volume

$$e_1 + \lambda q = e_2 \tag{12.3}$$

where "1" is the initial state and "2" is the state associated with the wave, e_i is the internal energy associated with state "i", and λ is the percent reaction complete and discussed in Chap. 11.

Or for an ideal gas

$$c_v T_1 + \lambda q = c_v T_2^V \tag{12.4}$$

And when $\lambda = 100\%$, then T_2^V is the adiabatic flame temperature, constant volume and now as the wave moves to the right, the wave transitions through Mach 1 and becomes a shock wave with $Ma_1 > 1$ and the combustion is now a detonation.

Please note that given $c_p > c_v \rightarrow T_2^V > T_2^P$ for a given λq.

This chapter is solely about waves traveling much less than Mach 1 ($Ma_1 < 0.01$), deflagrations. When the deflagration involves a mixture of air and fuel that has been premixed, the rate of consumption is orders of magnitude greater than when it is not premixed. An example [6] is ethylene-oxygen (see Table 12.1).

Table 12.1 Comparison of fuel/O_2 consumption based on mixing [5]

Mixing	Fuel/oxidizer	Rate of consumption [Mole/cm³−s]	Relative rate
Premixed	Ethylene-oxygen	4.0	66,667
Non-premixed	Ethylene-oxygen	6E-5	1

12.3 Premixed Deflagration (Laminar Flames)

In this section, three different formulations for the laminar flame speed, which is defined as a premixed combustible gas and oxidizer with a wave traveling well below Mach 1, are given. The first is attributed to Mallard and Le Chatelier [10], the second is attributed to Spalding [10] and the third is a correlation from Metghalachi and Keck based on several hydrocarbon and air mixtures [7].

Given below in Fig. 12.1 is a premixed laminar flame where h represents the height of the premixed flame and H represents the height of the non-premixed (diffusion) flame.

The planar conservation principles, to be used in this section, and that include transport considerations derived in Chap. 8 are given below [6, 10].

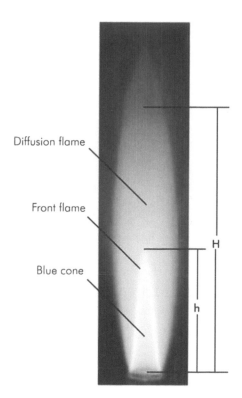

h= Height of blue cone   H= Height of diffusion flame

Fig. 12.1 Premixed laminar flame [1]

Conservation of Mass, Planar Coordinates

$$\frac{\partial(\rho v_x)}{\partial x} = 0 \tag{12.5}$$

or

$$\dot{m}'' = \rho v_x = \text{constant} \tag{12.6}$$

Conservation of Species, Planar Coordinates

$$\frac{d\dot{m}_i''}{dx} = \dot{m}_i''' \tag{12.7}$$

And in terms of Fick's law [10]

$$\frac{d}{dx}\left[\dot{m}''Y_i - \rho D\frac{dY_i}{dx}\right] = \dot{m}_i''' \tag{12.8}$$

And for the following chemical reaction

$$1\,\text{kg fuel} + S\,\text{kg Oxidizer} \rightarrow (S+1)\,\text{kg product} \tag{12.9}$$

Thus,

$$\dot{m}_{\text{Fuel}}''' = \frac{1}{S}\dot{m}_{\text{oxidizer}}''' = \frac{-1}{S+1}\dot{m}_{\text{Product}}''' \tag{12.10}$$

And Eq. 12.7 for each component is
Fuel

$$\dot{m}'''\frac{dY_{\text{Fuel}}}{dx} - \frac{d}{dx}\left[\rho D\frac{dY_{\text{Fuel}}}{dx}\right] = \dot{m}_{\text{Fuel}}''' \tag{12.11}$$

Oxidizer

$$\dot{m}'''\frac{dY_{\text{Oxidizer}}}{dx} - \frac{d}{dx}\left[\rho D\frac{dY_{\text{Oxidizer}}}{dx}\right] = Sm_{\text{Fuel}}''' \tag{12.12}$$

Product

$$\dot{m}'''\frac{dY_{\text{Product}}}{dx} - \frac{d}{dx}\left[\rho D\frac{dY_{\text{Product}}}{dx}\right] = -(1+S)\dot{m}'''_{\text{Fuel}} \tag{12.13}$$

Conservation of Energy, Planar Coordinates
Starting from Eq. 8.128,

$$\dot{m}''C_p\frac{dT}{dx} + \frac{d}{dx}\left(-\rho Dc_p\frac{dT}{dx}\right) = -\sum h^0_{f,i}\dot{m}'''_i \tag{12.14}$$

where

$$-\sum h^0_{f,i}\dot{m}'''_i$$
$$= -\left[h^0_{f,\text{Fuel}}\dot{m}'''_{\text{Fuel}} + h^0_{f,\text{Oxidizer}}S\dot{m}'''_{\text{Oxidizer}} - h^9_{f,\text{Product}}(S+1)\dot{m}'''_{\text{Fuel}}\right] = -\dot{m}'''_{\text{Fuel}}\Delta h_c \tag{12.15}$$

Substitution of Eq. 12.14 into Eq. 12.13 results in

$$\dot{m}'''\frac{dT}{dx} - \frac{1}{c_p}\frac{d\left[k\frac{dT}{dx}\right]}{dx} = -\frac{\dot{m}'''_{\text{Fuel}}\Delta h_c}{c_p} \tag{12.16}$$

where the intent of this section is to determine the laminar flame speed, S_L, which is related to conservation of mass through

$$\dot{m}'' = \rho_u S_L \tag{12.17}$$

where ρ_u is the density of unburned fuel/oxidizer.

12.3.1 Mallard and Le Chatelier's Laminar Flame Speed

In this section, a model is developed for the speed of unburned gas (at T_0), S_L, moving through the combustion zone (Zone II) normal to the wavefront where Zone II is where the reaction occurs (at T_i) and the burned gases are at T_f. This type of combustion is a deflagration and the Mach numbers are below 1 and the values of SL are typically below 1 m/s (Fig. 12.2).

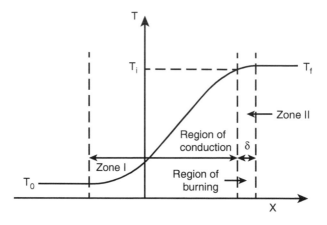

Fig. 12.2 Mallard and Le Chatelier's conceptual model

The approach taken is that of Mallard and Le Chatelier [10] where the energy balance into and out of the control volume for Zone II is

$$\dot{m}C_p\left(T_i - T_0\right) = \lambda \frac{\left(T_f - T_i\right)}{\delta} A \tag{12.18}$$

where λ is the heat convection, ρ is density, C_p is the specific heat, T_f is the final temperature, T_0 is the initial temperature, T_i is the ignition temperature, δ is the width of the reaction zone, and $\dot{m}$ is the mass rate into the control volume.
    And

$$\dot{m} = \rho A u = \rho S_L A \tag{12.19}$$

Substitution of Eq. 12.18 into 12.17 results in

$$\rho S_L C_p\left(T_i - T_0\right) = \lambda\left(T_f - T_i\right)/\delta \tag{12.20}$$

And solving for S_L results in

$$S_L = \frac{\lambda}{\rho C_p} \frac{\left(T_f - T_i\right)}{\left(T_i - T_o\right)} \frac{1}{\delta} \tag{12.21}$$

And a mass balance equation through Zone II is

$$\frac{\dot{m}}{A} = \rho u = \rho S_L = \omega \delta \tag{12.22}$$

Equation 12.21 is solved for δ and substituted into Eq. 12.20, which result in

$$S_L^2 = \frac{\lambda}{\rho C_p} \frac{\left(T_f - T_i\right)}{\left(T_i - T_o\right)} \frac{\dot{\omega}}{\rho} \qquad (12.23)$$

A challenge with using Eq. 12.22 is that T_i is not properly known and another form of Eq. 12.22 is determined in the next section that addresses this issue among others.

12.3.2   Spalding's Laminar Flame Speed Theory

Beginning with the energy equation, Eq. 12.15 [10],

$$\dot{m}''' \frac{dT}{dx} - \frac{1}{c_p} \frac{d\left[k\dfrac{dT}{dx}\right]}{dx} = -\frac{\dot{m}'''_{\text{Fuel}} \Delta h_c}{c_P} \qquad (12.24)$$

And integrating both sides of Eq. 12.23 where the boundary conditions far upstream are

$$T\left(-\infty\right) = T_u \text{ and } \frac{dT}{dx}\left(-\infty\right) = 0 \qquad (12.25)$$

And far downstream are

$$T\left(\infty\right) = T_b \text{ and } \frac{dT}{dx}\left(\infty\right) = 0 \qquad (12.26)$$

The results of integration Eq. 12.23 with the given boundary conditions results in

$$\dot{m}''\left(T_b - T_u\right) - \frac{k}{c_p} \frac{dT}{dx} \frac{\frac{dT}{dx}(\infty)}{\frac{dT}{dx}(-\infty)} = \frac{-\Delta h_c}{c_p} \int_{-\infty}^{\infty} \dot{m}'''_F dx \qquad (12.27)$$

or

$$\dot{m}''\left(T_b - T_u\right) = \frac{-\Delta h_c}{c_p} \int_{-\infty}^{\infty} \dot{m}'''_F dx \qquad (12.28)$$

Where the assumed temperature profile from T_u to T_b is linear and the differential appearing in the above integration and the endpoints of integration can be redefined in terms of temperatures by using

$$\frac{dT}{dx} = \frac{T_b - T_u}{\delta} \text{ or } dx = \frac{\delta}{T_b - T_u} dT \tag{12.29}$$

And Eq. 12.27 becomes

$$\dot{m}'' \left(T_b - T_u \right) = \frac{-\Delta h_c}{c_p} \frac{\delta}{T_b - T_u} \int_{T_u}^{T_b} \dot{m}_F''' \, dT \tag{12.30}$$

where

$$\frac{\delta}{T_b - T_u} \int_{T_u}^{T_b} \dot{m}_F''' \, dT = \delta \overline{\dot{m}}_F''' \tag{12.31}$$

which is the average reaction rate and

$$\dot{m}'' \left(T_b - T_u \right) = \frac{-\Delta h_c}{c_p} \delta \overline{\dot{m}}_F''' \tag{12.32}$$

Equation 12.31 can be thought of as an algebraic expression with two unknowns $\{\dot{m}'', \delta\}$ and we need another equation to solve for these two unknowns.
Resorting again to Eq. 12.29 but now the boundary conditions are
Far downstream

$$T(-\infty) = T_u \text{ and } \frac{dT}{dx}(-\infty) = 0 \tag{12.33}$$

And $x = \dfrac{\delta}{2}$

$$T\left(\frac{\delta}{2}\right) = \frac{T_b + T_u}{2} \text{ and } \frac{dT}{dx}\left(\frac{\delta}{2}\right) = \frac{T_b - T_u}{\delta} \tag{12.34}$$

Additionally, in the interval $x \epsilon \left(-\infty, \dfrac{\delta}{2} \right) \rightarrow \overline{\dot{m}}_F''' = 0$

Integration of Eq. 12.29 with the above boundary conditions results in

$$\dot{m}'' \left(T_{\delta/2} - T_u \right) - \frac{k}{c_p} \frac{dT}{dx} \frac{\frac{dT}{dx}(\delta/2)}{\frac{dT}{dx}(-\infty)} = 0 \tag{12.35}$$

or

$$\dot{m}'' \left(\frac{T_b + T_u}{2} - T_u \right) - \frac{k}{c_p} \left(\frac{T_b - T_u}{\delta} \right) = 0 \tag{12.36}$$

and

$$\dot{m}'' \left(\frac{T_b - T_u}{2} \right) - \frac{k}{c_p} \left(\frac{T_b - T_u}{\delta} \right) = 0$$

or

$$\frac{\dot{m}''}{2} - \frac{k}{c_p} \frac{1}{\delta} = 0 \tag{12.37}$$

The linear system to solve is

$$\dot{m}'' \left(T_b - T_u \right) + \frac{\Delta h_c}{c_p} \delta \overline{\dot{m}}_F''' = 0 \tag{12.38}$$

and

$$\frac{\dot{m}''}{2} - \frac{k}{c_p} \frac{1}{\delta} = 0 \tag{12.39}$$

Solving Eq. 12.39 for δ

$$\delta = \frac{k}{c_p} \frac{2}{\dot{m}''} \tag{12.40}$$

And substituting Eq. 12.40 into Eq. 12.38 results in

$$\dot{m}'' \left(T_b - T_u \right) = - \frac{\Delta h_c}{c_p} \overline{\dot{m}}_F''' \left[\frac{k}{c_p} \frac{2}{\dot{m}''} \right] \tag{12.41}$$

or

$$\left(\dot{m}'' \right)^2 \left(T_b - T_u \right) = - \frac{\Delta h_c}{C_p^2} \overline{\dot{m}}_F''' \left[2k \right] \tag{12.42}$$

And solving for $\dot{m}''$ results in

$$\dot{m}'' = \sqrt{2\frac{k}{c_p^2}\frac{(-\Delta h_c)}{(T_b - T_u)}\overline{\dot{m}}_F'''}$$

(12.43)

By definition the laminar flame speed and thermal diffusivity are

$$S_L = \frac{\dot{m}''}{\rho_u}$$

(12.44)

and

$$\alpha = \frac{k}{\rho_u c_p}$$

(12.45)

Further,

$$\Delta h_c = (S+1)c_p(T_b - T_u)$$

(12.46)

where S is the mass air to fuel ratio.
Substitution of Eqs. 11.44, 11.45, and 11.46 into Eq. 11.43 results in

$$S_L = \sqrt{-2\alpha(S+1)\frac{\overline{\dot{m}}_F'''}{\rho_u}}$$

(12.47)

and

$$\delta = \sqrt{\frac{-2\rho_u\alpha}{(S+1)\overline{\dot{m}}_F'''}}$$

(12.48)

or

$$\delta = \frac{2\alpha}{S_L}$$

(12.49)

Example 12.1 Laminar Flame Speed
Find the laminar flame speed for a stoichiometric mixture of propane and air at STP.
     In order to determine the laminar flame speed for a stoichiometric mixture of propane and air, several distinct steps are required to include:

1. Stoichiometric considerations
2. Reaction rate equation
3. Specific heat determined

4. Determination of adiabatic flame temperature
5. Determination of heat diffusivity
6. Laminar flame speed calculations

Each step is discussed below.

Stoichiometric Considerations

A screenshot of this worksheet within the "Laminar Flame Speed" worksheet is provided and essentially this worksheet determines the concentrations for fuel, oxidizer (air), and products.

It should be noted that

$$T_\infty = 300\,\text{K},$$

$$T_{50\%} = \frac{1}{2}\left(T_\infty + T_{\text{Burnt}}\right) = 1{,}280\,\text{K}$$

$$T_{75\%} = 1{,}770\,\text{K}$$

$$T_{\text{Burnt}} = T_{\text{ad}} = 2{,}270\,\text{K},$$

where T_∞ is the ambient temperature, $T_{50\%}$ is the temperature at the beginning of the reaction zone, $T_{75\%}$ is the temperature in the middle of the reaction zone, and T_{Burnt} is the temperature at the end of the reaction zone where it is assumed fuel and oxidizer have been completely consumed.

Also note the mass concentrations of fuel, air, and oxidizer are given at the ambient conditions and within the reaction zone.

C	[kg/kmole]	12.01
H	[kg/kmole]	1.01
O	[kg/kmole]	16.00
N	[kg/kmole]	14.01
x		3
y		8
a		5
m(air)	[kg]	15.7
m(fuel)	[kg]	1
MW(oxygen)	[kg/kmole]	29.00
MW(fuel)	[kg/kmole]	44.10
(A/F)stoich		15.7
(A/F)		15.7
Phi		1.00
Y(air)		0.94
Y(fuel)		0.06
Y(Ox)		0.22
Y(f,fuel)		0.030
Y(f,Ox)		0.110

Hydrocarbon Names

- Saturated hydrocarbons (alkanes) have general formula $C_n H_{2n-2}$

- CH_4 methane   • $C_6 H_{14}$ hexane
- $C_2 H_6$ ethane   • $C_7 H_{16}$ heptane
- $C_3 H_8$ propane   • $C_8 H_{18}$ octane
- $C_4 H_{10}$ butane   • $C_9 H_{20}$ nonane
- $C_5 H_{12}$ pentane   • $C_{10} H_{22}$ decane

Nonpolar bonds, only London dispersion forces

$$(A/F)_{stoic} = \left(\frac{m_{air}}{m_{fuel}}\right)_{stoic} = \frac{4.76a}{1}\frac{MW_{air}}{MW_{fuel}} \qquad a = x + \frac{y}{4}$$

$$C_x H_y + a(O_2 + 3.76N_2) \rightarrow xCO_2 + \frac{y}{2}H_2O + 3.76aN_2 \qquad \Phi = \frac{(A/F)_{Stoich}}{(A/F)}$$

Reaction Rate Equation

This worksheet within the spreadsheet "Laminar Flame Speed" determines the parameters associated with a global reaction rate equation, which was discussed in the previous chapter.

A global reaction rate equation engineering purposes and derived for deflagration was given in Chap. 11 as Eq. 11.113

$$k_{av} = \dot{\omega} = AT^x \exp\left[-\frac{E_a}{RT}\right][\text{Fuel}]^m [\text{Oxidizer}]^n$$

where x is set to zero and $\{A, a, b\}$ are determined through a regression method that involves fitting the model for the laminar flame speed equation (the parameters particular to a given fuel are given in the table below) and for a fuel of propane, the coefficients are $\{8.60e11, 15098, 0.1, 1.65\}$.

Fuel	x	y	A	E/R(u)	m	n
CH4	1	4	1.30E+08	24,358	-0.3	1.3
CH4	1	4	8.30E+05	15,098	-0.3	1.3
C2H6	2	6	1.10E+12	15,098	0.1	1.65
C3H8	3	8	8.60E+11	15,098	0.1	1.65
C4H10	4	10	7.40E+11	15,098	0.15	1.6
C5H12	5	12	6.40E+11	15,098	0.25	1.5
C6H14	6	14	5.70E+11	15,098	0.25	1.5
C7H16	7	16	5.10E+11	15,098	0.25	0.15
C8H18	8	18	4.60E+11	15,098	0.25	0.15
C8H18	8	18	7.20E+12	15,098	0.25	0.15
C9H20	9	20	4.20E+11	20,131	0.25	0.15
C10H22	10	22	3.80E+11	15,098	0.25	0.15
CH3OH			3.20E+12	15,098	0.25	1.5
C2H5OH			1.50E+12	15,098	0.15	1.6
C6H6	6	6	2.00E+11	15,098	-0.1	1.85
C7H8	7	8	1.60E+11	15,098	-0.1	1.85
C2H4	2	4	2.00E+12	15,098	0.1	1.65
C3H6	3	6	4.20E+11	15,098	-0.1	1.85
C2H2	2	2	6.50E+12	15,098	0.5	1.25

Specific Heat Determination

A cubic relationship for specific heat constant pressure for air is utilized to determine the specific heat constant pressure at T equal to 1280 K (see below).

Substance	Beta(0)	Beta(1)	Beta(2)	Beta(3)
Air	1.05	−0.365	0.85	−0.39
T	[K]	1280		
Theta	[K]	1.28		
C_p	[kJ/kg-K]	1.16		

Adiabatic Flame Temperature

The adiabatic flame temperature constant pressure for several fuels is provided below to include propane [T(ad) = 2267 K].

Formula	Fuel	x	y	T(ad)
				[K]
CH4	Methane	1	4	2226
C2H2	Acetylene	2	2	2539
C2H4	Ethene	2	4	2369
C2H6	Ethane	2	6	2259
C3H6	Propene	3	6	2334
C3H8	Propane	3	8	2267
C4H8	1-Butene	4	8	2322
C4H10	n-Butane	4	10	2270
C5H10	1-Pentene	5	10	2314
C5H12	n-Pentene	5	12	2272
C6H6	Benzene	6	6	2342
C6H12	1-Hexene	6	12	2308
C6H14	n-Hexene	6	14	2273

Thermal Properties

This worksheet within the spreadsheet "Laminar Flame Speed" determines the thermal diffusivity based on the following equation evaluated at two temperatures (300 K and 1280 K).

$$\alpha = \frac{k(1280)}{\rho(300)c_p(1280)}$$

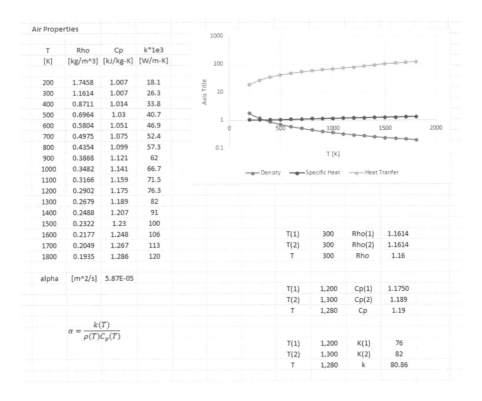

Air Properties

T	Rho	Cp	k*1e3
[K]	[kg/m^3]	[kJ/kg-K]	[W/m-K]
200	1.7458	1.007	18.1
300	1.1614	1.007	26.3
400	0.8711	1.014	33.8
500	0.6964	1.03	40.7
600	0.5804	1.051	46.9
700	0.4975	1.075	52.4
800	0.4354	1.099	57.3
900	0.3868	1.121	62
1000	0.3482	1.141	66.7
1100	0.3166	1.159	71.5
1200	0.2902	1.175	76.3
1300	0.2679	1.189	82
1400	0.2488	1.207	91
1500	0.2322	1.23	100
1600	0.2177	1.248	106
1700	0.2049	1.267	113
1800	0.1935	1.286	120

alpha	[m^2/s]	5.87E-05

$$\alpha = \frac{k(T)}{\rho(T)C_p(T)}$$

T(1)	300	Rho(1)	1.1614
T(2)	300	Rho(2)	1.1614
T	300	Rho	1.16

T(1)	1,200	Cp(1)	1.1750
T(2)	1,300	Cp(2)	1.189
T	1,280	Cp	1.19

T(1)	1,200	K(1)	76
T(2)	1,300	K(2)	82
T	1,280	k	80.86

Laminar Flame Speed Calculations

In this worksheet, the laminar flame speed is calculated from the following equation

$$S_L^2 = -2\alpha\left(A/_F + 1\right)\frac{\dot{m}_{\text{Fuel}}}{\rho_u}$$

The real challenge in this worksheet is to get the mass rate of fuel in the correct units. The following process was utilized to determine $\dot{m}_{\text{Fuel}}$

A	[gmole/cm³-s]	8.60E+11	Original units
A	[kmole/m³-s]	4.84E+09	New units
E/R(u)		15,098	
m		0.1	
n		1.65	
[Fuel]		0.00067901	Mass concentration
[Ox]		0.00342262	Mass concentration
k(G)		9.55E+05	
d(omega)/dt	[kmole/m³-s]	−2.43E+00	Units from reaction equation
dm(omega)/dt	[kg/m³-s]	−107.2	Appropriate units

The complete worksheet is given below where the solution is $S(L) = 43$ cm/s.

		Value	Comments
Rho(0)	[kg/m^3]	1.20	
Rho(75%)	[kg/m^3]	0.20	
Rho(50%)	[kg/m^3]	1.16	
C(p,50%)	[J/kg-K]	1190.00	
Lambda(50%)		0.0809	
Alpha	[m^2/s]	5.86E-05	
A	[gmole/cm^3-s]	8.60E+11	Original Units
A	[kmole/m^3-s]	4.84E+09	New Units
E/R(u)		15,098	
m		0.1	
n		1.65	
[Fuel]		0.00067901	Mass Concentration
[Ox]		0.00342262	Mass Concentration
k(G)		9.55E+05	
d(omega)/dt	[kmole/m^3-s]	-2.43E+00	Units from Reaction Equation
dm(omega)/dt	[kg/m^3-s]	-107.2	Appropriate Units
S(L)^2		0.17	
S(L)	[m/s]	0.42	
S(L)	[cm/s]	42	
delta(L)	[m]	2.81E-04	

12.3.3   *Metghalachi and Keck's Correlations for Laminar Flame Speed*

Metghalchi and Keck [7] have developed accurate correlations that allow us to determine the laminar flame speed for several fuels where the equations are given below and the fuel-specific parameters are given as Table 12.2.

Table 12.2   Parameters associated with a particular fuel

Fuel	Phi(M)	B(M) [cm/s]	B(2) [cm/s]
Methanol	1.11	36.92	−140.51
Propane	1.08	34.22	−138.65
Isooctane	1.13	26.32	−84.72
RMFD-303	1.13	27.58	−78.34

$$S_L = S_L^{\text{ref}} \left(\frac{T_u}{T_u^{\text{ref}}} \right)^{\gamma} \left(\frac{P}{P^{\text{ref}}} \right)^{\beta} \left(1 - 2.1 Y_{\text{dil}} \right) \tag{12.50}$$

And S_L^{ref} has the relationship

$$S_L^{\text{ref}} = B_M + B_2 \left(\phi - \phi_M \right)^2 \tag{12.51}$$

where

$$T_u^{\text{ref}} = 298 \text{ K}, \ P^{\text{ref}} = 1 \text{ atm} \tag{12.52}$$

And β has the relationship

$$\beta = -0.16 + 0.22 \left(\phi - 1 \right) \tag{12.53}$$

And γ has the relationship

$$\gamma = 2.18 - 0.8 \left(\phi - 1 \right) \tag{12.54}$$

Utilizing the correlations given above and assuming the substance is RMFD-303, a sensitivity analysis was performed to show the effects of increased temperature, increased pressure, and increased dilution had on $S(L)$. The results are given as Figs. 12.3, 12.4, and 12.5 where

$$T^* = \frac{T}{T_u^{\text{ref}}}, P^* = \frac{P}{P^{\text{ref}}} \tag{12.55}$$

It can be seen that increasing T^* has a drastic positive effect on $S(L)$ and that increasing P^* or $Y(\text{dil})$ has a slight negative effect on $S(L)$.

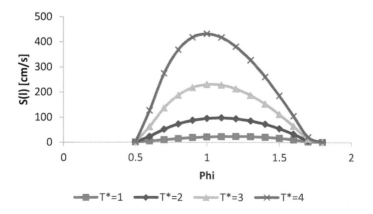

Fig. 12.3 $S(L)$ versus Phi for increasing T^*

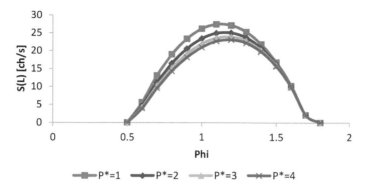

Fig. 12.4 $S(L)$ versus Phi for increasing P^*

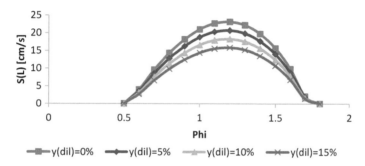

Fig. 12.5 $S(L)$ versus Phi for increasing dilution

12.4 Non-premixed Deflagration (Diffusion Flames)

In this section, the first attempt to model the height of a diffusion flame (non-premixed flame) was by Burke and Schumann [2], which has been shown to work well for circular burners. For burners of other geometries, buoyant forces become more important and Burke and Schumann [2] ignored these considerations. Roper [8, 9] addresses buoyant forces and the temperature dependence of diffusion.

The axi-symmetric conservation principles, to be used in this section, and that include transport considerations presented in Chap. 8 are given below.

Conservation of Mass, Axi-symmetric Coordinates

$$\frac{1}{r}\frac{\partial\left(r\rho v_r\right)}{\partial r}+\frac{\partial\left(\rho v_x\right)}{\partial x}=0 \tag{12.56}$$

Conservation of Axial Momentum, Axi-symmetric Coordinates

$$\frac{1}{r}\frac{\partial\left(r\rho v_x v_x\right)}{\partial x}+\frac{1}{r}\frac{\partial\left(\rho v_x v_r\right)}{\partial r}-\frac{1}{r}\frac{\partial\left(r\mu\dfrac{\partial v_x}{\partial r}\right)}{\partial r}=\left(\rho_\infty-\rho\right)g \qquad (12.57)$$

Conservation of Species, Axi-symmetric Coordinates

$$\frac{1}{r}\frac{\partial\left(r\rho v_r Y_i\right)}{\partial x}+\frac{\partial\left(r\rho v_x Y_i\right)}{\partial r}-\frac{1}{r}\frac{\partial}{\partial r}\left(r\rho D\frac{\partial Y_i}{\partial r}\right)=\dot{m}_i''' \qquad (12.58)$$

where

$$Y_{\mathrm{Pr}}=1-Y_F-Y_{\mathrm{ox}} \qquad (12.59)$$

12.4.1   Reacting, Constant Density Laminar Jet Flow (Burke and Schumann)

In Burke and Schumann [2], the coordinate system is axi-symmetric and two concentric cylinders are utilized (see Figs. 12.6 and 12.7) where the inner cylinder represents the fuel input and the outer cylinder represents the air (oxidizer); please

Fig. 12.6 Burke and Schumann, conceptual model [3]

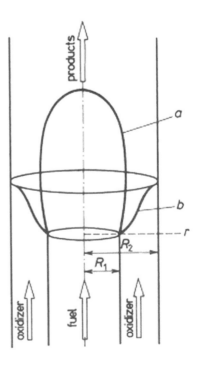

Fig. 12.7 Flame front [4]

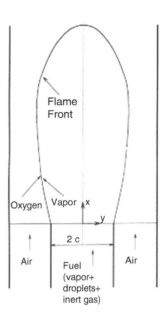

note the radius of each cylinder is carefully selected to insure the velocities of air and fuel are equal. Also shown in Fig. 12.6 is two flame configurations – "a" and "b" where a represents a situation where there is an over-abundance of air for combustion (over-ventilated) and b represents a situation where there is an under-abundance of air for combustion (under-ventilated). We see for over-ventilation the flame expands to a width equal to the inner cylinder and for an under-ventilated flame it expands to the outer cylinder.

One of the assumptions to be made for a Burke and Schumann flame is that the reaction is infinitely fast and that a "flame front" exists (see Fig. 12.7) where the equivalence ratio is 1 and, theoretically, inside the flame front the composition is fuel and outside the flame front the composition is air.

Consider two concentric cylinders where the outer cylinder (oxidizer cylinder) has radius b and the inner cylinder (fuel) has radius a. The cylinders end at $z = 0$ and this is the zone where mixing begins. Let the initial mass fraction of fuel be Y_{F0} and the initial mass fraction of oxygen be Y_{O0}. The following assumptions are made [2, 10]:

1. The average velocity is parallel to the z-axis of the ducts
2. The mass flux in the axial direction is constant
3. Axial diffusion is negligible when compared to transverse/radial diffusion
4. The flame is infinitely fast and occurs in a thin reaction sheet, Flame Front, where the equivalence ratio is one
5. Effects of gravity (buoyancy) are ignored

Consider a very simple chemical reaction

$$\text{Fuel} + sO_2 \rightarrow (1+s)\text{Products} + q \tag{12.60}$$

where s is the mass of oxygen and q is the heat released per unit mass of fuel consumed.

The following non-dimensional quantities are defined

$$y_F = \frac{Y_F}{Y_{F0}}, y_O = \frac{Y_O}{Y_{O0}}, S = \frac{sY_{F0}}{Y_{O0}} \tag{12.61}$$

The conservation of mass for the fuel and oxidizer are given as

$$\frac{\rho D}{r} \frac{\partial}{\partial r}\left(r\frac{\partial Y_F}{\partial r}\right) - \rho v \frac{\partial Y_F}{\partial z} = \frac{\omega}{Y_{F0}} \tag{12.62}$$

and

$$\frac{\rho D}{r} \frac{\partial}{\partial r}\left(r\frac{\partial Y_O}{\partial r}\right) - \rho v \frac{\partial Y_O}{\partial z} = S\frac{\omega}{Y_{F0}} \tag{12.63}$$

where ω is the number of moles of fuel burned per unit volume per time and, as previously stated, $\rho D = \text{constant}$.

The necessary boundary conditions are
Within the inner cylinder,

$$\text{at } z = 0, 0 < r < a, y_F = 1, y_O = 0 \tag{12.64}$$

Within the outer cylinder,

$$\text{at } z = 0, a < r < b, y_F = 0, y_O = 1 \tag{12.65}$$

and

$$\text{at } r = b, 0 < z < \infty, \frac{\partial Y_F}{\partial r} = 0, \frac{\partial Y_O}{\partial r} = 0 \tag{12.66}$$

Equations 12.62 and 12.63 can be combined by using a linear combination of the $\{y_F, y_O\}$ where the new variable, mixture fraction is

$$Z = \frac{Sy_F - y_O + 1}{S+1} \tag{12.67}$$

$$\text{for } 0 < r < a \rightarrow Z = 1 \text{ and } a < r < b \rightarrow Z = 0$$

Thus, assuming no fuel leaks outside of the flame front, nor any oxygen leaks inside of the flame front. Please note S is the molar stoichiometric oxidizer to fuel ratio.

Equations 12.62 and 12.63 where Z has been substituted reduces to

$$\frac{1}{r}\frac{\partial}{\partial r}\left(r\frac{\partial Z}{\partial r}\right) - \frac{\rho v}{\rho D}\left(\frac{\partial Z}{\partial z}\right) = 0 \tag{12.68}$$

Please note we have reduced two in-homogenous differential equations to one homogenous differential equation, which reduces the complexity of the solution considerably.

Going further, the following non-dimensional variables are introduced

$$\xi = \frac{r}{b}, \eta = \frac{\rho D}{\rho v}\frac{z}{b^2}, c = \frac{a}{b} \tag{12.69}$$

Substitutions of the equations given as Eq. 12.69 into Eq. 12.68 result in

$$\frac{1}{\xi}\frac{\partial}{\partial \xi}\left(\xi\frac{\partial Z}{\partial \xi}\right) - \frac{\partial Z}{\partial \eta} = 0 \tag{12.70}$$

With the following non-dimensional boundary conditions

$$\text{at } \eta = 0 \text{ and } 0 < \xi < c \rightarrow Z = 1 \tag{12.71}$$

$$\text{at } \eta = 0 \text{ and } c < \xi < 1 \rightarrow Z = 0 \tag{12.72}$$

and

$$\text{at } \xi = 1 \text{ and } 0 < \eta < \infty, \rightarrow \frac{\partial Z}{\partial \xi} = 0 \tag{12.73}$$

Equation 12.70 with boundary conditions 12.71, 12.72, and 12.73 has solution [6]

$$Z(\xi, \eta) = c^2 + 2c\sum_{n=1}^{\infty}\frac{1}{\lambda_n}\frac{J_1(c\lambda_n)}{J_0^2(\lambda_n)}J_0(c\lambda_n)e^{-\lambda_n^2\eta} \tag{12.74}$$

where J_0 and J_1 are the Bessel function of the first type and λ_n is the nth root of $J_1(\lambda) = 0$.

An approximate solution for Eq. 12.74 can be determined by noting that $\lambda_1 = 3.83$ and that only one term of the infinite series will be necessary because

$$e^{-\lambda_n^2} \text{ for } n > 1 \rightarrow e^{-\lambda_n^2} \approx 0 \tag{12.75}$$

where $\lambda_{n+1} > \lambda_n$ for $n = 1, 2, 3\ldots$

Approximate Solution for Burke Schumann Flames [11]

It has been assumed that the fuel and oxygen only co-exist into a thin shell known as the flame front and so

$$\text{outside the flame front} \rightarrow y_F y_O = 0 \tag{12.76}$$

and

$$\text{inside the flame front} \rightarrow S y_F = y_o \tag{12.77}$$

And thus Z inside the flame front takes the value

$$Z_{stoich} = Z_s = \frac{1}{S+1} \tag{12.78}$$

And substituting Eq. 12.78 into Eq. 12.74 results in

$$\frac{1}{S+1} = c^2 + 2c \sum_{n=1}^{\infty} \frac{1}{\lambda_n} \frac{J_1(c\lambda_n)}{J_0^2(\lambda_n)} J_0(c\lambda_n) e^{-\lambda_n^2 \eta} \tag{12.79}$$

where

$$S \gg 1 \rightarrow \text{under} - \text{ventilated case} \tag{12.80}$$

and

$$S \ll 1 \rightarrow \text{over} - \text{ventilated case} \tag{12.81}$$

And utilizing one term from Eq. 12.79 and setting $\xi = 0$ (over-ventilated case)

$$\eta_{over} = \frac{1}{\lambda_1^2} \ln \left[\frac{2c J_1(c\lambda_1)}{(Z_2 - c^2) \lambda_1 J_0(\lambda_1)} \right] \tag{12.82}$$

And utilizing one term from Eq. 12.79 and setting $\xi = 1$ (under-ventilated case)

$$\eta_{under} = \frac{1}{\lambda_1^2} \ln \left[\frac{2c J_1(c\lambda_1)}{(Z_2 - c^2) \lambda_1 J_0^2(\lambda_1)} \right] \tag{12.83}$$

Example 12.2 Burke Schumann Flame

What's the flame height for a circular burner with inner diameter of 10 cm, outer diameter of 20 cm, a fuel with axial velocity of 0.05 m/s, density of 1.2 kg/m^3, and diffusion of 1e-4 m^2/s?

The solution is given in the tables below from which we see the system is over-ventilated and the flame height is 0.48 m.

Variable	Units	Value	Comments
a	[m]	0.1	Inner diameter
b	[m]	0.2	Outer diameter
D	[m^2/s]	1.00E-04	Mass diffusion
v	[m/s]	0.05	Velocity
ρ	[kg/m^3]	1.2	Density
r	[m]	0.05	Distance from center line
z	[m]	0.1	Distance above release
S		0.1	s*Y(F,0)/Y(O,0)

ξ		0.25	
η		0.005	
c		0.5	
S		0.1	
$Z(s)$		0.91	
Etta'		0	
Lambda(1)		3.83	
Ventilation		Over ventilated	
J1(c*Lamba1)		0.58	
J0(Lambda1)		−0.403	
Etta		#NUM!	Under-ventilated
Etta		0.02	Over-ventilated
Flame height	[m]	#NUM!	Under-ventilated
Flame height	[m]	0.48	Over-ventilated

12.4.2 Reacting, Buoyant Laminar Jet Flow (Roper)

For the Burke and Schumann flame and a circular burner, assuming no buoyancy and constant diffusion are assumptions that the errors tend to cancel each other out, but for other burner geometries this is no longer the case. Besides, it is intuitively obvious that the diffusion will change with location because of temperature as will the density.

Roper provides a conservation principle given as

$$\frac{\partial C}{\partial t} = \frac{D}{x_f^2}\frac{\partial^2 C}{\partial \eta^2} + \frac{D}{y_f^2}\frac{\partial^2 C}{\partial \xi^2} \qquad (12.84)$$

where C is concentration, D is diffusion, t is time, x_f and y_f are coordinates, and $\{\eta, \xi\}$ are defined below

$$t = \int_0^z \frac{dz}{v_z}, \eta = \frac{x}{x_f}, \xi = \frac{y}{y_f} \tag{12.85}$$

Roper, Circular Burner

For a circular port burner, a form of conservation principles was derived similar to the beginning equation in Burke and Schumann [2], which is given as

$$\frac{\delta C}{\delta y} = \frac{k}{v}\left(\frac{\delta^2 C}{\delta r^2} + \frac{1}{r}\frac{\delta C}{\delta r}\right) \tag{12.86}$$

where r and y are independent variables and k is the heat transfer coefficient and v is the velocity.

The main difference is in the Roper equation the independent non-dimensional variables have been defined as

$$\theta = \int \frac{dz}{v_x x_f^2}, \eta = \frac{x}{x_f}, \xi = \frac{y}{y_f} \tag{12.87}$$

These stretched variables allow for changes in density and diffusion with location.

A solution for a circular burner to the conservation principle posed in Roper [8] is given as Eqs. 12.88, 12.89, 12.90, and 12.91.

$$\theta = \frac{DzT_{\text{fuel}}}{r^2 v_{\text{fuel}} T_{\text{flame}}} \tag{12.88}$$

$$c_m = 1 - \exp\left(-\frac{1}{4\theta}\right) \tag{12.89}$$

$$\frac{D}{D_\infty} = \left(\frac{T_{\text{fuel}}}{T_\infty}\right)^{5/3} \tag{12.90}$$

and

$$c_m = \frac{1}{S+1}, \text{when } z = H \tag{12.91}$$

where D is the diffusion coefficient at T_{fuel}, D_∞ is the diffusion coefficient at T_∞, z is the distance above the opening, T_{fuel} is the temperature of the fuel, T_{flame} is the temperature of the flame, T_∞ is the ambient temperature, r is the radius of the circular opening, v_{fuel} is the velocity of the fuel, c_m is the concentration of fuel, and z is the height of the flame.

Combining Eqs. 12.91 and 12.89 results in

$$\frac{1}{S+1} = 1 - \exp\left(-\frac{1}{4\theta}\right) \tag{12.92}$$

and

$$\exp\left(-\frac{1}{4\theta}\right) = 1 - \frac{1}{S+1} = \frac{S}{S+1} \tag{12.93}$$

Further,

$$-\frac{1}{4\theta} = \ln\left(\frac{S}{S+1}\right), \theta = \frac{-1}{4\ln\left(\dfrac{S}{S+1}\right)} \tag{12.94}$$

And substituting in for θ Eq. 12.88 and Eq. 12.90 for D results in

$$\frac{H}{r^2 v_{\text{fuel}}} D \frac{T_{\text{fuel}}}{T_{\text{flame}}} = \frac{H}{r^2 v_{\text{fuel}}} D_\infty \left(\frac{T_\infty}{T_{\text{fuel}}}\right)^{5\!/\!3} \frac{T_{\text{fuel}}}{T_{\text{flame}}} = \frac{-1}{4\ln\left(\dfrac{S}{S+1}\right)} \tag{12.95}$$

And solving for $\dfrac{H}{Q}$ gives us

$$\frac{H}{Q_\infty} = \frac{\left(\dfrac{T_\infty}{T_{\text{flame}}}\right)^{2\!/\!3}}{4\pi D_\infty \ln\left(1 + \dfrac{1}{S}\right)} \tag{12.96}$$

where Q_∞ is the mixture flow rate reference to T_∞ and is equal to

$$Q_\infty = \pi v_{\text{fuel}} r^2 \frac{T_\infty}{T_{\text{fuel}}} \tag{12.97}$$

Roper, Square Burner

For a square burner, the H/Q model is

$$\frac{H}{Q_\infty} = \frac{\frac{1}{16D_\infty}\left(\frac{T_\infty}{T_{flame}}\right)^{2/3}}{\left[inverf\left\{\frac{1}{\sqrt{1+S}}\right\}\right]^2} \tag{12.98}$$

where Q_∞ equals

$$Q_\infty = 4r^2 v_{fuel}\frac{T_\infty}{T_{fuel}} \tag{12.99}$$

where now r is the length of a side of the square opening.

It needs to be noted that for the range of S from methane ($S = 9.52$) to pentane ($S = 38.08$) that the Eqs. 12.96 and 12.98 will essentially predict the same $\frac{H}{Q}$. Additionally, for circular and square geometries the "flow is unaffected by buoyancy forces [8]."

Example 12.3 Comparison of Flow Rates and Heat Rates for Propane and Methane

We want to operate a square port diffusion flame burner with a 100 mm high flame in a laboratory. Determine the volumetric flow rate required if the fuel is methane. What flow rate would be required if the fuel is propane?

In this problem, a SI version of the equation For square port burners will be utilized and is given as

$$\frac{H}{Q_{fuel}}\left[\frac{s}{m^2}\right] = \frac{1045\left(\frac{T_\infty}{T_{flame}}\right)}{\left(inverf\left(\frac{1}{S+1}\right)\right)^2} \tag{12.100}$$

where $D_\infty = 20\frac{mm^2}{s}, T_{fuel} = T_\infty = 300\,K, T_{flame} \approx 1500\,K$ [9], please note that the $\frac{H}{Q}$ determined using this equation is reference to T_{fuel}

The following equation was utilized to determine the heat rate

$$P = \dot{m}\Delta h_{combustion}$$

where $\Delta h_{combustion}$ was determined from [10].

The results of the analysis are given below and it's seen that even though propane is a much heavier hydrocarbon ($S = 23.80$ *versus* $S = 9.52$), the heat rates are similar (260 watts versus 243 watts).

Substance		Propane	Methane
x		3	1
y		8	4
S		23.80	9.52
$1/sqrt(S + 1)$		0.20	0.31
Inverf(arg)		0.18	0.28
H/Q(fuel)	[s/m²]	31874	13403
H	[m]	0.10	0.10
Q	[m³/s]	3.14E-06	7.46E-06
Delta(h)	[J/kg]	46,357,000	50,016,000
Density	[kg/m³]	1.79	0.65
Power	[watts]	260	243

For slot burners, the combustion system may be controlled by whether the flow is dominated by momentum considerations, buoyancy considerations, or a mixture of the two. In order to determine the regime of flow, an appropriate Froude number is calculated.

Diffusion Flame Froude Number

The Froude number for a laminar flow into a stagnant environment [10] is

$$\text{Fr} = \frac{\left[vIY_{F,\text{stoich}} \right]^2}{aH} \tag{12.101}$$

where v is the velocity, I is a constant associated with the shape of the velocity profile, $Y_{F,\text{sthoic}}$ is the mass fraction of fuel for a stoichiometric flow, a is an acceleration term (given below as Eq. 12.107), and H is the height of the flame.

When

$$\text{Fr} \gg 1 \rightarrow \text{Momentum Controlled}$$
$$\text{Fr} \approx 1 \rightarrow \text{Mixed Conditions} \tag{12.102}$$
$$\text{Fr} \ll 1 \rightarrow \text{Buoyancy Controlled}$$

Roper, Slot Burner, Momentum Controlled

A process similar to the process for the derivation of H/Q for the circular burner results for a slot burner where momentum controls and is given as

$$\frac{H_M}{Q_\infty} = \left(\frac{b}{L} \right) \phi^2 \frac{T_\infty}{T_{\text{fuel}}} \frac{I}{D_\infty} \left(\frac{T_{\text{flame}}}{T_\infty} \right)^{1/3} \tag{12.103}$$

where $\left(\dfrac{b}{L} \right)$ is the ratio of slot height to width, I is a momentum factor taken to be 1, and ϕ is given as

$$\phi = \frac{1}{4 \text{inverf}\left\{\dfrac{1}{S+1}\right\}}$$ (12.104)

For a slot burner Q equals

$$Q_\infty = v_{\text{fuel}} bL \frac{T_\infty}{T_{\text{fuel}}}$$ (12.105)

Roper, Slot Burner, Buoyancy Controlled
For a slot burner where buoyancy controls, the H/Q model is

$$\frac{H_B}{Q^{4/3}} = \left\{\frac{9}{8}\phi^4 D_\infty^2 aL^4\right\}^{1/3}\left(\frac{T_{\text{flame}}}{T_\infty}\right)^{.22}$$ (12.106)

where a is an acceleration term and takes the value

$$a \approx 0.6g\left[\frac{T_{\text{flame}}}{T_\infty} - 1\right]$$ (12.107)

Roper, Slot Burner, Transition Zone
And in the transition zone the H/Q model is

$$H_{\text{Trans}} = \left\{\frac{4}{9}H_M\right\}\left\{\frac{H_B}{H_M}\right\}^3\left[\left\{1+\frac{27}{8}\left(\frac{H_B}{H_M}\right)^3\right\}^{2/3} - 1\right]$$ (12.108)

In Roper's second paper [9], experimental validation for circular geometries occurred where the flame height was determined by two means for various hydrocarbons and it was showed that a linear relationship exists between $\dfrac{H}{Q}$ versus $\dfrac{1}{\ln\left\{1+\dfrac{1}{S}\right\}}$ where S is the ratio of moles of oxygen to moles of fuel. It can also be shown mathematically

$$\frac{1}{\ln\left\{1+\dfrac{1}{S}\right\}} \cong S + \frac{1}{2}$$ (12.109)

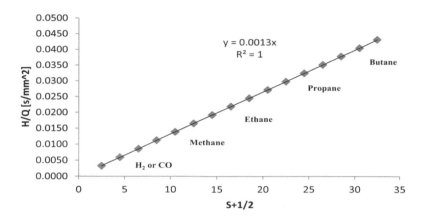

Fig. 12.8 H/Q versus $S + \dfrac{1}{2}$

For $S\epsilon(2, 30)$

The relationship developed in Roper [9] is given as

$$\frac{H}{Q}\left[\frac{s}{mm^2}\right] = 0.00133 * \frac{1}{\ln\left\{1 + \dfrac{1}{S}\right\}} \tag{12.110}$$

A relationship provided as Eqs. 12.96 and 12.110 is shown below as Fig. 12.8.

Experimental validation [9] was also conducted for slot geometries where either momentum or buoyancy controlled.

12.5   Problems

Problem 12.1 Determination of the laminar flame speed using Spalding's presentation

Using Example 12.1 vary the equivalence ratio from 0.5 to 4 and show the effect on the laminar flame speed.

Problem 12.2 Determine the laminar flame speed for propane using the correlation given in Sect. 12.3.3 where the equivalence ratio varies from 0.5 to 4 and compare against Problem 12.1.

Problem 12.3 Determine the flame height for a circular burner based on the approximate solution for the Burke and Schumann for the following conditions

A circular burner with inner diameter of 5 cm, outer diameter of 10 cm, a fuel with axial velocity of 0.1 m/s, density of 1.2 kg/m³, and diffusion of 1e-4 m²/s?

Problem 12.4 Determine the flow rate for a square flame using Roper's method for the following conditions

Ethane and assume $D_\infty = 20 \dfrac{mm^2}{s}, T_{fuel} = T_\infty = 300\,K, T_{flame} \approx 1500\,K$

For $Q = 10 \dfrac{mm}{s}$ what's the height [mm]?

Problem 12.5 Determine the flame height using Roper's method for a circular and square burner with the diameter equal to the square's length and show that the solutions are essentially the same.

Appendix 12.1: Laminar Flame Speed (website)

An Excel Spreadsheet was developed to illustrate a particular concept and is given on the companion website.

Appendix 12.2: Laminar Flame Speed Correlations (website)

An Excel Spreadsheet was developed to illustrate a particular concept and is given on the companion website.

Appendix 12.3: Burke and Schumann's Model (website)

An Excel Spreadsheet was developed to illustrate a particular concept and is given on the companion website.

Appendix 12.4: Roper's Model (website)

An Excel Spreadsheet was developed to illustrate a particular concept and is given on the companion website.

References

1. Amell, A. (2007). Influence of altitude on the height of blue cone in a premixed flame. *Applied Thermal Engineering, 27*(2-3), 408–412.
2. Burke, S. P., & Schumann, T. E. W. (1928). Diffusion flames. In *76th meeting of the American Chemical Society*.
3. Chomiak, J. (1990). *Combustion. A study in theory, fact and application*. Gordon and Breach Science Publishers.
4. Fukutani, S., Kunioshi, N., & Jinno, H. (1991). Flame structure of an axi-symmetric hydrogen-air diffusion flame. *Proceedings of the Combustion Institute, 23*, 567–573.
5. Glassman, I., & Yetter, R. A. (2008). *Combustion* (4th ed.). Academic Press.
6. Kuo, K. K. (2005). *Principles of combustion* (2nd ed.). Wiley.
7. Metghalchi, M., & Keck, J. C. (1982). Burning velocities of mixtures of air with methanol, iso-ocatne, and indolene at high pressure and temperature. *Combustion and Flame, 48*, 191–210.
8. Roper, F. G., Smith, C., & Cunnigham, A. C. (1977). The prediction of laminar jet diffusion flame sizes: Part I. *Combustion and Flame, 29*, 219–226.
9. Roper, F. G., Smith, C., & Cunnigham, A. C. (1977). The prediction of laminar jet diffusion flame sizes: Part II. *Combustion and Flame, 29*, 227–234.
10. Turns, S. R. (2012). *An introduction to combustion (concepts and applications)* (3rd ed.). Mc-Graw Hill.
11. Wikipedia contributors. (2018, December 9). Burke–Schumann flame. In *Wikipedia, The Free Encyclopedia*. Retrieved 15:08, June 6, 2020, from https://en.wikipedia.org/w/index.php?title=Burke%E2%80%93Schumann_flame&oldid=872753342

Chapter 13
Detonations

13.1 Preview

Please refer to Fig. 13.1 where the shock is moving from right to left and the initial states are at "1," the shock states are at "2," and the reaction is complete at "3."

We saw in Chap. 10 that using the conservation principles we can determine the states after the shock (2) or at the endpoint of the reaction (3) based on the initial conditions (1). This theory is known as the Rankine-Hugoniot theory and is based solely on considerations of physics. In order to understand changes in states between 2 and 3, we need to delve into the chemistry of the problem.

The heat is not released instantaneously, and a reaction rate equation added to the conservation principles will address this issue. When the heat release changes with time (distance), so do the other states. This is the basis to a dynamic detonation model or ZND model.

Development of the dynamic detonation model is a major theme of this chapter in that it addresses some of the features of what is changing within the "reaction zone." You'll notice in Fig. 13.1 there are actually two zones within the "reaction zone" and these are (1) induction zone and (2) reaction zone. In order to properly address these two regions a better dynamic detonation model would have two reaction rate equations where the induction zone reaction rate equation doesn't release heat but delays the onset of heat release and a second reaction rate equation for the reaction zone that releases all of the heat. Both forms of dynamic detonation model will be discussed – one reaction rate and two reaction rates.

Combustion has traditionally been considered as either a constant volume process or a constant pressure process. An example of constant pressure combustion is the Brayton cycle; an example of a constant volume process is the Humphrey's cycle. Both types of combustion will be discussed in a simplified manner.

Electronic Supplementary Material: The online version of this chapter (https://doi.org/10.1007/978-3-030-87387-5_13) contains supplementary material, which is available to authorized users.

H. C. Foust III, *Thermodynamics, Gas Dynamics, and Combustion*, https://doi.org/10.1007/978-3-030-87387-5_13

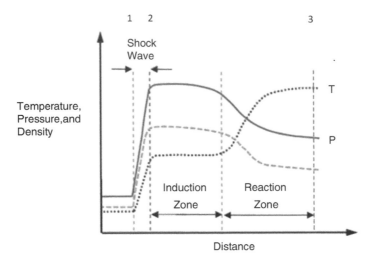

Fig. 13.1 States within the detonation structure [9]

The chapter will end by showing how the detonation structure is much more rich and strange than envisioned by the dynamic detonation models developed in this chapter; this section is entitled *Detonation Structure*.

13.2 Constant Volume Combustion

"Combustion in closed vessels is accompanied by a pressure rise. Whereas in open space the gas that is heated during combustion may expand freely and part of the heat released in the chemical reaction can go into the work of expansion, in closed vessels the walls prevent expansion of the gas and the heat of reaction goes solely into raising the internal energy of the gas. Because of the pressure rise, combustion in closed vessels involves a larger temperature rise than burning of the same mass of fuel of the same composition in open space at constant pressure." [13]

Theoretically, combustion chambers can be either constant pressure or constant volume; a third type, general combustion, will also be discussed in Sect. 13.4. Given below are the energy balance and entropy balance assuming homogenous flow and "uniform" reactions.

The entropy and energy balance for constant volume combustion are

$$ds = \frac{c_v(T)dT}{T} + Pd\upsilon \tag{13.1}$$

and (first law, closed system)

$$c_v T_1 + \lambda q = c_v T_2 \tag{13.2}$$

where ε is defined as

$$\lambda = \frac{c_v}{q}\left[T_2 - T_1\right]$$ (13.3)

For constant volume, Eq. 13.1 simplifies to

$$s_2 - s_1 = c_v \ln\left(\frac{T_2}{T_1}\right) = \frac{R}{\gamma - 1}\ln\left(\frac{T_2}{T_2}\right)$$ (13.4)

Additionally,

$$\frac{T_2}{T_1} = \frac{P_2 v_2}{P_1 v_1} = X'Y'$$ (13.5)

where $X' = \dfrac{v_2}{v_1}$ and $Y' = \dfrac{P_2}{P_1}$

Substituting Eq. 13.5 into Eq. 13.4 results in

$$s_2 - s_1 = \frac{R}{\gamma - 1}\ln\left(X'Y'\right)$$ (13.6)

A similar process can be utilized to define the energy equation (Eq. 13.3) in terms of X' and Y'

$$\epsilon q = \frac{R}{\gamma - 1} P_1 v_1\left[X'Y' - 1\right]$$ (13.7)

13.3 Constant Pressure Combustion

The energy (first law, open system) and entropy balance for a constant pressure combustor are

$$c_p T_1 + \lambda' q = c_p T_2$$ (13.8)

where

$$\lambda' = \frac{c_q}{q}\left[T_2 - T_1\right] = \gamma\lambda$$ (13.9)

and

$$s_2 - s_1 = c_p \ln\left(X'Y'\right) = \frac{R\gamma}{\gamma - 1}\ln\left(X'Y'\right)$$ (13.10)

13.4   Illustrated Example

The author of this book had the opportunity to work at the U.S. Air Force, Air Force Research Laboratory at Wright Patterson Air Force Base where he worked on determining the entropy increase associated with the combustion step of the rotational detonation engine [1], RDE (see Fig. 13.2). The RDE is a novel gas turbine that will likely be more fuel efficient in terms of the specific impulse versus a given Mach number.

The diagram the author was asked to create would look similar to what is given below (see Fig. 13.3) where 2 –>3''' is constant pressure combustion and 2->3' is constant volume compression. It will be shown there is a third distinct pathway associated with the RDE, 2->3.

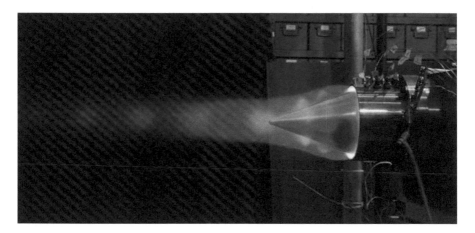

Fig. 13.2   RDE

Fig. 13.3   *ds* versus temperature [2]

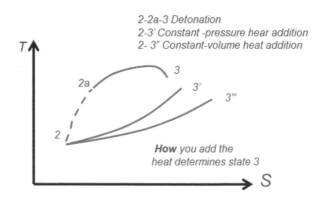

2-2a-3 Detonation
2-3' Constant -pressure hear addition
2- 3" Constant-volume heat addition

How you add the heat determines state 3

In order to do this work the following tasks were required:

- Task 1 – Develop first and second laws for constant volume and constant pressure combustion, which are given as Sects. 13.2 and 13.3
- Task 2 – Develop first and second laws for the RDE, which were done through a combination of using the Rankine-Hugoniot theory with partially combusted reactants and data from a dissertation on the thermodynamics associated with an RDE [4].
- Task 3 – Analysis of three scenarios

  - Task 3a – Temperature versus lambda
  - Task 3b – Δs versus lambda where $\lambda = 1 - \dfrac{[\text{Fuel}]_{lr}}{[\text{Fuel}]_0}$
  - Task 3c – $T/T(0)$ versus $\dfrac{\Delta s}{\overline{c}_p}$

Each task is discussed below. Additionally, the complete analysis is given as Appendix 13.1.

Task 1 – First and Second Law for Constant Volume and Constant Pressure Combustion

First Law, Constant Volume

$$c_v T_2 = c_v T_1 + \lambda [F]_0 Q \tag{13.11}$$

and

Second Law, Constant Volume

$$\Delta s = c_v \ln\left(\frac{T}{T_{\text{ref}}}\right) = \frac{R}{\gamma - 1} \ln\left(\frac{T}{T_{\text{ref}}}\right) \tag{13.12}$$

First Law, Constant Pressure

$$c_p T_2 = c_p T_1 + \lambda [F]_0 Q \tag{13.13}$$

and
Second Law, Constant Pressure

$$\Delta s = c_p \ln\left(\frac{T}{T_{\text{ref}}}\right) = \frac{R\gamma}{\gamma - 1} \ln\left(\frac{T}{T_{\text{ref}}}\right) \tag{13.14}$$

For hydrogen/air the following conditions prevail:
From the developed equations above and the parameters given in Table 13.1 graphs of lambda versus temperature and lambda versus ds are developed for both constant pressure and constant volume combustion (see Figs. 13.4 and 13.5).

Table 13.1 H$_2$/air parameters

$T(0)$	[K]	300
c_v	[cal/mole-deg]	5
c_p	[cal/mole-deg]	7
$Q*a(0)$	[cal/mole]	14,000

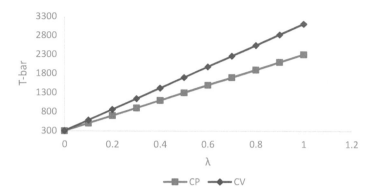

Fig. 13.4 Lambda versus temperature

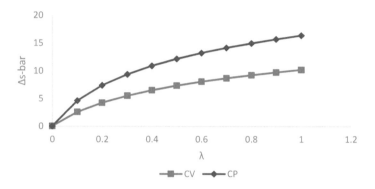

Fig. 13.5 Lambda versus *ds*

Task 2 – First and Second Law for RDE
First Law, RDE

The states for the RDE $\{T, P\}$ were determined through a combination of a run
of the RDE utilizing CFD modeling (see Fig. 13.6) with superposed Rankine lines
and Hugoniot curves.

From Fig. 13.6, a Rankine line was developed

From Fig. 13.7 and knowing the initial pressure-temperature (145 kPa and
226 K), the following table was developed for the states $\{P,T,\rho\}$.

Given the following equations from the Rankine-Hugoniot theory and Chapman-
Jouguet theory along with Table 13.2 above, a graph of X versus lambda was developed (see Fig. 13.8). This relationship was determined by solving for lambda when
$Y_{RL} = Y_{HC}$ for $X = \{0.2, 0.55\}$.

$$Y_{RL} - 1 = \gamma_1 \mathrm{Ma}_1^2 \left(1 - X\right) \tag{13.15}$$

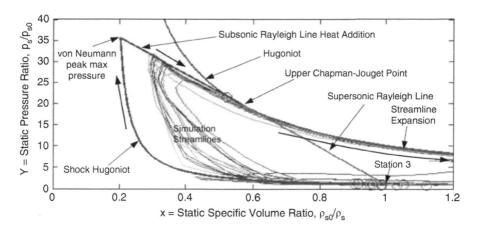

Fig. 13.6 Rankine line and Hugoniot curve for RDE [4]

Fig. 13.7 Rankine line for RDE

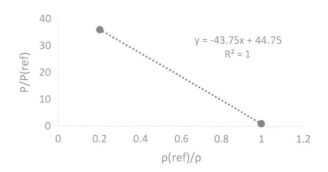

Table 13.2 States associated with Rankine line

X	Y	Rho [kg/m³]	Rho [kmoles/m³]	P [kPa]	T [K]
0.2	36	11.19	0.51	5220.00	1234.43
0.25	33.8125	8.95	0.41	4902.81	1449.27
0.3	31.625	7.46	0.34	4585.63	1626.62
0.35	29.4375	6.39	0.29	4268.44	1766.45
0.4	27.25	5.59	0.25	3951.25	1868.79
0.45	25.0625	4.97	0.23	3634.06	1933.62
0.5	22.875	4.48	0.20	3316.88	1960.94
0.55	20.6875	4.07	0.18	2999.69	1950.76

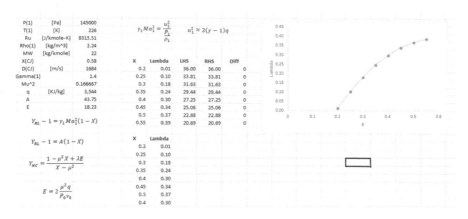

Fig. 13.8 Development of graph for X versus lambda

or

$$Y_{RL} - 1 = A(1 - X) \tag{13.16}$$

and

$$Y_{HC} = \frac{1 - \mu^2 X + \lambda E}{X - \mu^2} \tag{13.17}$$

where

$$E = 2\frac{\mu^2 q}{P_0 v_0} \tag{13.18}$$

where the detonation velocity $D(cj)$ is determined from the following equation

$$\gamma_1 \mathrm{Ma}_1^2 = \frac{u_1^2}{\dfrac{P_1}{\rho_1}} \tag{13.19}$$

And the heat release (q) is determined from the following equation

$$u_1^2 \approx 2(\gamma - 1)q \tag{13.20}$$

From the figure of X versus λ and the table of states associated with the Rankine line, the following figure was developed (Fig. 13.9).

Second Law, RDE

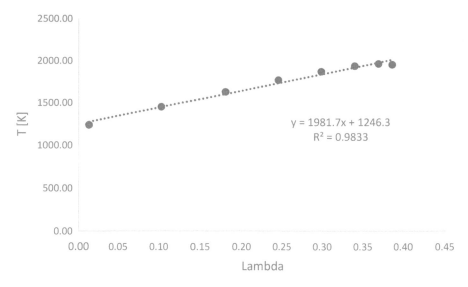

Fig. 13.9 T versus lambda

The entropy change can be determined from either Eq. 13.21 or 13.22.

$$\Delta s = \frac{R\gamma}{\gamma - 1} \ln\left(\frac{T}{T_{\text{ref}}}\right) - R \ln\left(\frac{P}{P_{\text{ref}}}\right) \tag{13.21}$$

or

$$\Delta s = \frac{R}{\gamma - 1} \ln\left(\frac{T}{T_{\text{ref}}}\right) + R \ln\left(\frac{V}{V_{\text{ref}}}\right) \tag{13.22}$$

Task 3 – Analysis of Three Scenarios

For the given parameters give in Table 13.3, graphs of λ versus T, λ versus Δs, and $T/T(0)$ versus $\Delta s/c_p$-bar were developed for constant volume, constant pressure, and RDE style combustion and given below as Figs. 13.10, 13.11, and 13.12.

Some of the parameters utilized in Table 13.3 come from an analysis for thermodynamic properties (see Appendix 13.1).

From Fig. 13.12, it can be seen that the RDE during the combustion step produces less entropy than the constant pressure and constant volume combustion and should provide all else being equal a gas turbine with higher overall thermal efficiency.

Table 13.3 Parameters for three scenarios [2]

C_P	[kJ/Kg-K]	1.58
Gamma		1.32
c_V	[kJ/Kg-K]	1.20
R	[kJ/Kg-K]	0.37
Rho(0)	[kg/m^3]	2.24
$P(0)$	[Pa]	145000.00
T(ad,CP)	[K]	2348.48
T(ad,CV)	[k]	3100.00

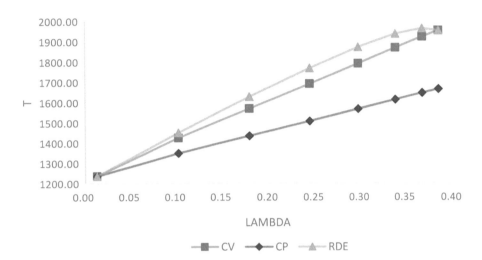

Fig. 13.10 λ vs. T for H$_2$/air for constant volume, constant pressure, and general combustion (RDE)

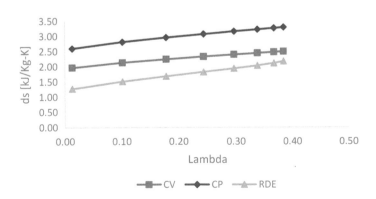

Fig. 13.11 λ vs. Δs for H$_2$/air for constant volume, constant pressure, and general combustion (RDE)

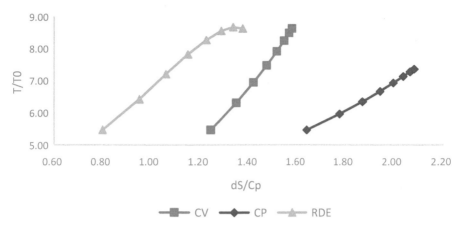

Fig. 13.12 ds/c_p versus T/T_0

13.5 Dynamic Detonation Models

13.5.1 Introduction

We've already seen one model for the detonation structure, which is defined as the modeling of states $\{T, \rho, P\}$ between the initial conditions and the final conditions where the reaction is 100% complete (also known as the CJ point) and this model is the Rankine-Hugoniot model (Fig. 13.13).

The Rankin-Hugoniot system allows us to determine the states after the shock (2) and at the CJ plane (3) for a given initial point (1); using the Chapman-Jouguet theory allows us to determine states associated with the CJ plane (2) such as $x(cj)$, $y(cj)$, and $Ma(cj)$. The dynamic model developed in this section will allow us to determine the states $\{T, \rho, P\}$ between the von Neumann point (1') to the CJ point (2) by coupling the conservation principles with an equation for the chemical reaction (Arrhenius type reaction) where the chemical reaction is irreversible

$$A \rightarrow B \tag{13.23}$$

Another model will then be developed around two irreversible chemical reactions

$$\begin{aligned} A &\rightarrow B \\ B &\rightarrow C \end{aligned} \tag{13.24}$$

where the first reaction (R_{AB}) is exothermic and the second reaction (R_{BC}) is endothermic

The model developed with double reactions (Eq. 13.24) will show that the reaction complete plane ($\lambda = 100\%$) no longer coincides with the CJ point and these types of detonations are known as pathological detonations.

Fig. 13.13 States within
the detonation structure
(Rouser KP, April 4, 2021,
personal communication)

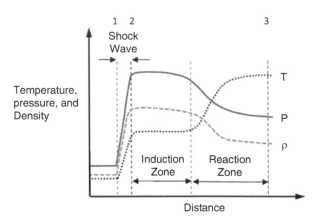

13.5.2   Reaction Rates

In this chapter the reaction rate equation will be of the form $r = k\left(1-\lambda\right)\exp\left(-\dfrac{E_a}{RT}\right)$.

13.5.3   Derivation of Dynamic Detonation Model

In Sect. 8.2, a more general form of the conservation principles is presented that
includes considerations of

1. Shear stress at the wall
2. Area changes
3. Heat release
4. Unsteady flow

In this chapter, we will assume no shear stress, no area changes, and steady flow.
Given Eqs. 8.7, 8.18, and 8.47

$$u\frac{d\rho}{dx} + \rho\frac{du}{dx} = 0 \tag{13.25}$$

$$\rho u\frac{du}{dx} + \frac{dP}{dx} = 0 \tag{13.26}$$

$$u\frac{dP}{dx} - uc^2\frac{d\rho}{dx} = \rho c^2 \sigma r \tag{13.27}$$

We'll pursue two approaches to developing a dynamic detonation model:

1. The first approach is to reduce the three equations into one equation in terms of

$$\frac{du}{dx} = f(X) \tag{13.28}$$

   And couple Eq. 13.28 with an appropriate equation for the reaction rate and equations to relate the states.

2. The second approach is to reduce Eqs. 13.25, 13.26, and 13.27 into three equations of the form

$$\frac{d\rho}{dx} = f(X) \tag{13.29}$$

$$\frac{du}{dx} = g(X) \tag{13.30}$$

$$\frac{dP}{dx} = h(X) \tag{13.31}$$

And couple Eqs. 13.29, 13.30, and 13.31 with an appropriate equation for the reaction rate.

First Approach – Single Dynamic Equation [14]

Multiplying both sides of Eq. 13.25 by dx and dividing each term by $\rho u A$ we can get to

Conservation of Mass

$$\frac{d\rho}{\rho} + \frac{du}{u} = 0 \tag{13.32}$$

Multiplying Eq. 13.26 by dx results in
Conservation of Momentum

$$dP + \rho u du = 0 \tag{13.33}$$

Conservation of Energy
Given

$$d\left(h + \frac{1}{2}u^2\right) = \Delta q d\lambda \tag{13.34}$$

or

$$dh + udu - \Delta q d\lambda = 0 \tag{13.35}$$

And substituting $dh = c_p dT$ for a perfect gas into Eq. 13.35 results in

$$c_p dT - \Delta q d\lambda + udu = 0 \tag{13.36}$$

Additionally, for an ideal, perfect gas the equation of state is
Equation of State

$$\frac{dP}{P} = \frac{d\rho}{\rho} + \frac{dT}{T} \tag{13.37}$$

Solving Eq. 13.36 for dT and diving by $c_p T$ results in

$$\frac{dT}{T} = \frac{\Delta q d\lambda}{c_p T} - \frac{udu}{c_p T} \tag{13.38}$$

And substituting Eq. 13.38 into Eq. 13.37 results in

$$\frac{dP}{P} = \frac{d\rho}{\rho} + \frac{\Delta q d\lambda}{c_p T} - \frac{udu}{c_p T} \tag{13.39}$$

Further, using Eq. 13.33

$$\frac{dP}{P} = -\frac{\rho udu}{P} \tag{13.40}$$

And substituting Eq. 13.40 into Eq. 13.39 results in

$$-\frac{\rho udu}{P} = \frac{d\rho}{\rho} + \frac{\Delta q d\lambda}{c_p T} - \frac{udu}{c_p T} \tag{13.41}$$

Substituting Eq. 13.32 into Eq. 13.41 results in

$$-\frac{\rho udu}{P} = -\frac{du}{u} + \frac{\Delta q d\lambda}{c_p T} - \frac{udu}{c_p T} \tag{13.42}$$

And solving in terms of $\dfrac{du}{u}$ results in

$$\frac{-udu}{P\big/\rho} + \frac{du}{u} + u\frac{du}{c_p t} = \frac{\Delta q d\lambda}{c_p T} \tag{13.43}$$

$$\left[-\frac{u^2}{\left(P\big/\rho \right)} + \frac{u^2}{c_p T} + 1 \right] \frac{du}{u} = \frac{\Delta q d\lambda}{c_p T} \tag{13.44}$$

and

$$\left[u^2 \left\{ \frac{-1}{RT} + \frac{1}{c_p T} \right\} + 1 \right] \frac{du}{u} = \frac{\Delta q d\lambda}{c_p T} \tag{13.45}$$

and

$$\left[u^2 \left\{ \frac{-c_p + R}{c_p RT} \right\} + 1 \right] \frac{du}{u} = \frac{\Delta q d\lambda}{c_p T} \tag{13.46}$$

And using $c_p - c_v = R$ results in

$$\left[\frac{-u^2}{\gamma RT} + 1 \right] \frac{du}{u} = \frac{\Delta q d\lambda}{c_p T} \tag{13.47}$$

or

$$\left[1 - \mathrm{Ma}^2 \right] \frac{du}{u} = \frac{\Delta q d\lambda}{c_p T} \tag{13.48}$$

And finally solving for $\dfrac{du}{u}$ results in

$$\frac{du}{u} = \frac{\dfrac{\Delta q d\lambda}{c_p T}}{1 - \mathrm{Ma}^2} \tag{13.49}$$

Additionally,

$$\dot{\lambda} = k (1 - \lambda) \exp\left(\frac{-E_a}{RT} \right) \tag{13.50}$$

And Eq. 13.50 can be related to $\dfrac{d\lambda}{dx}$ through the following equation

$$\dot{\lambda} = \frac{d\lambda}{dt} + u \frac{d\lambda}{dx} \tag{13.51}$$

And in a wave fixed frame of reference $\dfrac{d\lambda}{dt} = 0$ so that Eq. 13.49 becomes

$$\frac{du}{dx} = \frac{\dfrac{\Delta q}{c_p T}\dot{\lambda}}{1 - \text{Ma}^2} \tag{13.52}$$

In order to use Eq. 13.52 and assuming a constant c_p, the following additional models are required

1. Model for reaction rates, $\dot{\lambda}$, which is Eq. 13.50
2. Models for $\{T(x), P(x)\}$, $\{T(x), \rho(x)\}$ or $\{\rho(x), P(x)\}$ in order to determine $T(x)$ and $\text{Ma}(x)$

Second Approach – Three Dynamic Equations [6, 7]
The second approach is more efficient mathematically and provides four equations that relate $\{\rho, P, u, \lambda\}$.

Starting from Eqs. 8.7, 8.19, and 8.47 along with the reaction rate equation

$$u\frac{d\rho}{dx} + \rho\frac{du}{dx} = 0 \tag{13.53}$$

$$\rho u\frac{du}{dx} + \frac{dP}{dx} = 0 \tag{13.54}$$

$$u\frac{dP}{dx} - uc^2\frac{d\rho}{dx} = \rho c^2 \sigma r \tag{13.55}$$

$$u\frac{d\lambda}{dx} = r \tag{13.56}$$

Equations 13.53, 13.54, 13.55, and 13.56 have a matrix form, which is

$$\begin{bmatrix} u & \rho & 0 & 0 \\ 0 & u & \dfrac{1}{\rho} & 0 \\ -uc^2 & 0 & u & 0 \\ 0 & 0 & 0 & u \end{bmatrix} \begin{bmatrix} \dfrac{d\rho}{dx} \\ \dfrac{du}{dx} \\ \dfrac{dP}{dx} \\ \dfrac{d\lambda}{dx} \end{bmatrix} = \begin{bmatrix} 0 \\ 0 \\ \rho c^2 \sigma r \\ r \end{bmatrix} \tag{13.57}$$

The determinant of this system is $|A| = u^2(u^2 - c^2)$
And utilizing Kramer's rule we can solve for $\left\{\dfrac{d\rho}{dx}, \dfrac{du}{dx}, \dfrac{dP}{dx}, \dfrac{d\lambda}{dx}\right\}$ and these equations are

$$\frac{d\rho}{dx} = \frac{|A_1|}{|A|} = \frac{\begin{vmatrix} 0 & \rho & 0 & 0 \\ 0 & u & \dfrac{1}{\rho} & 0 \\ \rho c^2 \sigma r & 0 & u & 0 \\ r & 0 & 0 & u \end{vmatrix}}{u^2\left(u^2 - c^2\right)} \tag{13.58}$$

$$\frac{du}{dx} = \frac{|A_2|}{|A|} = \frac{\begin{vmatrix} u & 0 & 0 & 0 \\ 0 & 0 & \dfrac{1}{\rho} & 0 \\ -uc^2 & \rho c^2 \sigma r & u & 0 \\ 0 & 0 & 0 & u \end{vmatrix}}{u^2\left(u^2 - c^2\right)} \tag{13.59}$$

$$\frac{dP}{dx} = \frac{|A_3|}{|A|} = \frac{\begin{vmatrix} u & \rho & 0 & 0 \\ 0 & u & 0 & 0 \\ -uc^2 & 0 & \rho c^2 \sigma r & 0 \\ 0 & 0 & 0 & u \end{vmatrix}}{u^2\left(u^2 - c^2\right)} \tag{13.60}$$

$$\frac{d\lambda}{dx} = \frac{|A_4|}{|A|} = \frac{\begin{vmatrix} u & \rho & 0 & 0 \\ 0 & u & \dfrac{1}{\rho} & 0 \\ -uc^2 & 0 & u & \rho c^2 \sigma r \\ 0 & 0 & 0 & 0 \end{vmatrix}}{u^2\left(u^2 - c^2\right)} \tag{13.61}$$

Equations 13.58, 13.59, 13.60, and 13.61 simplify to

$$\frac{d\rho}{dx} = \frac{\rho c^2 \sigma r}{u\left(u^2 - c^2\right)} \tag{13.62}$$

$$\frac{du}{dx} = \frac{-c^2 \sigma r}{\left(u^2 - c^2\right)} \tag{13.63}$$

$$\frac{dP}{dx} = \frac{u\rho c^2 \sigma r}{\left(u^2 - c^2\right)} \tag{13.64}$$

$$\frac{d\lambda}{dx} = \frac{r}{u} \tag{13.65}$$

And using the definition of Mach number, Eqs. 13.62, 13.63, 13.64, and 13.65 further simplifies to

$$\frac{d\rho}{dx} = -\frac{\rho \sigma r}{u\left(1 - \mathrm{Ma}_1^2\right)} = -\frac{\rho}{u} K\left(\sigma r, \mathrm{Ma}\right) \tag{13.66}$$

$$\frac{du}{dx} = \frac{\sigma r}{1 - \mathrm{Ma}_1^2} = K\left(\sigma r, \mathrm{Ma}\right) \tag{13.67}$$

$$\frac{dP}{dx} = -\frac{\rho u \sigma r}{1 - \mathrm{Ma}_1^2} = -\rho u K\left(\sigma r, \mathrm{Ma}\right) \tag{13.68}$$

$$u\frac{d\lambda}{dx} = r = k\left(1 - \lambda\right)\exp\left(-\frac{E_a}{RT}\right) \tag{13.69}$$

Please note in *Approach 1* the system that is developed is an algebraic-differential system and notoriously difficult to solve; in *Approach 2*, the system is purely differential and we'll use *Approach 2* in the next two sections.

Given below are two sections that work through examples for a single-reaction dynamic detonation model (one-step model) and a double-reaction dynamic detonation model (two-step model). These two sections will also explore some issues associated with each approach.

13.5.4   Dynamic Detonation Model with Single Reaction

In this section, Eqs. 13.66, 13.67, and 13.68 along with a model for the reaction rate (Eq. 13.69) are solved numerically.

The procedure to be utilized will be as follows

1. Determine the states $\{\rho, u, P, \text{Ma}\}$ associated with the shocked conditions. These states will be where $\lambda = 0$
2. Set the following constraints

   (a) $\dfrac{E_a}{RT} = \text{constant}$

   (b) $k = 1$

3. Determine Δx and solve for $\lambda(x + \Delta x)$, $\rho(x + \Delta x)$, $u(x + \Delta x)$, $P(x + \Delta x)$

4. Solve for $\text{Ma}(x + \Delta x) = \dfrac{u(x + \Delta x)}{c(x + \Delta x)} = \dfrac{u(x + \Delta x)}{\sqrt{\gamma \dfrac{P(x + \Delta x)}{\rho(x + \Delta x)}}}$

5. Repeat until λ is near 1

Example 13.5 – Single Reactions

Given in Turns [12] is the discussion of the following detonation system (see Table 13.4 and Fig. 13.1), which is a stoichiometric mixture of C_2H_2 and air with the following states at the initial conditions (state 1), von Neumann point (state 2'), and CJ conditions (state 2). The value of $\dfrac{E}{R_u T}$ was set to 10 and k was 1.

Given the initial conditions, the states at 1' and 2 were determined through the Rankine-Hugoniot theory. Using the theory associated with dynamic detonation systems (Eqs. 13.66, 13.67, 13.68, and 13.69) and the following initial conditions (see Table 13.5), results of solving the system of equations are given as Figs. 13.14, 13.15, 13.16, 13.17, and 13.18 where it is noted that the x-axis has not been properly scaled and the values are approaching asymptotically the values at the CJ point.

Table 13.4 Detonation states for stoichiometric methane-air

Property	Units	State 1	State 1'	State 2
Rho	[kg/m³]	1.17	6.34	2.11
Pressure	[kPa]	101.3	3910	2087
T	[K]	298	2119	3531
Ma		5.78	0.4	1
C	[m/s]	345	845	1091
u	[m/s]	1997	338	1091
$C(P)$	[kJ/kg-K]	1.057	1.443	1.443
$R(g)$	[kJ/kg-K]	0.290	0.273	0.273
Gamma		1.379	1.233	1.233
dq	[kJ/kg]			3399

Table 13.5 Initial conditions

State	Initial value	Units
Density	6.34	[kg/m³]
Shock velocity	338	[m/s]
Pressure	3,910,000	[Pa]
Lambda	0	

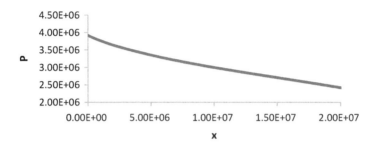

Fig. 13.14 *P* versus *x*

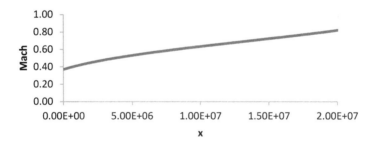

Fig. 13.15 Mach versus *x*

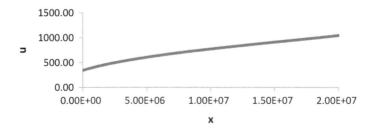

Fig. 13.16 *u* versus *x*

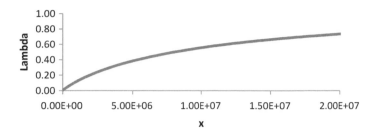

Fig. 13.17 Lambda versus *x*

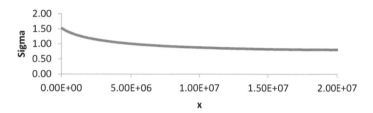

Fig. 13.18 Sigma versus *x*

13.5.5 Dynamic Detonation Model with Double Reactions

If we go back to Fig. 13.1, we see that the reaction zone is actually two separate zones – an induction zone followed by the actual reaction zone where the heat is released.

In this section, a detonation structure will be discussed that includes two reaction rate equations and the effect of two reaction rate equations on the determined final conditions. Another form of Eqs. 13.66, 13.67, 13.68, and 13.69 [7] that includes two reaction rate equations is given below.

$$\frac{d\rho}{dx} = -\frac{\rho\left(\sigma_1 r_1 + \sigma_2 r_2\right)}{u\left(1-\mathrm{Ma}_1^2\right)} \tag{13.70}$$

$$\frac{du}{dx} = \frac{\left(\sigma_1 r_1 + \sigma_2 r_2\right)}{1-\mathrm{Ma}_1^2} \tag{13.71}$$

$$\frac{dP}{dx} = -\frac{\rho u\left(\sigma_1 r_1 + \sigma_2 r_2\right)}{1-\mathrm{Ma}_1^2} \tag{13.72}$$

$$u\frac{d\lambda_1}{dx} = r_1 = \left(1-\lambda_1\right)k_1 \exp\left(-\frac{E_1}{RT_1}\right) \tag{13.73}$$

$$u\frac{d\lambda_2}{dx} = r_2 = \left(\lambda_1-\lambda_2\right)k_2 \exp\left(-\frac{E_2}{RT_2}\right) \tag{13.74}$$

If we were to solve this system of equations where r_1 is slightly endothermic (induction zone) and r_2 is exothermic, then we'd see the final states are no longer at the classic CJ conditions of $\lambda = 100\ \%$ and Mach = 1, but at some other values [7]. Another situation presented in the detonation literature is where the r_1 is exothermic and r_2 is endothermic and again the final state are no longer at the classic CJ conditions. The point of these exercises would be to show that there are detonation structures where the final conditions don't converge to the classic CJ conditions, but that the Rankine-Hugoniot and Chapman-Jouguet theory provide a reasonable estimate of the final states for most engineering analysis.

More exact solutions can be determined through incorporating more combustion chemistry with a more accurate numerical integration scheme and utilizing software that is readily available at [10, 11], which is known as the Caltech's Shock and Detonation Toolbox.

13.6   Detonation Structures

We saw in Sect. 13.5 how states within the detonation structure change with distance, but even the dynamic detonation model is only an approximation to the reality of a three-dimensional, transient detonation structure with instabilities (see Fig. 13.19, 13.20, and 13.21). It was found in 1957 by [3] that putting soot on the

Fig. 13.19 Detonation cell pattern

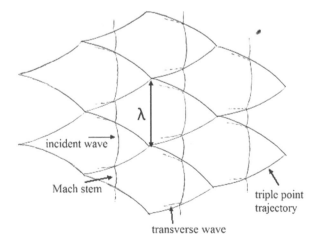

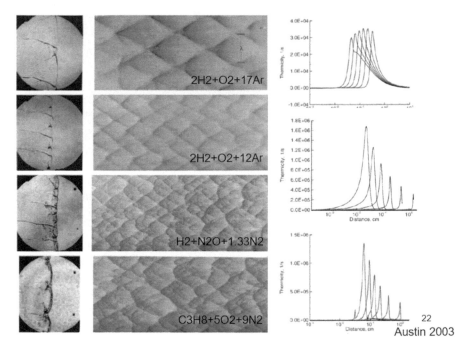

Fig. 13.20 Delineating regular and irregular detonation cell patterns [1]

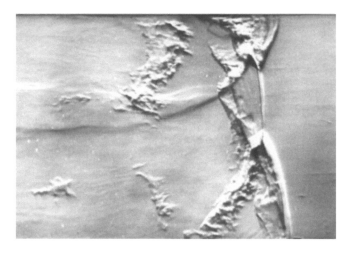

Fig. 13.21 Experimental results of a detonation structure [8]

inside of shock tubes and detonating the enclosed fuel/air mixture resulted in a particular fish scale pattern shown in Fig. 13.19 where λ is a measure of the detonation cells and related to other features of the detonation structure such as the intensity of transverse waves, the distance between the shock and CJ plane, etc.

It was also found that adding various amounts of inert gases such as argon to the fuel/air mixture resulted in a more regular detonation cell pattern (see Fig. 13.20) where the top image is an example of a regular detonation cell pattern and the bottom image is an example of an irregular detonation cell pattern. Please note all detonations are inherently unstable and the literature will classify regular detonations as (stable) and irregular detonations as (unstable).

Detonation stability is defined as the ability to estimate the states for a particular time and location; as the detonation becomes more unstable, this estimation becomes more unpredictable. There is a more exact, mathematical definition [2, 5, 8, 14], but it's beyond the scope of this book.

Given in Fig. 13.21 is an image of experiments within a shock tube where various features of the detonation structure are given in the carton (Fig. 13.22).

As the Mach shock moves from right to left through the detonation cell the Mach shock (Mach stem) velocity, $D(cj)$, is not constant but cycles through a range of values in a periodic fashion [8]. Transverse waves orthogonal to the Mach shock sweep through the detonation cell and toward the end of the detonation that occurs in each detonation cell a secondary combustion occurs as transverse waves collide. The strength of the transverse waves can be measured by the distance between waves and the detonation cell length (λ) is also a measure of the strength of the transverse waves. Triple points are the intersection of Mach stems and transverse waves.

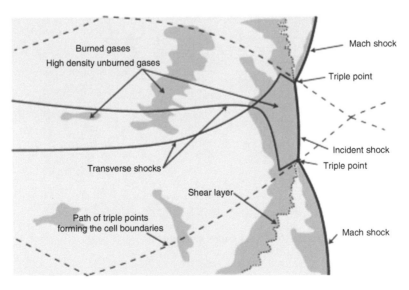

Fig. 13.22 Cartoon of detonation structure [8]

Regions of Stability

It has been found [3, 4, 10] that the ratio of the induction length (Δ_I) divided by the reaction length (Δ_R) is a measure of how stable a detonation structure is and Ng [4] utilized this observation among others to define a non-dimensional stability parameter given as

$$\chi = \epsilon_I \frac{\Delta_I}{\Delta_R} \tag{13.75}$$

where ϵ_I is "essentially the normalized activation energy of the induction reaction with respect to the shock temperature [3]", Δ_I is the length of the induction zone, and Δ_R is the length of the reaction zone.

In Fig. 13.23, detonations above the bold line are regular and those below irregular.

When the detonation is regular, the effort to model various features of the detonation structure or artifacts of the detonation is easier to accomplish. One example of this is as follows. It was found that detonations in shock tubes can be affected by boundary conditions when the inside diameter of the tube is below some critical diameter (D_c) and this phenomenon was the impetus to Fay's work [2]. Graphs of d_c/d versus V/V_{cj} for a regular and irregular detonation are given in Fig. 13.24 where the solid line is Fay's theory. It obvious that Fay's theory works for a regular detonation and not an irregular detonation.

This section has provided the briefest of abstracts for a very active and very complicated area within combustion – detonation structures and detonation shock dynamics. The following resources take this discussion much further:

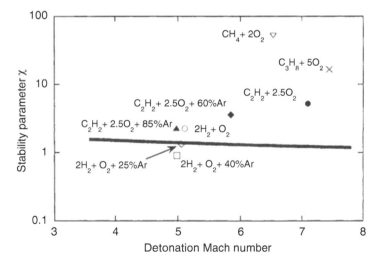

Fig. 13.23 Regions of stability versus detonation Mach [4]

Fig. 13.24 Detonation velocity deficit for a stable (**a**) and unstable (**b**) mixture [2]. (**a**) Stable mixture ($C_2H_2 + 2.5O_2 + xAr$). (**b**) Unstable mixture [$0.5(C_2H_2 + 5 N_2) + 0.5Ar$]

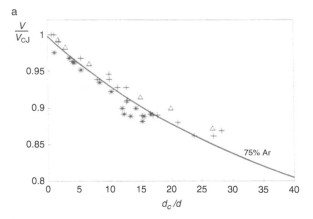

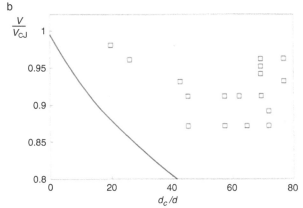

- Austin (2003). The role of instabilities in gaseous detonations; Ph.D. Thesis, California Institute of Technology.
- Ng (2005). The effect of chemical reaction kinetics on the structure of gaseous detonations; Ph.D. Thesis, McGill University.
- Oran, Weber, Steaniw, Lefebvre, and Anderson (1998). "A Numerical Study of a Two-Dimensional H2-O2-Air Detonation Using a Detailed Chemical Reaction Model"; *Combustion and Flame*, v. 113, pp. 147 to 163.
- Radulescu, Sharpe, Law and Lee (2007). "The hydrodynamic structure of unstable cellular detonations"; *Journal of Fluid Mechanics*; v. 580, n. 10, June, pp. 31 to 81.

13.7   Problems

Problem 13.1 ZND Model, Single Reaction
Given a H_2-air detonation system [7]

Parameter	Units	Value
γ		1.4
MW	Kg/kmole	20.91
R	J/kg-K	397.58
$P(0)$	Pa	1.01325e5
$T(0)$	K	298
$\rho(0)$	Kg/m^3	0.85521
$\nu(0)$	M3/kg	1.1693
q	J/kg	1.89566e6
E	J/kg	8.29352e6
a	1/s	5e9
β		0

Where the reaction rate equation is

$$r = aT^{\beta} \exp\left[-\frac{E}{RT}\right](1-\lambda)$$

Model this system to determine how the following states change with distance within the reaction/induction zone

- Pressure
- Density
- Mach
- Temperature
- Λ

Problem 13.2 ZND Model, Double Reactions

Results for the simple analysis (Problem 13.1) and detailed analysis [7] are given below.

Parameter	Simple analysis	Detailed analysis
L(rxn)	0.01 m	0.01 m
L(induction)	0.001 m	0.0001 m
$D(cj)$	1991.1 m/s	1979.7 m/s
$P(s)$	2.80849e6 Pa	2.8323e6 Pa
$P(cj)$	1.4553d6 Pa	1.6483e6 Pa
$T(s)$	1664.4 K	1542.7 K
T(cj)	2570.86 K	2982.1 K
Rho(s)	4.244 kg/m^3	4.618 kg/m^3
Rho(cj)	1.424 kg/m^3	1.5882 kg/m^3
$M(0)$	4.88887	4.8594
$M(s)$	0.41687	0.40779
M(cj)	1	0.93823

Repeat Problem 13.1, but now there are two reactions where the first is exothermic and the second is endothermic [6]. It is assumed that

$$k = k_1 = k_2$$

$$MW = MW_A = MW_B = MW_c$$

And all other reaction rate parameters are the same $\{a, \beta, R\}$
Given

Parameter	Units	Value
γ		1.4
MW	Kg/kmole	20.91
R	J/kg-K	397.58
$P(0)$	Pa	1.01325e5
$T(0)$	K	298
$\rho(0)$	Kg/m³	0.85521
$\nu(0)$	M3/kg	1.1693
$q(1)$	J/kg	7.58265e6
$Q(2)$	J/Kg	−5.68698e6
E	J/kg	8.29352e6
a	1/s	5e9
β		0

Where the reaction rate equation is

$$r = aT^{\beta} \exp\left[-\frac{E}{RT}\right](1-\lambda)$$

Model this system to determine how the following states change with distance within the reaction/induction zone

- Pressure
- Density
- Mach
- Temperature
- Λ

Appendix 13.1: Illustrated Example (website)

An Excel Spreadsheet was developed to illustrate a particular concept and is given on the companion website.

Appendix 13.2: Dynamic Detonation Model – One Step Model (website)

An Excel Spreadsheet was developed to illustrate a particular concept and is given on the companion website.

References

1. Foust, H. (2015). *Delineating entropy change and temperature for a rotating detonation engine*; Final Report for U.S. Air Force Summer Fellowship at Wright Patterson Air Force Base.
2. Lee, J. H. S. (2008). *The detonation phenomena.* Cambridge University Press.
3. Ng, H. D., Radulesu, M. I., Higgins, A. J., Nikiforakis, N., & JHS, L. (2007). Numerical investigation of the instability for one-dimensional Chapman-Jouguet detonations with chain-branching kinetics. *Combustion Theory and Modeling, 9*(3), 385–401.
4. Nordeen, C. (2013). Thermodynamics of a rotating detonation engine; Ph.D. Thesis, University of Connecticut.
5. Oran, E., Weber, J. W., Stefaniw, E. I., Lefebvre, M., & Anderson, J. (1998). A numerical study of a two-dimensional H2-O2-Ar detonation using a detailed chemical reaction model. *Combustion and Flame, 113,* 147–163.
6. Powers, J. (2014). *Lecture notes for AME 60636, fundamentals of combustion.* Notre Dame University.
7. Powers, J. (2020). *Lecture notes for AME 60635, intermediate fluid mechanics.* Notre Dame University.
8. Radulescu, M. I., Sharpe, G. J., Law, C. K., & Lee, J. H. S. (2007). The hydrodynamic structure of unstable cellular detonations. *Journal of Fluid Mechanics, 580*(10), 31–81.
9. Rouser, K. P. (2021) April 4, personal communication.
10. Sheppard, J. E. (2009). Detonation in gases. *Proceedings of the Combustion Institute, 32*(1), 83–98.
11. Shepherd. (2018). Shock and detonation toolbox. https://shepherd.caltech.edu/EDL/PublicResources/sdt/. Accessed 06/09/2020.
12. Turns, S. R. (2012). *An introduction to combustion (concepts and applications)* (3rd ed.). Mc-Graw Hill.
13. Zeldovich, I. B., Barenblatt, G. I., Librovich, V. B., & Makhviladze, G. M. (1985). *Mathematical theory of combustion and explosions.* Springer.
14. Zhang, F. (2012). *Shock wave science and technology reference library* (Vol. 6). Springer.

Chapter 14
Blast Waves

14.1 Preview

G.I. Taylor was one of the greatest physicists of the twentieth century and worked in classical mechanics. One area of interest for Mr. Taylor was blast waves associated with explosions and he had the privilege to work on the Manhattan Project that developed the atomic bomb. Two classic papers [8, 9] he has written form the basis to a theoretical model that estimated the energy released and states across the blast wave associated with a very intense explosion.

Mr. Taylor reduced the partial differential equations (representing the conservation principles) into a system of ordinary differential equations (ODEs) and solved these ODEs both numerical and provided approximate forms (algebraic equations) for the non-dimensional states. He also utilized a dimensional analysis argument to find a critical parameter, $K(\gamma)$, and showed that

$$\frac{E}{\rho_1 R^5} t^2 = K(\gamma) \approx O(1) \tag{14.1}$$

where E is energy release, ρ_1 is the density of air before the blast wave, R is the radius of the spherical blast, and t is time after detonation (see Fig. 14.1).

Mr. Taylor's work in this area will be discussed in this chapter. Mr. Taylor's work was in a particular coordinate system and a more general coordinate system associated with blast waves is the subject of J.H.S. Lee's work [3] and will be reviewed. As will the work of three undergraduate students, who applied the tools of G.I. Taylor to solve a blast wave associated with certain stellar explosions and note other systems exist [1, 2, 5, 10].

Electronic Supplementary Material: The online version of this chapter (https://doi.org/10.1007/978-3-030-87387-5_14) contains supplementary material, which is available to authorized users.

H. C. Foust III, *Thermodynamics, Gas Dynamics, and Combustion*, https://doi.org/10.1007/978-3-030-87387-5_14

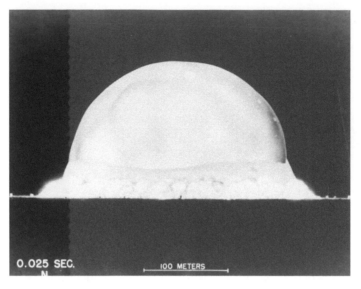

Fig. 14.1 Trinity test site detonation

14.2 Euler's Reactive Flow Equations

The Euler reactive flow equations given as Eqs. 14.2, 14.3, and 14.4 provided differential forms for the conservation principles and are a set of partial differential equations. These equations represent the conservation principles across the shock wave that includes blast waves. A *blast wave* is defined as "a shock wave whose strength decreases as it propagates away from the source [8]."

$$\frac{\partial \rho}{\partial t} + u \frac{\partial \rho}{\partial r} + \rho \frac{\partial u}{\partial r} + j \frac{\rho u}{r} = 0 \tag{14.2}$$

$$\frac{\partial u}{\partial t} + u \frac{\partial u}{\partial r} + \frac{1}{\rho} \frac{\partial P}{\partial r} = 0 \tag{14.3}$$

$$\frac{DP}{Dt} + \gamma P \frac{\partial u}{\partial r} + j\gamma \frac{Pu}{r} = 0 \tag{14.4}$$

where ρ is density, t is time, u is velocity, r is the radius of the shock sphere, j is a coefficient, P is pressure, and γ is the thermodynamic constant particular to the gas.

Partial differential equations are not easily solved numerically and would have been especially onerous in a time before computers and so are often reduced to ordinary differential equations through various means where one method is the similarity argument. Once Eqs. 14.2, 14.3, and 14.4 are reduced to ODEs, this new system can be solved numerically, which will be explored further in Sect. 14.4.

14.3   G.I. Taylor's Blast Theory (G.I. Taylor)

14.3.1   Overview

The equation of motion, continuity, and equation of state are given as Eqs. 14.5, 14.6, and 14.7. These equations are partial differential equations (PDE) and difficult to solve. G.I. Taylor used similarity arguments to reduce the set of PDEs into a set of ordinary differential equations (ODE).

$$\frac{\partial u}{\partial t} + u\frac{\partial u}{\partial r} = \frac{-1}{\rho}\frac{\partial P}{\partial r} \tag{14.5}$$

$$\frac{\partial \rho}{\partial t} + u\frac{\partial \rho}{\partial r} + \rho\left(\frac{\partial u}{\partial r} + \frac{2u}{r}\right) = 0 \tag{14.6}$$

$$\left(\frac{\partial}{\partial t} + u\frac{\partial}{\partial r}\right)\left(\frac{P}{\rho}\right) = 0 \tag{14.7}$$

where t is time after detonation, u is the one-dimensional velocity, r is the radius of the spherical shock wave, P is pressure, and ρ is density.

The resulting ODEs are

$$\phi'(\eta - \phi) = \frac{1}{\gamma}\frac{f'}{\psi} - \frac{3}{2}\phi \tag{14.8}$$

$$\frac{\psi'}{\psi} = \frac{\phi' + 2\phi/\eta}{\eta - \phi} \tag{14.9}$$

$$3f + \eta f' + \gamma\frac{\psi'}{\psi}(\phi - \eta)f - \phi'f' = 0 \tag{14.10}$$

With the following final conditions (come from the Rankine-Hugoniot theory)

$$\phi(1) = \frac{u_1}{U} = \frac{2}{\gamma + 1} \tag{14.11}$$

$$\psi(1) = \frac{\rho_1}{\rho_0} = \frac{\gamma + 1}{\gamma - 1} \tag{14.12}$$

$$f(1) = \frac{U^2}{a^2 y_1} = \frac{\gamma + 1}{2\gamma} \tag{14.13}$$

where γ is a thermodynamic constant, u_1, ρ_1, y_1 represent the values of u, ρ, y immediately behind the shock wave, and U is the radial velocity of the shock wave.

Variables utilized are defined below (the choice of certain non-dimensional groups is determined by group theory)

$$\eta = \frac{r}{R} \tag{14.14}$$

$$\psi = \frac{\rho}{\rho_0} \tag{14.15}$$

$$f = f_1 \frac{a^2}{A} \tag{14.16}$$

$$a^2 = \gamma \frac{P_0}{\rho_0} \tag{14.17}$$

where r is the distance from the point of detonation, R is the current radius of the detonation, ρ is air density, A is a constant to be determined, and P is pressure.

You'll notice in Eqs. 14.8, 14.9, and 14.10 there is only one independent variable, η and one parameter, γ. G.I. Taylor numerically integrated these equations (for $\gamma = 1.4$) and the results are given in Fig. 14.2.

GI Taylor then derived an equation for the energy release, which is

$$E = B\rho_0 A^2 \tag{14.18}$$

and B is defined as

$$B = 2\pi \int_0^1 \psi \phi^2 \eta^2 d\eta + \frac{4\pi}{\gamma(\gamma-1)} \int_0^1 f \eta^2 d\eta = 2\pi I_1 + \frac{4\pi}{\gamma(\gamma-1)} I_2 \tag{14.19}$$

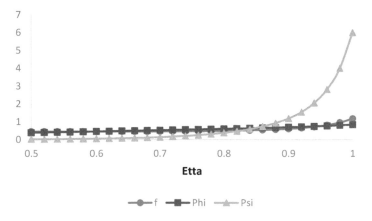

Fig. 14.2 η versus f, φ and ϕ

where A is a parameter determined experimentally from data of $\{t, R\}$ given in Fig. 14.3 and A is defined as

$$\frac{dR}{dt} = AR^{-1.5} \tag{14.20}$$

G.I. Taylor then developed algebraic expressions for f, ϕ, and ψ, which are given below.

$$\frac{f'}{f} \approx a\eta^b \tag{14.21}$$

$$\phi \approx \frac{\eta}{\gamma} + a\eta^b \tag{14.22}$$

$$\psi \approx a\eta^b \tag{14.23}$$

where the parameters $\{a, b\}$ were determined from Eqs. 14.11, 14.12, and 14.13 and other considerations.

In Sect. 14.3.2, we show how Eqs. 14.5, 14.6, and 14.7 were reduced to Eqs. 14.8, 14.9, and 14.10. In Sect. 14.3.3, more details of the numerical integration will be discussed to include where the final values (Eqs. 14.11, 14.12, and 14.13 come from). It will be shown how the approximate forms were derived in Sect. 14.3.4 and what the values of the parameters $\{a, b\}$ are for Eqs. 14.20, 14.21, and 14.22. In Sect. 14.3.5, we will discuss the form of the energy equation (Eqs. 14.18 and 14.19), how A is estimated from experimental data for $\{t, R\}$ and an estimate for the energy released at the Trinity test site.

In Sect. 14.4, numerical solutions are provided for a more general set of ODEs applicable to three coordinate systems – planar, cylindrical, and spherical.

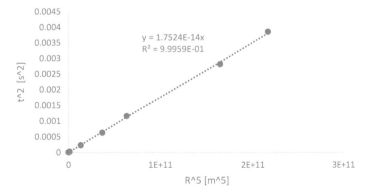

Fig. 14.3 t^2 versus R^5

G.I. Taylor developed two approaches to solving for states associated within very intense blast waves: (a) numerical methods and (b) approximate solutions (algebraic equations). In Sect. 14.5, these two tools are applied to stellar explosions in cylindrical coordinates.

14.3.2   Similarity Arguments

In this section, it will be shown the process of reducing one of the partial differential equations (Eqs. 14.5, 14.6 and 14.7) into an ordinary differential Equation (Eqs. 14.8, 14.9, and 14.10). This section will also derive the final conditions given as Eqs. 14.11, 14.12, and 14.13.

Much of this section comes from the excellent lecture notes of Dr. Joseph Powers [4].

..... Transformations

The independent variables $\{r, t\}$ are transformed to $\{\eta, \tau\}$

$$\{r,t\} \rightarrow \{\eta,\tau\} \tag{14.24}$$

And the following dependent variables are defined

$$\frac{P_2}{P_1} = y = R^{-3} f_1(\eta) \tag{14.25}$$

$$\frac{\rho_2}{\rho_1} = \psi(\eta) \tag{14.26}$$

$$u = R^{-1.5} \phi_1(\eta) \tag{14.27}$$

where

$$\eta = \frac{r}{R} \text{ and } U = \frac{dR}{dt} = AR^{-3/2} \tag{14.28}$$

The transformation also transforms the differential operators such as

$$\{\partial r, \partial t\} \rightarrow \{d\eta, d\tau\} \tag{14.29}$$

where

$$\frac{\partial}{\partial t} = \frac{\partial \eta}{\partial t} \frac{\partial}{\partial \eta} + \frac{\partial \tau}{\partial t} \frac{\partial}{\partial \tau} \tag{14.30}$$

and

$$\frac{\partial \eta}{\partial t} = \frac{-r}{R^2}\frac{dR}{dt} = -\frac{\eta}{R}\frac{dR}{dt} = -\frac{\eta}{R}AR^{-3/2} = -\frac{A\eta}{R^{2.5}}$$

(14.31)

and

$$\frac{\partial \tau}{\partial t} = 1$$

(14.32)

And substituting Eqs. 14.31 and 14.32 into Eq. 14.30 results in

$$\frac{\partial}{\partial t} = -\frac{A\eta}{R^{2.5}}\frac{\partial}{\partial \eta} + \frac{\partial}{\partial \tau}$$

(14.33)

where $\dfrac{\partial}{\partial \tau} = 0$, therefore

$$\frac{\partial}{\partial t} = -\frac{A\eta}{R^{2.5}}\frac{\partial}{\partial \eta}$$

(14.34)

Also,

$$\frac{\partial}{\partial r} = \frac{\partial \eta}{\partial r}\frac{\partial}{\partial \eta} + \frac{\partial \tau}{\partial r}\frac{\partial}{\partial \tau} = \frac{1}{R}\frac{d}{d\eta}$$

(14.35)

..... Transform one of the conservation principles
Starting with

$$\frac{\partial \rho}{\partial t} + u\frac{\partial \rho}{\partial r} + \rho\frac{\partial u}{\partial r} = -2\frac{\rho u}{r}$$

(14.36)

We want to transform each term

$$\frac{\partial \rho}{\partial t} = \left\{ -\frac{A\eta}{R^{2.5}}\frac{\partial}{\partial \eta} \right\}\{\rho_1\psi\}$$

(14.37)

$$u\frac{\partial \rho}{\partial r} = \left\{ R^{-3/2}\phi_1 \right\}\left\{ \frac{1}{R}\frac{d}{d\eta} \right\}\{\rho_1\psi\}$$

(14.38)

$$\rho\frac{\partial u}{\partial r} = \{\rho_1\psi\}\left\{ \frac{1}{R}\frac{d}{d\eta} \right\}\{R^{-3/2}\phi_1\}$$

(14.39)

and

$$-2\frac{\rho u}{r} = -\frac{2}{r}\{\rho_1\psi\}\{R^{-3/2}\phi_1\}$$

(14.40)

Substituting Eqs. 14.37, 14.38, 14.39, and 14.40 into Eq. 14.36 results in

$$\frac{-A\eta}{R^{5/2}}\frac{d\psi}{d\eta}+\frac{\phi_1}{R^{5/2}}\frac{d\psi}{d\eta}+\frac{\psi}{R^{5/2}}\frac{d\phi_1}{d\eta}=-\frac{2}{\eta}\frac{\psi\phi_1}{R^{5/2}} \tag{14.41}$$

Simplifying results in

$$-A\eta\psi'+\phi_1\psi'+\psi\left[\phi_1'+\frac{2}{\eta}\phi_1\right]=0 \tag{14.42}$$

With the substitution $\phi=\dfrac{\phi_1}{A}$ into Eq. 14.42 results in

$$-\eta\psi'+\phi\psi'+\psi\left(\phi'+\frac{2}{\eta}\phi\right)=0 \tag{14.43}$$

Which is also Eq. 14.9
….. States associated with a large blast wave
Another form of the Hugoniot curve is

$$\left(y+\mu^2\right)\left(x-\mu^2\right)=2\lambda\mu^2q'+1-\mu^4 \tag{14.44}$$

where $\mu^2=\dfrac{\gamma-1}{\gamma+1}$

And it is assumed the thermodynamic properties are the same on both sides of the shock.

If we substitute the Rankine line into Eq. 14.44, we get

$$\left(\left\{\left(1+\gamma\mathrm{Ma}_1^2\right)-\gamma\mathrm{Ma}_1^2x\right\}+\mu^2\right)\left(x-\mu^2\right)=2\lambda\mu^2q'+1-\mu^4 \tag{14.45}$$

Or

$$-\gamma\mathrm{Ma}_1^2x^2+\left[\left(1+\gamma\mathrm{Ma}_1^2\right)+\mu^2+\gamma\mathrm{Ma}_1^2\mu^2\right]x-\left(1+\gamma\mathrm{Ma}_1^2\right)\mu^2-2\lambda q'^{\mu^2}=0 \tag{14.46}$$

which is a quadratic in x and

$$A=-\gamma\mathrm{Ma}_1^2,B=1+\gamma\mathrm{Ma}_1^2\left(1+\mu^2\right)+\mu^2,C=-\mu^2\left(1+\gamma\mathrm{Ma}_1^2+2\lambda q'\right) \tag{14.47}$$

and has solution

$$x=\frac{1+\gamma\mathrm{Ma}_1^2\left(1+\mu^2\right)+\mu^2}{2\gamma\mathrm{Ma}_1^2}$$
$$\mp\frac{\sqrt{\left(1+\gamma\mathrm{Ma}_1^2\left(1+\mu^2\right)+\mu^2\right)^2-4\gamma\mathrm{Ma}_1^2\mu^2\left(1+\gamma\mathrm{Ma}_1^2+2\lambda q'\right)}}{2\gamma\mathrm{Ma}_1^2} \tag{14.48}$$

Either before or after the shock and attached reaction zone, $\lambda = 0$ and the solution for Eq. 14.48 for these conditions is either

$$x = 1 \text{ or } x = \mu^2 + \frac{1}{\gamma \text{Ma}_1^2}\left(1 + \mu^2\right) \tag{14.49}$$

where the second solution is appropriate [9]
And as $\text{Ma}_1^2 \rightarrow \infty$ implies that Eq. 14.49 goes to

$$x = \frac{\rho_0}{\rho} = \mu^2 = \frac{\gamma - 1}{\gamma + 1} \tag{14.50}$$

and

$$\psi\left(1\right) = \frac{1}{x} = \frac{\gamma + 1}{\gamma - 1} \tag{14.51}$$

Substituting $x = \mu^2$ into the Rankine Line results in

$$y = \frac{P}{P_0} = \left(1 + \gamma \text{Ma}_1^2\right) - \gamma \text{Ma}_1^2 \mu^2 = 1 + \gamma \text{Ma}_1^2\left(1 - \mu^2\right) = 1 + \gamma \text{Ma}_1^2\left(\frac{2}{\gamma + 1}\right) \tag{14.52}$$

And as $\text{Ma}_1^2 \rightarrow \infty$ implies that Eq. 14.52 goes to

$$\frac{P}{P_0} = \gamma \text{Ma}_1^2 \frac{2}{\gamma + 1} \tag{14.53}$$

Further,

$$\gamma \text{Ma}_1^2 \frac{2}{\gamma + 1} = R^{-3} f_1 = R^{-3} A^2 \frac{f}{a^2} \tag{14.54}$$

and

$$\left(\frac{dR}{dt}\right)^2 = A^2 R^{-3} = u^2 \tag{14.55}$$

where u is the shock speed.
Substituting Eq. 14.55 into Eq. 14.54 results in

$$\gamma \text{Ma}_1^2 \frac{2}{\gamma + 1} = f \frac{u^2}{a^2} \tag{14.56}$$

and

$$\frac{2\gamma}{\gamma+1} = f(1) \tag{14.57}$$

From the conservation of mass and in a laboratory frame of reference

$$\frac{u-D}{D} = \frac{-v}{v_0} = -\mu^2 - \frac{1}{\gamma \mathrm{Ma}_1^2}\left(1+\mu^2\right) \tag{14.58}$$

where

$$u - D = \text{particle velocity and is in laboratory frame of reference} \tag{14.59}$$

And as $\mathrm{Ma}_1^2 \to \infty$ implies that Eq. 14.58 goes to

$$\frac{u-D}{D} = \frac{-v}{v_0} = -\mu^2 \tag{14.60}$$

and

$$u = R^{-1.5}\phi_1 = R^{-1.5}A\phi = u\phi \tag{14.61}$$

And as such

$$\frac{u}{D} = 1 - \mu^2 = \frac{2}{\gamma+1} = \phi(1) \tag{14.62}$$

14.3.3   Numerical Solutions

In order to numerically integrate the system of ordinary differential equations given as Eqs. 14.8, 14.9, and 14.10, we need to get the system in the form

$$\frac{f'}{f} = f(f,\phi,\psi) \tag{14.63}$$

$$\frac{\phi'}{\phi}(\eta-\phi) = g(f',\phi,\psi) \tag{14.64}$$

$$\frac{\psi'}{\psi}(\eta-\phi) = h(\phi') \tag{14.65}$$

Substituting Eq. 14.9 into Eq. 14.10 results in

$$3f + \eta f' + \gamma \left[\phi' + 2\frac{\phi}{\eta} \right](-f) - \phi f' = 0 \tag{14.66}$$

and

$$(\eta - \phi) f' = \gamma \left[\phi' + 2\frac{\phi}{\eta} \right] f - 3f \tag{14.67}$$

$$(\eta - \phi) f' = \gamma \left[\frac{1}{\eta - \phi} \left(\frac{1}{\gamma} \frac{f'}{\psi} - \frac{3}{2}\phi \right) + 2\frac{\phi}{\eta} - 3 \right] f \tag{14.68}$$

And multiplying both sides by $(\eta - \phi)$ results in

$$(\eta - \phi)^2 f' = \gamma \left[\left(\frac{1}{\gamma} \frac{f'}{\psi} - \frac{3}{2}\phi \right) + \left(2\frac{\phi}{\eta} - 3 \right)(\eta - \phi) \right] f \tag{14.69}$$

and

$$f' \left[(\eta - \phi)^2 - \frac{f}{\psi} \right] = f \left[-3\eta + \phi \left(3 + \frac{1}{2}\gamma \right) - 2\gamma \frac{\phi^2}{\eta} \right] \tag{14.70}$$

or

$$\frac{f'}{f} = \frac{-3\eta + \phi \left(3 + \frac{1}{2}\gamma \right) - 2\gamma \dfrac{\phi^2}{\eta}}{(\eta - \phi)^2 - \dfrac{f}{\psi}} \tag{14.71}$$

And from Eq. 14.8

$$(\eta - \phi)\frac{\phi'}{\phi} = \frac{1}{\gamma} \frac{f'}{\psi} \frac{1}{\phi} - \frac{3}{2} \tag{14.72}$$

And from Eq. 14.9

$$(\eta - \phi)\frac{\psi'}{\psi} = \phi' + 2\frac{\phi'}{\eta} \tag{14.73}$$

Equations 14.71, 14.72, and 14.73 are solved using Euler's method and a small time step with the final conditions given as Eqs. 14.11, 14.12, and 14.13. As shown in [8, 9], Euler's method is a rather crude numerical integration technique, but in the figures below shows good agreement with G.I. Taylor's results (Figs. 14.4, 14.5, and 14.6).

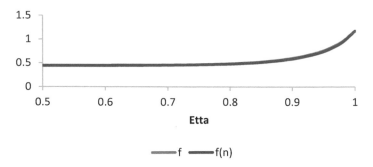

Fig. 14.4 Etta versus *f*

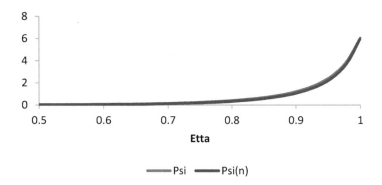

Fig. 14.5 Etta versus Psi

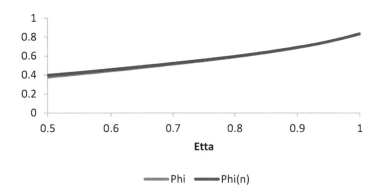

Fig. 14.6 Etta versus Phi

14.3.4   Approximate Forms

In this section, approximate forms are developed from the ordinary differential equations (Eqs. 14.8, 14.9, and 14.10); parameters of these algebraic equations are determined from the final conditions given as Eqs. 14.11, 14.12, and 14.13.

.... Deriving approximate form for φ

From Eq. 14.67,

$$(\eta - \phi)\frac{f'}{f} = \gamma\phi' - 3 + 2\gamma\frac{\phi}{\eta} \tag{14.74}$$

If we were to graph $\dfrac{f'}{f}$ versus η, then we'd see for $\eta < 0.6$ the left-hand side of Eq. 14.74 is essentially zero and thus

$$0 = \gamma\phi' - 3 + 2\gamma\frac{\phi}{\eta} \tag{14.75}$$

or

$$\frac{d\phi}{d\eta} + \frac{2}{\eta}\phi = \frac{3}{\gamma} \tag{14.76}$$

Which is a first-order, differential equation [6] with solution provided via an integrating factor. Equation 14.76 is of the form

$$\frac{dy}{dx} + P(x)y = Q(x) \tag{14.77}$$

With integrating factor

$$\mu = e^{\int P(x)dx} \tag{14.78}$$

And for Eq. 14.76 μ is

$$\mu = \eta^2 \tag{14.79}$$

And multiplying both sides of Eq. 14.76 by Eq. 14.79 results in

$$\frac{d\phi}{d\eta}\eta^2 + 2\eta\phi = \frac{3}{\gamma}\eta^2 \tag{14.80}$$

or

$$\frac{d}{d\eta}\left[\eta^2\phi\right] = \frac{3}{\gamma}\eta^2 \tag{14.81}$$

With solution

$$\phi = \frac{\eta}{\gamma} \tag{14.82}$$

Looking at the graph of ϕ versus η (see Fig. 14.4) there is a second portion to the right-hand side of Eq. 14.82 to account for the nonlinear changes in ϕ as a function of η

$$\phi = \frac{\eta}{\gamma} + \alpha \eta^n \tag{14.83}$$

Utilizing the final condition $\phi(1)$ and Eq. 14.83, α is determined

$$\phi(1) = \frac{2}{\gamma + 1} = \frac{1}{\gamma} + \alpha \tag{14.84}$$

or

$$\alpha = \frac{1}{\gamma} \frac{\gamma - 1}{\gamma + 1} \tag{14.85}$$

Determining n is more involved.

Using Eq. 14.74 and substituting in for ϕ and ϕ' the approximate forms just determined one gets to

$$\frac{f'}{f}(\eta - \phi) = \left[1 + \alpha n \gamma \eta^{n-1}\right] - 3 + 2 \frac{\gamma}{\eta} \left[\frac{\eta}{\gamma} + \alpha \eta^n\right] \tag{14.86}$$

Solving the right-hand side of Eq. 14.86 when $\eta = 1$ results in

$$\frac{f'}{f}(\eta - \phi) = \left[1 + \alpha n \gamma\right] - 3 + 2\gamma \left[\frac{1}{\gamma} + \alpha\right] = \alpha \gamma (n + 2) = \frac{\gamma - 1}{\gamma + 1}(2 + n) \tag{14.87}$$

Solving for $\eta - \phi$ when $\eta = 1$ results in

$$\eta - \phi = 1 - \frac{1}{\gamma} - \frac{1}{\gamma} \frac{\gamma - 1}{\gamma + 1} = \frac{\gamma - 1}{\gamma + 1} \tag{14.88}$$

And substituting Eqs. 14.88 into Eq. 14.17 results in

$$\frac{\gamma - 1}{\gamma + 1} \frac{f'}{f} = \frac{\gamma - 1}{\gamma + 1}(2 + n) \tag{14.89}$$

or when $\eta = 1$

$$\frac{f'}{f} = 2 + n \tag{14.90}$$

Additionally,

$$\frac{f'}{f} = \frac{-3\eta + \phi\left(3 + \frac{1}{2}\gamma\right) - 2\gamma\dfrac{\phi^2}{\eta}}{(\eta - \phi)^2 - \dfrac{f}{\psi}} \tag{14.91}$$

And when $\eta = 1$

$$\psi = \frac{\gamma + 1}{\gamma - 1}, f = \frac{2\gamma}{\gamma + 1}, \phi = \frac{2}{\gamma + 1} \tag{14.92}$$

Substituting Eq. 14.90 and Eq. 14.92 into Eq. 14.91 results in

$$\frac{f'}{f} = n + 2 = \frac{-3 + \dfrac{2}{\gamma + 1}\left(3 + \dfrac{\gamma}{2}\right) - 2\gamma\left(\dfrac{2}{\gamma + 1}\right)^2}{\left(\dfrac{\gamma - 1}{\gamma + 1}\right)^2 - \dfrac{\dfrac{2\gamma}{\gamma + 1}}{\dfrac{\gamma + 1}{\gamma - 1}}} \tag{14.93}$$

and

$$\frac{f'}{f} = n + 2 = \frac{-3 + \dfrac{6 + \gamma}{\gamma + 1} - \dfrac{8\gamma}{(\gamma + 1)^2}}{\left(\dfrac{\gamma - 1}{\gamma + 1}\right)^2 - 2\dfrac{\gamma(\gamma - 1)}{(\gamma + 1)^2}}$$

$$= \frac{-3(\gamma + 1)^2 + (\gamma + 6)(\gamma + 1) - 8\gamma}{1 - \gamma^2} = \frac{-2\gamma^2 - 7\gamma + 3}{1 - \gamma^2} \tag{14.94}$$

and

$$2 = \frac{2(1 + \gamma)(1 - \gamma)}{(1 + \gamma)(1 - \gamma)} = \frac{2 - 2\gamma^2}{1 - \gamma^2} \tag{14.95}$$

Therefore,

$$n = \frac{-7\gamma+1}{1-\gamma^2} = \frac{7\gamma-1}{\gamma^2-1} \tag{14.96}$$

……….. Deriving approximate form for f
Using Eq. 14.91,

$$\frac{f'}{f}(\eta-\phi) = \gamma\phi' - 3 + 2\gamma\frac{\phi}{\eta} = \left[1+\alpha n\gamma\eta^{n-1}\right] - 3$$

$$+ 2\gamma\left[\frac{1}{\gamma}+\alpha\eta^{n-1}\right] = (n+2)\alpha\gamma\eta^{n-1} \tag{14.97}$$

and

$$\eta-\phi = \eta - \frac{\eta}{\gamma} - \alpha\eta^n \tag{14.98}$$

Substitution of Eqs. 14.98 into Eq. 14.97 and solving for $\dfrac{f'}{f}$ results in

$$\frac{f'}{f} = \frac{(n+2)\alpha\gamma\eta^{n-1}}{\eta\left(1-\dfrac{1}{\gamma}\right)-\alpha\eta^n} = \frac{(n+2)\alpha\gamma^2\eta^{n-2}}{(\gamma-1)-\alpha\gamma\eta^{n-1}} \tag{14.99}$$

With solution

$$\int_\eta^1 \frac{df}{f} = (n+2)\gamma\int_\eta^1 \frac{\alpha\gamma\eta^{n-2}\,d\eta}{(\gamma-1)-\alpha\gamma\eta^{n-1}} \tag{14.100}$$

Set $b = \gamma-1, u = \alpha\gamma\eta^{n-1}$, and $\dfrac{du}{n-1} = \alpha\gamma\eta^{n-2}d\eta$ results in

$$\int\frac{df}{f} = \frac{n+2}{n-1}\gamma\int\frac{du}{b-u} \tag{14.101}$$

And set $x = b - u$ and $dx = -du$ results in

$$\int\frac{df}{f} = -\frac{n+2}{n-1}\gamma\int\frac{dx}{x} \tag{14.102}$$

Equation 14.102 has solution

$$\int_\eta^1 \frac{df}{f} = -\frac{n+2}{n-1}\gamma\ln\left[b-\alpha\gamma\eta^{n-1}\right]\Bigg|_\eta^1 \tag{14.103}$$

and

$$\ln\left(\frac{2\gamma}{\gamma+1}\right) - \ln(f) = \frac{n+2}{n-1}\gamma \ln\left[\frac{b-\alpha\gamma\eta^{n-1}}{b-\alpha\gamma}\right] \tag{14.104}$$

or

$$\ln(f) = \ln\left(\frac{2\gamma}{\gamma+1}\right) - \frac{n+2}{n-1}\gamma \ln\left[\frac{b-\alpha\gamma\eta^{n-1}}{b-\alpha\gamma}\right] \tag{14.105}$$

.... Deriving approximate form for ψ

Given

$$\frac{\psi'}{\psi} = \frac{\phi' + 2\dfrac{\phi}{\eta}}{\eta - \phi} \tag{14.106}$$

And approximate forms for ϕ and ϕ' results in

$$\frac{\psi'}{\psi} = \frac{\dfrac{1}{\gamma} + \alpha n\eta^{n-1} + 2\left(\dfrac{1}{\gamma} + \alpha\eta^{n-1}\right)}{\eta\left(\dfrac{\gamma-1}{\gamma}\right) - \alpha\eta^{n}} \tag{14.107}$$

and

$$\frac{\psi'}{\psi} = \frac{\dfrac{3}{\eta} + \alpha\gamma n\eta^{n-2}(n+2)}{(\gamma-1) - \alpha\gamma\eta^{n-1}} = \frac{\dfrac{3}{\eta} + \alpha\gamma\eta^{n-2}(n+1)}{b - \alpha\gamma\eta^{n-1}} \tag{14.108}$$

Equation 14.108 won't be solved but the solution is given by G.I. Taylor [8] as

$$\ln(\psi) = \ln\left(\frac{\gamma+1}{\gamma-1}\right) + \frac{3}{\gamma-1}\ln(\eta) - 2\frac{\gamma+5}{7-\gamma}\ln\left(\frac{\gamma+1-\eta^{n-1}}{\gamma}\right) \tag{14.109}$$

14.3.5   Energy Released

In this chapter, the initiation of the explosion and the geometry of the container have been ignored. It can be shown that quickly after ignition, a very intense explosion can be treated as a point explosion developing into a sphere. As such, the infinitesimal volume is given as

$$dV = \{\text{Surface Area}\} \times \{\text{thickness}\} = 4\pi r^2 dr \qquad (14.110)$$

And the differential mass of this shell is

$$dm = \rho dV = 4\pi \rho r^2 dr \qquad (14.111)$$

The differential energy of the system includes internal energy plus kinetic energy

$$dE = \left(e + \frac{1}{2}u^2\right)dm = \left(e + \frac{1}{2}u^2\right)\rho r^2 dr = 4\pi \rho \left(e + \frac{1}{2}u^2\right)r^2 dr \quad (14.112)$$

And the total energy is

$$E = \int_0^{R(t)} 4\pi \rho \left(e + \frac{1}{2}u^2\right)r^2 dr = \int_0^{R(t)} 4\pi \left(\frac{1}{\gamma-1}\frac{P}{\rho} + \frac{1}{2}u^2\right)r^2 dr$$

$$= \frac{4\pi}{\gamma-1}\int_0^{R(t)} Pr^2 dr + 2\pi \int_0^{R(t)} \rho u^2 r^2 dr \qquad (14.113)$$

$$= \text{Total Thermal Energy} + \text{Total Kinetic Energy}$$

Equation 14.113 is now put into a non-dimensional form based on Eqs. 14.14, 14.15, 14.16, and 14.17 where the transformations are given below.

$$P = P_0 R^{-3} f_1, r = R\eta, dr = Rd\eta, \rho = \rho_0 \psi, u^2 = R^{-3}\phi_1^2 \qquad (14.114)$$

Substitution of Eq. 14.114 into Eq. 14.113 results in

$$E = \frac{4\pi}{\gamma-1}\int_0^1 \{P_0 R^{-3} f_1\}\{R^2\eta^2\}\{Rd\eta\} + 2\pi \int_0^1 \{\rho_0\psi\}\{R^{-3}\phi_1^2\}\{R^2\eta^2\}\{Rd\eta\} \qquad (14.115)$$

and

$$E = \frac{4\pi}{\gamma-1}\int_0^1 P_0 f_1 \eta^2 d\eta + 2\pi \int_0^1 \rho_0 \psi \phi_1^2 \eta^2 d\eta \qquad (14.116)$$

Further

$$E = \frac{4\pi}{\gamma-1}\int_0^1 \frac{P_0 A^2}{c_0^2} f\eta^2 d\eta + 2\pi \int_0^1 \rho_0 \psi A^2 \phi^2 \eta^2 d\eta \qquad (14.117)$$

and

$$E = 4\pi A^2 \left[\frac{P_0}{c_0^2(\gamma-1)}\int_0^1 f\eta^2 d\eta + \frac{\rho_0}{2}\int_0^1 \psi \phi^2 \eta^2 d\eta \right] \qquad (14.118)$$

or

$$E = 4\pi A^2 \rho_0 \left[\frac{1}{\gamma(\gamma-1)} \int_0^1 f\eta^2 d\eta + \frac{1}{2} \int_0^1 \psi\phi^2\eta^2 d\eta \right]$$ (14.119)

Additionally,

$$E = \rho_0 A^2 B$$ (14.120)

where B is

$$B = 2\pi \int_0^1 \psi\phi^2\eta^2 d\eta + \frac{4\pi}{\gamma(\gamma-1)} \int_0^1 f\eta^2 d\eta = 2\pi I_1 + \frac{4\pi}{\gamma(\gamma-1)} I_2$$ (14.121)

Please note than I_1 and I_2 have only one parameter (γ) and can be solved numerically using mathematical software.

The crux of the problem is to determine A, which can be surmised from how it was originally defined

$$\frac{dR}{dt} = AR^{-1.5}$$ (14.122)

The solution to Eq. 14.122 is

$$\frac{2}{5} R^{5/2} = At$$ (14.123)

In G.I. Taylor [9], experimental results of time after ignition [t] versus radius of the spherical shock blast [R] are given. Graphing t versus $\frac{2}{5} R^{5/2}$ provides an estimate of A.

A graph of R^5 versus t^2 is shown below (Fig. 14.7).

Given below is a reprint of G.I. Taylor's Table 3 [8].

Table 14.1   Energy release from Trinity test site [8]

γ	1.2	1.3	1.4	1.667
I_1	0.259	0.221	0.185	0.123
I_2	0.175	0.183	0.187	0.201
K	1.727	1.167	0.856	0.487
Ex1e-20 [erg]	14.4	9.74	7.14	4.06
TNT	34,000	22,900	16,800	9,500

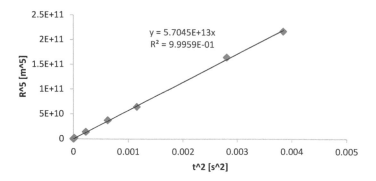

Fig. 14.7 R^5 versus t^2 for Trinity test site

$$K = \frac{4}{25}\left\{2\pi I_1 + \frac{4\pi}{\gamma(\gamma-1)}I_2\right\}$$ (14.124)

and

$$E = K\rho_0\frac{R^5}{t^2}$$ (14.125)

G.I. Taylor went on to prove that the likely gamma of the system is 1.4 and so the energy released was near 16,800 T.N.T.!

14.4   More General Theory (JHS Lee)

The equations given in below [3] are a more general form of the equations given in the sections on G.I. Taylor's work given as Eqs. 14.8, 14.9, and 14.10.

14.4.1   Reduced Forms

The reduced form of the PDE's given as Eqs. 14.2, 14.3, and 14.4 are

$$\psi' = \frac{-\left[(\phi-\xi)^2\,\dfrac{j\phi\psi}{\xi} + 2\theta f - \theta\phi\psi\,(\phi-\xi)\right]}{(\phi-\xi)\left[(\phi-\xi)^2 - \gamma\dfrac{f}{\psi}\right]}$$ (14.126)

$$\phi' = \frac{-\theta\phi(\phi-\xi)+\dfrac{f}{\psi}\left(2\theta+\gamma j\dfrac{\phi}{\xi}\right)}{(\phi-\xi)^2-\gamma\dfrac{f}{\psi}}$$

(14.127)

$$f' = \frac{-(\phi-\xi)\left(2\theta f+\gamma j\dfrac{\phi f}{\xi}\right)+\gamma f\phi\theta}{(\phi-\xi)^2-\gamma\dfrac{f}{\psi}}$$

(14.128)

where

$$\psi=\frac{\rho}{\rho_1},\phi=\frac{u(t)}{\dot{R}_s(t)},f=\frac{P}{\rho_1\dot{R}_s^2(t)},\xi=\frac{r}{R_s(t)}$$

(14.129)

With final conditions

$$\psi(1)=\frac{\gamma+1}{\gamma-1}$$

(14.130)

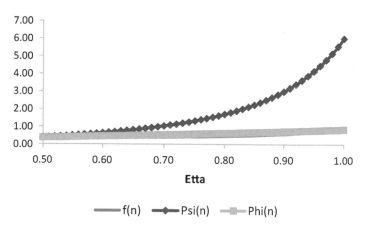

Fig. 14.8 Etta versus non-dimensional states ($j=0$)

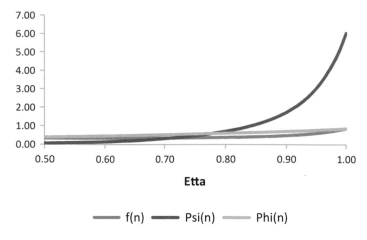

Fig. 14.9 Etta versus non-dimensional states ($j = 1$)

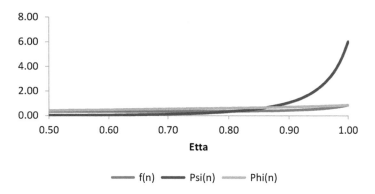

Fig. 14.10 Etta versus non-dimensional states ($j = 2$)

$$\phi(1) = \frac{2}{\gamma + 1} \tag{14.131}$$

$$f(1) = \frac{2}{\gamma + 1} \tag{14.132}$$

14.4.2   Numerical Solutions

Utilizing Eqs. 14.126, 14.127, and 14.128 and an Euler's scheme with $h = 0.01$, the following graphs were created for j equal to 0, 1, and 2 (Figs. 14.8, 14.9, and 14.10).

14.5   Illustrated Example (Explosions Associated with Rotating Stars)

The author of this book had the pleasure of teaching a numerical analysis course where we essentially worked through understanding G.I. Taylor's two papers [6, 7] and solved the system of ODEs using package differential equation solvers. This author has also had the opportunity as a mentor for undergraduate research experience to lead three undergraduate researchers through these two papers and applying the two developed tools (numerical integration and approximate forms) to a similar system [10]. It was quite pleasing how well these tools applied to the particular system chosen and the results of this work are given below.

The system discussed below is the conservation principles for an adiabatic flow behind a cylindrical shock propagating in a rotational axi-symmetric flow of a perfect gas with initial density and velocity as a function of distance from the axis of symmetry [10], which would model a very intense explosion from a rotating star.

14.5.1   Reduce Form

Given from [10]

$$U' = \frac{1}{\gamma P - (\eta - U)} \left[\frac{v^2 g (\eta - U)}{\eta} - \frac{P \gamma U}{\eta} - b(\eta - U)Ug \quad Pd - 2bP \right] \tag{14.133}$$

$$V' = \frac{V}{\eta - U} \left(b + \frac{U}{\eta} \right) \tag{14.134}$$

$$W' = \frac{bW}{\eta - U} \tag{14.135}$$

$$g' = \frac{g}{\eta - U}\left[g' + d + \frac{U}{\eta}\right] \tag{14.136}$$

$$P' = \frac{V'g}{\eta} - bUg + (\eta - U)gU' \tag{14.137}$$

where

$$u = \dot{R}U, v = \dot{R}V, w = \dot{R}W, \rho = \rho_0, p = \rho_0\dot{R}^2P \tag{14.138}$$

And final conditions are

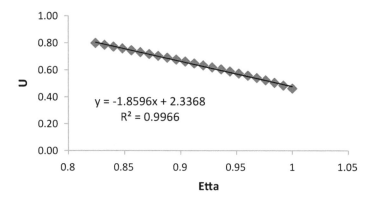

Fig. 14.11 Etta versus U

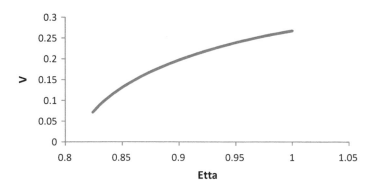

Fig. 14.12 Etta versus V

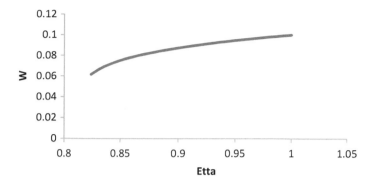

Fig. 14.13 Etta versus *W*

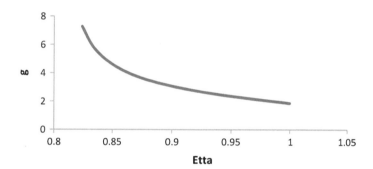

Fig. 14.14 Etta versus *g*

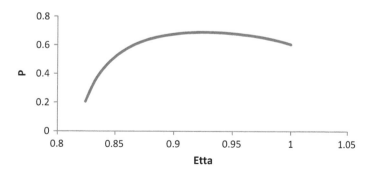

Fig. 14.15 Etta versus *P*

$$U(1) = 1 - \beta \tag{14.139}$$

$$V(1) = \frac{1}{M}\sqrt{\frac{2b+d}{\gamma}} \tag{14.140}$$

$$W(1) = \frac{A}{Q} \tag{14.141}$$

$$g(1) = \frac{1}{\beta} \tag{14.142}$$

$$P(1) = \frac{1}{\gamma M^2} + 1 - \beta \tag{14.143}$$

14.5.2   Numerical Solutions

Solutions for the five non-dimensional quantities are given below as Figs. 14.11, 14.12, 14.13, 14.14, and 14.15.

14.5.3   Approximate Forms

Referring to Fig. 14.11, there is a linear relationship between η and U; this linear relationship will be exploited below.

Another form of Eqs. 14.133, 14.134, 14.135, 14.136, and 14.137 is

$$(\eta - U)\frac{g'}{g} = U' + \left(d + \frac{U}{\eta}\right) \tag{14.144}$$

$$(\eta - U)\frac{U'}{U} = b + \frac{P'}{gU} - \frac{V^2}{\eta U} \tag{14.145}$$

$$(\eta - U)\frac{V'}{V} = b + \frac{U}{\eta} \tag{14.146}$$

$$(\eta - U)\frac{W'}{W} = b \tag{14.147}$$

and

$$(\eta - U)\frac{P'}{P} = (\eta - U)\gamma\frac{g'}{g} + d + 2b - \gamma d \qquad (14.148)$$

.... Deriving approximate form for U

It has been shown numerically that the relationship between ξ and U is linear for the admissible range of parameters and as such

$$U = m\eta + k \qquad (14.149)$$

Given Eq. 14.149 the following relationships result

$$\eta - U = (1 - m)\eta - k \qquad (14.150)$$

$$U' = m \qquad (14.151)$$

and

$$\frac{U}{\eta} = m + \frac{k}{\eta} \qquad (14.152)$$

$$(\eta - U)\frac{W'}{W} = b \qquad (14.153)$$

And subtracting Eq. 14.147 from Eq. 14.146 results in

$$\frac{V'}{V} - \frac{W'}{W} = \frac{1}{\eta - U}\frac{U}{\eta} \qquad (14.154)$$

Further, another form of Eq. 14.148 is

$$(\eta - U)\frac{P'}{P} = \gamma U' + \gamma\frac{U}{\eta} + s(1 - b) - 2 \qquad (14.155)$$

.... Deriving approximate form for W

Given

$$\frac{W'}{W} = \frac{b}{\eta - U} = \frac{b}{(1 - m)\eta - k} \qquad (14.156)$$

and

$$\frac{1}{W}\frac{dW}{d\eta} = \frac{b}{(1-m)\eta - k} \tag{14.157}$$

With solution

$$\ln(W) = \frac{b}{1-m}\ln\left[(1-m)\eta - k\right] \tag{14.158}$$

or

$$W = W_0\left[(1-m)\eta - k\right]^{b/1-m} \tag{14.159}$$

..…..... Deriving approximate form for V
Given

$$\frac{V'}{V} - \frac{b}{(1-m)\eta - k} = \frac{1}{(1-m)}\frac{U}{\eta} \tag{14.160}$$

and

$$\frac{V'}{V} = \frac{1}{(1-m)\eta - k}\left[b + \frac{U}{\eta}\right] \tag{14.161}$$

and

$$\frac{V'}{V} = \frac{1}{(1-m)\eta - k}\left[b + m + \frac{k}{\eta}\right] \tag{14.162}$$

The solution to Eq. 14.162 is

$$\ln(V) = \frac{b}{1-m}\ln\left[(1-m)\eta - k\right] + \frac{m}{1-m}\ln\left[(1-m)\eta - k\right] + \int\frac{k/\eta}{(1-m)\eta - k}d\eta \tag{14.163}$$

where

$$X = (1-m)\eta - k, dX = (1-m)d\eta, \text{and}\frac{U+k}{1-m} = \eta \tag{14.164}$$

Substituting Eq. 14.164 into the integral within Eq. 14.163 results in

$$\frac{1}{1-m}\int\frac{\overline{\frac{k(1-m)}{U+k}}}{u}\,du = k\int\frac{du}{u^2+ku} = k\int\frac{du}{u(k+u)} \tag{14.165}$$

Which has solution [7]

$$\int\frac{dx}{x(a+bx)} = \frac{1}{a}\ln\left|\frac{x}{a+bx}\right| \tag{14.166}$$

.... Deriving approximate form for P

Given

$$(\eta-U)\frac{P'}{P} = \gamma U' + \gamma\frac{U}{\eta} + s(1-b) - 2 \tag{14.167}$$

and

$$U' = m, \frac{U}{\eta} = m + \frac{k}{\eta}, \eta - U = (1-m)\eta - k \tag{14.168}$$

Substituting the relationships from Eq. 14.168 into Eq. 14.167 results in

$$\left[(1-m)\eta - k\right]\frac{P'}{P} = \gamma m + \gamma\left[m+\frac{k}{\eta}\right] + s(1-b) - 2 \tag{14.169}$$

and

$$\frac{P'}{P} = \frac{\gamma m + \gamma\left[m+\dfrac{k}{\eta}\right] + s(1-b) - 2}{\left[(1-m)\eta - k\right]} \tag{14.170}$$

and

$$X = (1-m)\eta - k, \frac{dX}{1-m} = d\eta \text{ and } m + \frac{k}{\eta} = m + \frac{1-m}{X+k}k \tag{14.171}$$

Such that

$$\ln(P) = \int\frac{\gamma m + \gamma\left[m+\dfrac{1-m}{X+k}k\right] + s(1-b) - 2}{X(1-m)}\,dX \tag{14.172}$$

and

$$\ln(P) = \frac{\gamma m}{1-m}\int\frac{dX}{X} + \frac{\gamma}{1-m}\int\frac{m+\dfrac{1-m}{X+k}k}{X}dX + \frac{s(1-b)}{1-m}\int\frac{dU}{U} - \frac{2}{1-m}\int\frac{dU}{U} \qquad (14.173)$$

The second integral can be simplified as

$$\frac{\gamma}{1-m}\int\frac{m+\dfrac{1-m}{X+k}k}{X}dX = \frac{m\gamma}{1-m}\int\frac{dU}{U} + \gamma k\int\frac{dX}{X^2+kX} \qquad (14.174)$$

.... Deriving approximate form for g

$$(\eta-u)\frac{g'}{g} = u' + \left(d+\frac{u}{\eta}\right) \qquad (14.175)$$

and

$$\left[(1-m)\eta-k\right]\frac{g'}{g} = m+d+m+\frac{k}{\eta} = 2m+d+\frac{k}{\eta} \qquad (14.176)$$

or

$$\frac{1}{g}\frac{dg}{d\eta} = \frac{2m+d+\dfrac{k}{\eta}}{(1-m)\eta-k} \qquad (14.177)$$

With solution

$$\ln(g) = \int\frac{2m+d}{(1-m)\eta-k}d\eta + \int\frac{\dfrac{k}{\eta}}{(1-m)\eta-k}d\eta \qquad (14.178)$$

and

$$x = (1-m)\eta-k, \frac{dx}{m-1} = d\eta \qquad (14.179)$$

Substituting Eq. 14.179 into 14.178 results in

$$\ln(g) = \frac{1}{1-m}\int\frac{2m}{x}dx + \frac{1}{1-m}\int\frac{d}{x}dx + \int\frac{\dfrac{k}{\eta}}{(1-m)\eta-k}d\eta \qquad (14.180)$$

14.6 Problems

Problem 14.1 Reduce Eq. 14.6 to an ordinary differential equation.

Problem 14.2 Reduce Eq. 14.7 to an ordinary differential equation.

Problem 14.3 Using Matlab's ODE solvers, solve Eqs. 14.8, 14.9, and 14.10 with the final conditions given as Eqs. 14.11, 14.12, and 14.13.

Problem 14.4 Using Matlab's ODE solvers, solve Eqs. 14.144, 14.145, 14.146, 14.147, and 14.148 with the final conditions given as Eqs. 14.139, 14.140, 14.141, 14.142, and 14.143.

Problem 14.5 Using Matlab develop a program to determine the value of I_1 and I_2 for a given value of γ and compare against Table 14.1.

Problem 14.6 Develop approximate form (algebraic solution) for Eq. 14.128.

Problem 14.7 Reduce Eq. 14.2 to an ordinary differential equation.

Problem 14.8 Reduce Eq. 14.3 to an ordinary differential equation.

Appendix 14.1: Similarity Arguments (website)

An Excel Spreadsheet was developed to illustrate a particular concept and is given on the companion website.

Appendix 14.2: GI Taylor's Work (website)

An Excel Spreadsheet was developed to illustrate a particular concept and is given on the companion website.

Appendix 14.3: JHS Lee's Work (website)

An Excel Spreadsheet was developed to illustrate a particular concept and is given on the companion website.

Appendix 14.4: Fisk, Tjandra and Vaughan's Work (website)

An Excel Spreadsheet was developed to illustrate a particular concept and is given on the companion website.

References

1. Jones, J. (1991). The spherical detonation. *Advances in Applied Mathematics, 12*, 147–186.
2. Kandula, M., & Freeman, R. (2008). On the interaction and coalescence of spherical blast waves. *Shock Waves, 18*(1), 21–33.
3. Lee, J. H. S. (2016). *The gas dynamics of explosions*. Cambridge University Press.
4. Powers, J. (2008). *Lecture notes for introduction to combustion*. Notre Dame University.
5. Rogers, M. H. (1957). Similarity flows behind strong shock waves. *Journal of Mechanics and Applied Mechanics, 11*(4), 411–422.
6. Ross, S. (1980). *Introduction to ordinary differential equations* (3rd ed.). Wiley.
7. Stewart, J. (2015). *Calculus* (8th ed.). Cengage Learning.
8. Taylor, G. I. (1950). The formation of a blast wave by a very intense explosion, part I. *Proceedings of the Royal Society of London. Series A, Mathematical and Physical Sciences, 201*(1065), 159–174.
9. Taylor, G. I. (1950). The formation of a blast wave by a very intense explosion, part II. *Proceedings of the Royal Society of London. Series A, Mathematical and Physical Sciences, 201*(1065), 175–186.
10. Vishwakarama, J. P., & Pathak, P. (2012). Similarity solutions for a cylindrical shock wave in a rotational axisymmetric gas flow. *Journal of Theoretical and Applied Mechanics, 2*, 563–575.

Index

© Springer Nature Switzerland AG 2022
H. C. Foust III, *Thermodynamics, Gas Dynamics, and Combustion*,
https://doi.org/10.1007/978-3-030-87387-5